Wildlife Conservation in Managed Woodlands and Forests

SECOND EDITION

FORESTRY SERIES

Series Editor: **Esmond H. M. Harris,** BSc., Dip.For., FICFor., CBiol., MIBiol.

*Forthcoming

Wildlife Conservation in Managed Woodlands and Forests
SECOND EDITION

Esmond Harris
BSc (Forestry and Botany)
Chartered Biologist, Fellow of the Institute of Chartered Foresters
formerly Director of the Royal Forestry Society, Woodland Owner

and

Jeanette Harris
BSc (Zoology), BSc (Agriculture)
Affiliate of the Institute of Chartered Foresters, Farmer and Woodland Owner

RESEARCH STUDIES PRESS LTD.
Taunton, Somerset, England

JOHN WILEY & SONS INC.
New York • Chichester • Toronto • Brisbane ª Singapore

RESEARCH STUDIES PRESS LTD.

24 Belvedere Road, Taunton, Somerset, England TA1 1HD

First edition 1991, Basil Blackwell Ltd.

Marketing and Distribution:

Australia and New Zealand:
Jacaranda Wiley Ltd.
Sydney Office, Suite 4A, 113 Wicks Road, North Ryde, NSW 2113, Australia

Canada:
JOHN WILEY & SONS CANADA LIMITED
22 Worcester Road, Rexdale, Ontario, Canada

Europe, Africa, Middle East and Japan:
JOHN WILEY & SONS LIMITED
Baffins Lane, Chichester, West Sussex, UK, PO19 1UD

North and South America:
JOHN WILEY & SONS INC.
605 Third Avenue, New York, NY 10158, USA

South East Asia:
JOHN WILEY & SONS (ASIA) PTE. LIMITED
2 Clementi Loop #02-01
Jin Xing Distripark, Singapore 129809

Library of Congress Cataloging-in-Publication Data

Harris, Esmond.
    Wildlife conservation in managed woodlands and forests / Esmond
Harris and Jeanette Harris. - - 2nd ed.
        p.   cm. - - (Forestry series ; 4)
    Includes bibliographical references (p.   ) and index.
    ISBN 0-471-96932-X (pbk. : alk. paper)
    1. Wildlife conservation - - Great Gritain.   2. Forest management -
- Great Britain.   I. Harris, Jeanette, 1929-   . II. Title.
QL84.4.G7H37   1997
639.9(0941 - - dc21
                                96-48744
                                    CIP

British Library Cataloguing in Publication Data

A catalogue record for this book is available from the British Library.

ISBN 0 86380 206 0 (Research Studies Press Ltd.) *[Identifies the book for orders except in America.]*
ISBN 0 471 96932 X (John Wiley & Sons Inc.) *[Identifies the book for orders in USA.]*

Typeset by Bernie Nicholls Typesetting, Aylesbury, UK
Printed in Great Britain by SRP Ltd., Exeter

Contents

Foreword

Nothing could give me greater pleasure than to be associated in this small way with the second edition of this particular book. Both publisher and Esmond and Jeanette Harris are to be congratulated for updating it and re-issuing the result. It can fill an otherwise unoccupied niche in the forester's library.

My own interest in woodland and its management began when I was sent to a magical place called Slapton in South Devon to create a field centre and manage Slapton Ley Nature Reserve, then an SSSI, now an NNR. Here were two long-neglected woods. France Wood, planted in the nineteenth century, and Slapton Wood, an ancient semi-natural woodland, offered so much scope to an over-energetic young naturalist suddenly charged with their management. They were already rich in wildlife, but beginning a 100 year management plan, written by one Eric (Robbie) Roberts of the then Nature Conservancy, enhanced wildlife opportunity almost immediately. Much needed income was realised and ploughed back into the Reserve. I thus began my first lessons in combining conservation and cash production. Involvement in the woodland estates of a National Park, the Forestry Commission, and the Countryside Council for Wales since, has increased that interest.

I thus have no hesitation, as you might imagine, in commending this book and particularly its updating. Chapter 3 brings the reader bang up to date on all those national and international nature conservation recent developments, which provide the statutory context of wildlife protection and biodiversity enhancement within which the woodland manager can and should work.

It is to be hoped that government will soon come round to the other relevant component of that context - a forestry strategy for the UK. After all, 'amenity' woodland is extending by the year via 'National Forest', community forest, community woods and farm woodlands. Welsh ministers have welcomed proposals for the extension of productive woodland cover in Wales on lower slopes. Despite agriculture's fluctuating fortunes, a gentle shift from food towards fibre production on farms, alongside the growing stewardship of wildlife by farmers, must be a sensible order of the day. Pulling all these together into a strategy must soon become essential. At the other end of the spectrum this book can be a primer for managers changing direction, a guide to a thinking forester, and a reassurance to the worried forestry observer.

In all respects therefore it is more than welcome at this time.

Ian Mercer
Professor of Rural Conservation Practice
School of Agriculture and Forestry
University of Wales

Preface to the Second Edition

We are pleased to have been invited to prepare a new edition of this book by Research Studies Press for their Forestry Series because there have been significant developments in British forestry since the first edition was published, in particular following from the Rio summit in 1992, which led to commitments in Britain to sustainability and biodiversity as cornerstones of future UK forestry policy. These have been outlined in chapter 9 and we discuss and demonstrate how these principles can be applied. At about the same time there came into being several small woodland initiatives with objectives specific to their region but with the common aim of improving Britain's small, unmanaged woodlands. All these to some extent have wildlife conservation as one of their objectives.

Thus there has also been increasing interest in wildlife conservation, particularly as part of revitalized small woodlands. Early retirement has offered to many the opportunity to move to the countryside and buy a small piece of woodland of their own but soon there is a desire to make it productive alongside other objectives. Whilst we first wrote this book mainly with large-scale forestry in mind, the same principles apply to smaller woods. We are often asked where information can be obtained concerning productive woodland and wildlife. As it is scattered throughout the literature and not readily available to the practising forest manager and small woodland owner, we have summarized it here. The book also gives guidance on how to incorporate wildlife conservation objectives (provision for wildlife) in forests and small woods, and indeed the main purpose of the book is to show how this can be done *alongside other objectives*. (We do not see wildlife conservation very often best served when it is the sole objective.)

Even in our British woods with the limited range of native species that we have, as explained in chapter 1, there is potential for a vast range of animals, birds, plants and invertebrates. We therefore discuss general principles with some specific examples and have referenced the text fully so that particular interests can be followed up in the published literature with regard to particular species. This is in Part II but in

Part I we have also referenced the statements we have made to show they are not just our own observations but are supported by objective research carried out by field scientists. This, then, is a book for the 'general practitioner' and the amateur woodland manager as well as for the professional.

In the first edition we expressed enthusiasm for and confidence in Britain's young forests for wildlife and made positive suggestions about their potential but they were controversial views at the time and we are aware that not all we said was widely accepted then. In the ten years since we did the research for that edition many scientific papers in this area have appeared and we have been impressed by the many examples they provide of species numbers and diversity increasing in productive plantations. This has softened attitudes, and a more objective approach to the compatibility of wildlife conservation and wood production is emerging. We have added many of these references to this new edition for those who want to follow them up and we welcome the opportunity that current attitudes and practices in British forestry can provide for applying the principles agreed at Rio.

Throughout the book we emphasize that sustained yield for wood production is the best basis for woodland wildlife conservation and that the development of our maturing forests in Britain has significant potential in this respect. It is of the greatest importance therefore to conservation and the present-day concern for the 'sustainability' of all resources that throughout Europe forest resources are being not only maintained but increased. To quote Dr Hummel, one-time Head of the Forestry Division of the European Commission, writing in the Institute of Chartered Foresters *News* (2/96): 'For the past 50 years the forest area, volume of growing stock and increment have been rising steadily throughout Europe and fellings have consistently been less than the increment. In this respect our forests are, therefore, sustainable.' There could be no better basis upon which to build a wide range of wildlife habitats.

Esmond and Jeanette Harris
Calstock, Cornwall 1996

Preface to the First Edition

In Britain we have two million hectares of forest managed primarily for commercial purposes. Much of it, created in the last sixty years, is composed primarily of introduced species, principally conifers, due to the low fertility of the ground available for planting. Until now, such forests and woodlands have been largely dismissed as a wildlife resource but their potential, through appropriate management, for providing a wide range of habitats for both common and rare species, is considerable. Managed forests and woodlands, therefore, have a vital part to play in what has been termed 'extensive conservation', i.e. broadening the base for conservation by extending it beyond sites of special importance.

In writing this book we had two purposes: firstly to show that commercial management of forests and woodlands is not inconsistent with, and indeed can assist, a thriving and varied population of fauna and flora; secondly to provide some guidance on what can be done to encourage and sustain wildlife in productive forest environments. Imaginative and practical forestry can greatly enhance Britain's natural heritage and we want to draw attention to the opportunities provided, both by maturing first generation forests and by other productive woodlands, for broadening the spectrum of wildlife habitats.

Feeling that our own young forests in Britain are too often considered in isolation, and their conservation value overlooked, we have drawn widely on overseas examples of both natural and long-established managed forests, as these clearly indicate the way in which our commercial forests could offer similar valuable habitats in the future.

Since we started researching for this book the public has become aware of the devastating exploitation of forests worldwide. If this means more timber should be grown at home there need be no alarm about sterility of the forests and woods that will be required, as the encouraging accumulation of evidence from plantations shows that they can support a wide range of wildlife while providing a renewable source of timber.

E.H.M. Harris
J.A. Harris
Tring, Hertfordshire

Acknowledgements

We are grateful to the following for reading the manuscript of this book and for their helpful and constructive comments:

Peter Garthwaite, a forester with a lifetime's experience of forestry both at home and abroad and a keen naturalist.

Bill Copland, who has had a varied career as an educationalist and when with the Nature Conservancy was involved in organizing the Duke of Edinburgh's 'Countryside in 1970' Conference. He has been Senior Education Adviser to Dorset County Council, responsible for environmental education, and is currently an educational consultant to the National Trust.

Stephen Ball, for his many useful suggestions and for his professional editing of the text.

In addition to the literature cited, much relevant information has been obtained, both at home and abroad, during the study tours of the Royal Forestry Society of England, Wales and Northern Ireland. We are indebted to all the landowners, foresters and naturalists who shared their knowledge with us.

Many relevant observations were made during judging for the Duke of Cornwall's Award for Forestry and Conservation in the period 1985-9 when the Award was being established and we were fortunate to accompany the judges, Dick Steele, Director General of the Nature Conservancy, Rod Hewitt, former Conservator, Forestry Commission, and Dr Bill Wright, Adviser on Forestry and Conservation to the National Trust. We are grateful to all the owners who entered the competition and whose woodlands were visited, and for the time they took to explain and demonstrate their ideas.

Additionally, for the second edition, our grateful thanks are due to Roger Lines for pointing out several minor errors in the text of the first edition and to Dr Keith Kirby for drawing our attention to some now out-of-date botanical names, all of

which have been corrected. Simon Humphreys guided us on the recently introduced 'woodmark' certification scheme outlined in chapter 9. To Veronica Wallace
of Research Studies Press we are particularly grateful. She made several suggestions for the new edition and patiently improved the grammatical presentation of
our text, as well as bringing together the second edition in its present and
improved form.

Introduction

Forests of one type or another form the vegetation cover of at least 31 per cent of the land surface of the world. More than half of land-dwelling species of animals and plants depend on these forests: some live, feed or grow solely within the forest, whilst many animals, both vertebrates and invertebrates, feed outside but use the forests for shelter, protection and breeding. A considerable number of animals and plants are edge species, taking advantage of the conditions created at the marginal zones between the forest and other neighbouring habitats (ecotones) to satisfy their requirements. Finally, the importance of tree cover for plant and animal life, for erosion control and windbreaks, and to satisfy our economic needs for timber, fuel and other forest products, is generally accepted throughout the world.

In some countries considerable effort is now being directed at halting the ravages perpetrated by man in the world's remaining naturally occurring forests; also to rehabilitate forests in areas that have been felled, especially where erosion has become a serious problem. This is achieved both by afforestation and by allowing the natural regrowth of secondary forest. China, for instance, has a constructive programme but in many countries deforestation continues without thought for the future. Complete removal of the original forest cover usually initiates soil degradation and erosion by wind and water, so that the choice of species for reafforestation is often limited. Although in many situations a local natural 'colonizer' species can be and often is used as a first stage of reafforestation, experience has shown that the same species from a different climatic region, or a similar species from another country with similar conditions, grows more quickly and fulfils the aim of providing ground cover more rapidly.

In areas such as tropical forest, where selected trees of prime timber species are felled, leaving the rest, the forest will recover from the disturbance relatively quickly but the resulting forest composition is changed by human intervention. Felling and species selection have operated in Britain over many centuries. The so-called natural or semi-natural forests and woodlands that remain with us today result form human intervention not just on one occasion, as is still the case in many other countries, but repeatedly over the centuries. The great extent of deforestation in Britain so reduced the area of natural tree cover that it reached one of the lowest levels in the world, at its nadir early in the twentieth century.

Because the pressures of population dictate land use in Britain, priority is given to urbanization, industry and agriculture; reafforestation programmes have therefore concentrated on the poorer soils, which are mainly in the uplands. Here, due to the long-term loss of tree cover, the already infertile soil has become degraded. As we have noted, this tends to restrict the choice of native trees available for reafforestation, but early in the twentieth century the discovery of the advantages offered by many of the North American conifers changed the whole concept of re-establishing tree cover in Britain.

Public reaction to the use of such introduced species has been emotive in recent years, both economic and conservation arguments being put forward to condemn their use. The economic case has been based on the assertion that the timber grown is of poor quality and that anyway it is not required. The conservation movement has stated that the alien habitats that result are both shunned by wildlife and detrimental to the well-being of native species. These views are misleading simplifications and fail to take account of the complex factors involved; a better understanding of these factors is one for the purposes of this book.

Similar climates tend to produce similar forest types, although some of the tree species are different. These similar forest types support similar plant and animal associations, with many genera and even some species in common. Many of the species found in Britain occur in similar forests elsewhere in the world, where familiar species and genera can be found thriving in habitats different to those of man-modified British countryside. By learning how plants and animals thrive in these less disturbed situations, and by creating similar conditions in new forests, we can manage productive woodlands to enhance the wildlife potential of Britain. This cannot be achieved by considerations of timber production alone, but modifications compatible with good forest management can make woodlands and forests financially viable and at the same time provided new ecosystems attractive to a wide spectrum of wildlife. To learn how these objectives can be achieved we need to look at more natural environments than those that now exist in Britain, including comparable situations in Europe and further afield.

In this book we give an account of how present-day managed forests have become what they are and discuss ways of improving their conservation potential. It is of note that no native British tree species is under threat of extinction (unless Plymouth pear is regarded as a truly British species), in marked contrast to some native animals and plants that are. In Part II individual animal and plant species are therefore dealt with and for those whose numbers are at a low level we suggest appropriate conservation measures. As there is no comparable threat to trees they have been treated differently: in Part I we consider the management of commercial forests as a conservation resource.

The Evolution and Management of Forests

CHAPTER 1

Forests of the World

It has been estimated that if conditions throughout the world had not been modified by the influence of human beings, forests of one kind or another would form the natural *climax* (that is, the vegetation that becomes dominant in particular climatic conditions) vegetation over at least 40 per cent of the world's surface (Hora 1981). The remaining 60 per cent would be frozen wastes, mountain tops, natural grassland, savannah and desert; how much of these last three represent truly natural states is not certain. Much of the original forest cover has been cleared for agriculture, particularly in the Far East, Western Europe and the United States of America, and some of this land has been built on. In Britain, urban and industrial areas now equal the area of forest at 9 per cent each, although none of the latter is the original natural forest. Clearing forests for agriculture still takes place today, especially in the tropics, so that the present area of the world's surface covered by forest, natural and man-made, is 31 per cent (Kallio et al. 1987). A considerable proportion of this is stunted boreal (mainly coniferous) and alpine forest, or secondary scrub growth from previous clearance (see figure 1.1).

CLIMATE

Climatic influences are the predominant natural factors limiting tree growth and thus determine the type of vegetation present. The most important of these are temperature, water availability, waterlogging, exposure, and the length and strength of daylight during the growing season; the last is particularly important to ground vegetation. Soil fertility is less important but soil depth is significant. The overall influence of one or a complex interaction of a number of these factors determines the type of vegetation in any particular place. A change in the balance, even over small areas, can dramatically influence tree growth and the type of trees present.

Broadly speaking, forest types vary with latitude because conditions become cooler and less favourable for growth as the distance from the equator increases. Areas on different continents but in the same hemisphere with similar climates

6

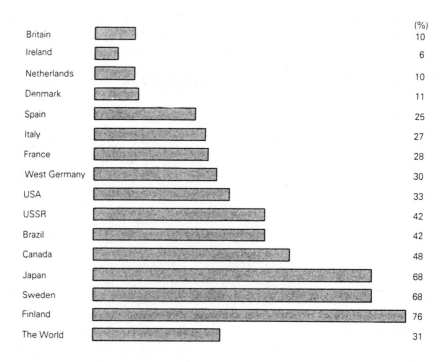

	(%)
Britain	10
Ireland	6
Netherlands	10
Denmark	11
Spain	25
Italy	27
France	28
West Germany	30
USA	33
USSR	42
Brazil	42
Canada	48
Japan	68
Sweden	68
Finland	76
The World	31

Figure 1.1 Proportion of land occupied by forest of all types Sixty per cent of the world's forests are broadleaved, but in the northern hemisphere the majority are coniferous.
(From FAO figures)

produce forests of similar structure, appearance and species type. As well as the broad change caused by latitude, a similar influence is exerted by altitude in mountainous regions. Numerous classifications of forest types have been proposed for various countries, based mainly on temperature. The first was that of Mayr in 1906 who related growth to the temperature in the growing season, the length of the growing season and the lowest annual temperature. Later, a simple classification was worked out by the Arnold Arboretum in America. This is based on average winter temperature and defines Tree Hardiness Zones which indicate regions in which particular species can grow; it was developed mainly for gardeners. In the USA these zones are latitudinal but similar zones worked out for tree hardiness in Western Europe show a longitudinal pattern resulting from the influence of the Atlantic Gulf Stream (Johnson 1984). Classifications in Britain include those of Fairburn (1968) and the Institute for Terrestrial Ecology (ITE). The latter designates 32 land classes based on minimum January temperature, days when snow is falling, daily sunshine, altitude, distance from the west coast and geology (Last et al. 1986). However, in addition to limiting winter temperatures, summer temperature in the growing season is important to tree growth and so is critical from the point of the production of good timber. This is why there are limits to the places where broadleaved trees can be grown economically in Britain. Climatological maps produced by ITE (White and Smith 1982) provide data on summer temperatures as well as rainfall, sunshine, wind and snow cover in Britain, and methods are given which show how to calculate conditions in a particular area (see table 1.1 and figures 1.2, 1.3a, 1.3b and 1.4).

Recent predictions of temperature rises resulting from the 'greenhouse effect' have received attention and there have been speculations about the effects on vegetation, trees and forests. Records from 1851 to 1984 suggest a *global* rise of 0.5 to 0.7°C but British records show a cooling of at least 2.5°C in the 20 years from 1970 (Cannell et al. 1989). British temperatures are related to those of the north Atlantic ocean surface, which fell by 0.3°C from the middle of the nineteenth to the early twentieth century, followed by a rise of 0.5°C between the 1920s and 1930s and a further cooling by 0.3°C in recent years. These changes are within normal fluctuations and 'it is not yet possible to predict future trends in our climate with any confidence' (Parker 1988). It seems likely that many of the models predicting temperature rise have overstated likely increases but only continued and careful monitoring of actual change will eventually give a more realistic picture. As far as tree growth is concerned, it will not be a rise in average temperature that will be significant but any increase in temperature during the growing season.

SOILS

Soils play an important part in the local distribution of vegetation. The interaction of soils and the plants growing upon them is complex and has been the subject of extensive investigation. The classification of soils themselves is difficult and a number of systems exist, including the FAO/ UNESCO classification of 1968, the United States Department of Agriculture system, the Canadian classification and various Russian classifications (Russell 1957; Whitmore 1975; Eyre 1984; Larsen 1980; Knystautas 1987).

The origin of soil is important because old, acidic rocks produce soils poor in plant nutrients, whilst softer sedimentary deposits produce more fertile soils. In general, however, climatic factors exert the main influence on soil properties irrespective of soil origin. Where precipitation exceeds evaporation, soil minerals and plant nutrients are leached out to varying degrees. The extreme case is where waterlogging is severe in soils over permafrost, impervious rocks or soil pans, resulting in minimal leaching and anaerobic conditions. This inhibits soil microbial activity which in turn, together with low temperatures in high latitude or at altitude, allows undecomposed plant remains (raw humus) to accumulate on the surface. This leads to the formation of peat which can build up to considerable depths. Peat is not limited to high latitudes, however, as anaerobic

Table 1.1 Forest zones and factors controlling tree growth

Forest zone	Soil type	Dominant trees	Secondary trees	Mean January temperatures	Mean July temperatures	Mean annual temperatures	Annual rainfall	Number of days available for growth
7 Taiga and Mountain Forest	Podzol	Larch Pine Spruce Fir Birch	Willow Aspen	−12 to −30°C (11 to −22 °F)	12 to 20°C (53 to 68 °F)	2 to 6°C (35 to 42 :F)	500mm upwards (20″ upwards)	60 to 145
6 Mixed Forest	Brown earth	Larch Pine Spruce Fir Birch Oak Ash Lime[a] Beech[a]	Willow Aspen Rowan Poplar Maple Elm Alder Cherry Hazel	−4 to −8°C (25 to 18 °F)	14 to 20°C (57 to 68 °F)	2 to 8°C (35 to 47 °F)	400mm upwards (16″ upwards)	145 to 170
5.2 Northern Temperate Deciduous Forest	Brown earth	Birch Oak Ash Lime[a] Beech[a]	Willow Aspen Rowan Poplar Maple	−4 to −8°C	14 to 20°C	2 to 8°C	400mm upwards	145 to 170

Zone	Soil	Dominant species	Associated trees	Temp	Temp	Temp	Rainfall	Growing season (days)
		Pine	Elm, Alder, Cherry, Hazel	(25 to 18 °F)	(57 to 68 °F)	(35 to 47 °F)	(16" upwards)	170 to 240
5.1 Southern Temperate Deciduous Forest	Brown earth	Oak[a] Beech[a]	Ash, Hornbeam[a], Sweet chestnut, Walnut, Horse chestnut, Magnolia	0 to 14 °C (32 to 57 °F)	21 to 29 °C (70 to 84 °F)	9 to 14 °C (48 to 57 °F)	500 to 750mm 20 to 27"	
Height above sea level 561m (1840 ft) the altitude above which timber production cannot be expected in Britain — Moorhouse (Yorkshire)				0 °C (32 °F)	12 °C (53 °F)	6 °C (42 °F)	1800mm (70")	125 to 145
Represents climatic conditions in south-east Britain, the northern limit of zone 5:1 — London				3 to 5 °C (37 to 41 °F)	14 to 18 °C (57 to 64 °F)	8 to 11 °C (47 to 52 °F)	600 to 800mm 25 to 30"	200 to 240

Source: data from Manley 1953; Wang 1961; Barry and Chorley 1976; FAO 1982; White and Smith 1982; Parker 1988.

[a] These trees are at the northern extreme of their range in Britain.

The table details the forest zones represented in Britain and the main factors determining their occurrence in the northern hemisphere. The distribution of rainfall is particularly important, the amount available in the growing season being the most significant. This, together with summer temperature, determines rate of growth. These forest zones merge gradually into one another except where climatic conditions change abruptly.

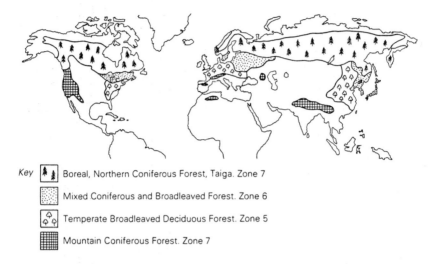

Key
| | |
Boreal, Northern Coniferous Forest, Taiga. Zone 7

Mixed Coniferous and Broadleaved Forest. Zone 6

Temperate Broadleaved Deciduous Forest. Zone 5

Mountain Coniferous Forest. Zone 7

Figure 1.2 Northern hemisphere forest zones of relevance to Britain The indicated regions are those occupied by forest Zones 5, 6 and 7 in the northern hemisphere. Zone 7 includes both taiga and coniferous forest that sustain similar coniferous species. The range and distribution should be compared with figure 1.3, which illustrates winter and summer temperatures.
(Data from USDA 1949; DFS 1949; Kuchler 1964; FAO 1982; Flint 1984; Polunin and Walters 1985; Knystautas 1987)

conditions also occur in some parts of the tropics. In northern latitudes peat formed from sphagnum moss is often found as raised bogs which are higher than the surrounding land and are highest in the centre; it is thought that in the initial stages these formed rapidly with steep sides, becoming flatter as the rate of peat formation declined.

Where conditions are not so extreme and soil is well aerated, soil microbes and fungi are more active, especially where there are warmer temperatures, so some recycling of nutrients takes place allowing arboreal vegetation to survive. Trees in such conditions are often shallow rooting and do not demand a high level of nutrients. On soils of poor nutrient status, such as those derived from acidic rocks, many trees have alternative methods of improving nutrient uptake: alder uses bacteria to supply nitrogen and conifers use mycorrhizal fungi to assist in the uptake of nutrients, particularly phosphates (Kormanik 1979). In high-rainfall areas and on freely draining soils, minerals are leached down through the soil profile and distinct layers are formed. Such soils are known as podzols. In areas of very high rainfall, minerals are completely removed from the soil and because it is these minerals which prevent the soil from becoming acidic, the soil is degraded and can only support undemanding vegetation. (The acidity of soil is measured on the pH scale, which

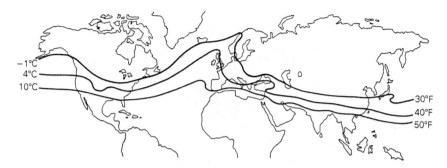

Figure 1.3a **Temperatures in the northern hemisphere: winter isotherms, January**
The isotherms connect places with the same mean January temperature. Their pattern
over Britain illustrates the ameliorating influence of the Gulf Stream.

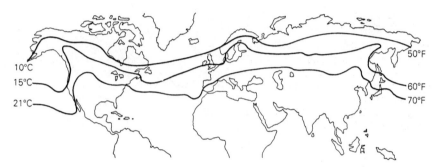

Figure 1.3b **Temperatures in the northern hemisphere: summer isotherms, July**
The isotherms connect places with the same mean July temperature. The 10 °C isotherm
marks the northern limit of tree growth, the 15 °C isotherm marks the centre of the
mixed forest zone (Zone 6), and the 21 °C isotherm marks the boundary between
southern temperate deciduous forest (Zone 5.1) and the northern temperate deciduous
forest (Zone 5.2). At a July mean temperature of 21 °C plant growth is twice as rapid as at
14 °C. In Britain, small rises in altitude cause relatively large falls in temperature, mean
temperatures falling by 1 °C for every 84 m (270 ft) increase in altitude.

indicates the concentration of free hydrogen ions contained in it. The scale is not
arithmetic but logarithmic, so that pH 5 is ten times more acid than pH 6; lower figures indicate the most acidic soils, a pH of 7 is neutral and pH figures above 7 indicate basic soils.)

Depending on the amount of leaching, different soil types are formed. The variation is further influenced by temperature, which increases the rate of nutrient
cycling as it rises, and also by increased weathering. In regions with tropical rain
forest, nutrients are recycled three times faster than in temperate regions. Although

12

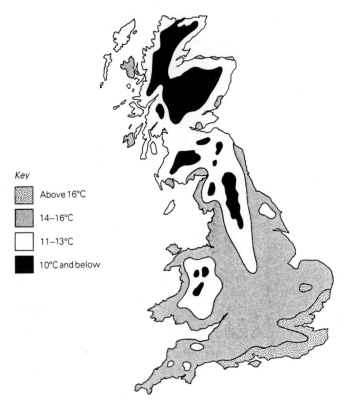

Key

Above 16°C

14–16°C

11–13°C

10°C and below

Figure 1.4 Mean summer temperatures in Britain The temperature contrast demonstrates why satisfactory economic growth of broadleaves is confined to southern and eastern Britain. Although British winters are mild and the growing season is long, growth of broadleaves is everywhere slow compared to that on the European mainland. (Data from White and Smith 1982)

the rainfall is high in the tropics, recycling is so rapid that nutrients remain on the surface (Whitmore 1975; May 1979b). When such forests are cleared for agriculture the soil is inherently poor, so once humus has been used by the first crops the remaining soil is infertile. In temperate regions the warm conditions encourage broadleaves which, with their deeper roots, are able to exploit greater soil depth than conifers. The depth of nutrient cycling is therefore increased. On soils from less acidic rocks which are not leached excessively and in regions where the temperature is higher with a pH of 5.5 to 6, both conifers and broadleaves produce similar soils, called brown earths. Podzolization is most pronounced on acidic and free-draining soils, occurring under broadleaves as well as conifers (Dimbleby and Gill 1957; Miles 1986). Soils of pH 5.5 upwards contain sufficient basic (alkaline) minerals to exert a buffering effect in the soil and thus 'cushion' the acidic effects. Weathering and deposition from the atmosphere are also continually adding soil nutrients.

As the amount of precipitation decreases, a stage is reached where evaporation exceeds precipitation and soils are then formed where leaching hardly occurs or is reversed. Such soils have little moisture for plant growth and support only stunted trees, arid grasslands or desert. In deserts where no leaching at all occurs, salts accumulate on the surface, severely restricting tree growth.

WORLD FOREST ZONES

The forest zones of the world can be briefly summarized as follows in relation to temperature, rainfall and soils. Information has been derived mainly from: Wang 1961; Kuchler 1964; Washburn 1973; Whitmore 1975; Larsen 1980; Hora 1981; Mitchell Beazley 1981; FAO 1982; Johnson 1984; Eyre 1984; Flint 1984; Chabot and Mooney 1985; Polunin and Walters 1985; Harris and Harris 1987 and Knystautas 1987.

Zone 1 Tropical Rain Forest

This is of various types and is found in the Amazon region, Central Africa and Malaysia. These forests occur in areas with high temperatures, high rainfall, constant day-length and with the sun high in the sky. All these factors allow growth and fruiting all the year round. The tree species are all evergreen. This zone includes the following types of forest:

1.1 *Mountain tropical forest.* This type occurs on mountains where rainfall and temperature are sufficient for abundant tree growth.
1.2 *Tropical forest on beaches and in mangrove swamps.* These forests are coastal and the trees can tolerate salt and brackish conditions.
1.3 *Tropical forests growing on peat and fresh water swamps in waterlogged conditions.* The trees have special mechanisms for overcoming the adverse conditions.
1.4 *Tropical forests growing on poor soils,* for example on limestone or very basic rocks and soils with a low level of plant nutrients, such as podzols formed on freely draining soils. The trees are adapted to grow in conditions of water stress and mineral deficiency.
1.5 *Tropical lowland, evergreen forest with* either a permanently humid climate or a humid climate alternating with dry periods. These forests are the most species-rich and are those usually described as 'tropical rain forest'. The underlying soils are often poor in nutrients as the high temperatures and high humidity produce rapid recycling.

Zone 2 Monsoon and Savannah Forest

These types of forest occur in India, Africa and South America. They depend on the seasonal availability of moisture. Fruiting is seasonal and the trees are semi-evergreen or deciduous so damage to the leaves does not occur during the dry season.

Zone 3 Dry Evergreen Forest
These forests occur in the Mediterranean, Chile, South Africa, southwest Australia and southern California. They are adapted to hot summers with drought conditions and warm, dry winters. The leaves of trees are leathery and the bark thick to avoid water loss in summer. The trees are evergreen as the winter is mild.

Zone 4 Warm Temperate Evergreen Forest
This zone consists of two types:

4.1 *Regions with warm, dry winters and hot summers but with sufficient moisture for growth.* These forests occur in the USA (California), the former USSR (West Georgia) and South America (Chile).

4.2 *Regions with warm winters and hot summers but rain all the year round.* These forests occur in central China, Korea, the USA (Florida) and New Zealand.

Zone 5 Temperate Broadleaved Deciduous Forest
These forests occur mainly in the northern hemisphere, particularly in Western Europe, the eastern USA, China, Japan and the Soviet Far East. Small areas of *Nothofagus* forest represent this type in South America.

The trees in this zone lose their leaves in winter because a marked cold season restricts the availability of water. Bud scales are present in winter to protect the young leaves and flowers from frost and desiccation during this period. Broadleaves, typical of this type of forest, give way to conifers at about the 55th parallel in the northern hemisphere.

Forests in this zone fall into two categories:

5.1 *Southern temperate broadleaved deciduous forest.* Forests in this category are found in the southern part of the Temperate Broadleaved Deciduous Zone, and in Europe consist normally of one dominant species, usually oak or beech, with some ash or hornbeam. In the USA oak forest contains hickory as a main component. Mean annual temperatures within the zone range from 9 to 14°C. Mean summer temperatures range from 21 to 29°C, giving 170 to 240 days annually available for growth. These forests are most extensive on the dry, eastern sides of continents.

5.2 *Northern temperate broadleaved deciduous forest: mixed hardwoods and northern hardwoods.* These forests are found in the northern part of the temperate broadleaved deciduous zone, and the species that compose them consist of broadleaves which tolerate reasonably harsh conditions. They contain a mixture of tree and shrub species with no single species dominant over large areas. Oak is a component but only forms pure stands on the warmer, drier soils. Elm, birch, lime and maple are also important, together with a range of minor species, such as cherry, poplar and alder. The mean annual temperature ranges from 2 to 8°C, with summer temperatures from 14 to 20°C. A minimum of 145 days annually available for growth is required. This zone gives way southwards to zone 5.1. Species such as lime occur in the south part of the zone (see chapter 10). Hardwoods tolerating exacting conditions are termed 'northern hardwoods'; in the USA, beech, birch, aspens and maples are the main species.

Zone 6 Mixed Coniferous and Broadleaved Forest
Between the Temperate Broadleaved Deciduous Zone and the Coniferous Zones (both latitudinally and altitudinally), a mixed transitional zone of conifers and broadleaves occurs where climatic changes are not abrupt, so that one group gradually gives way to the other (Walter 1973; Lowe 1977). The mean annual temperature ranges from 2 to 8°C, with a summer mean from 14 to 20°C. This zone and Zone 5 have probably lost more to agriculture than any other forest type throughout the world. Mixed forests in western Europe, including Britain, are not as rich in species as those of the USA and Asia where fewer species were destroyed in the Pleistocene ice ages. The great variety of tree species in the latter forests led to varied structure with a correspondingly richer fauna and flora. Extensive stands of a single species are not found, except where the topography is limiting. In the least-disturbed mixed forests of Europe and Asia, intimate mixtures of species occur, the proportion of conifer to broadleaves varying with the conditions. Small groups of a particular species do occur but single trees of various species are also scattered through the forest. High on mountains and at higher latitudes, more conifers are found, and vice versa. The mixed forests of east Asia are the richest in both fauna and flora; species were not destroyed during the ice ages because they could move south, unimpeded by mountains.

Zone 7 The Boreal, Northern Coniferous Forest, or Taiga
This zone occurs on the North American continent and in northern Europe and Asia. The northern part of the zone is not so fully covered by trees as further south where growth conditions are better. This northern area alone is referred to by some authors as 'taiga', but in this book that term is used to cover the whole of the northern coniferous forest south of the open tundra.

The taiga comprises 29 per cent of all forest of the world and 73 per cent of all coniferous forest (Kuusela 1987). The winters are cold, long and dry. Although water is present as snow and ice, it is mainly unavailable. The summers are short with a growing season of about three months in the south of the zone and as little as 30 days in the north. Rainfall is low but ice and snow melt provides moisture and long summer day-length enhances growth. Waterlogging occurs with the wet conditions and produces raised sphagnum bog, providing breeding grounds for insect

Plate 1.1 Northern coniferous forest untouched by man at 1200m on Chang Bai Shan mountain in north-east China. *Abies nephrolepis, A. holophylla, Larix olgensis, Picea jezoensis, Pinus koraiensis, Acer, Betula, Sorbus, Prunus and Populus* species.
(Photo: E. H. M. Harris)

life such as midges, which in turn encourages many birds to migrate to this zone for breeding during the short summer. In taiga forest the mean annual temperature is 4°C and temperatures range from -50 to +35°C. Because of the low temperatures, vegetation litter accumulates as raw (mor) humus and podzolic conditions occur, forming an acid layer favourable to the growth of ericaceous plants and fungi, both of which are important food sources for wild life. In the north tree cover is sparse, gradually increasing southwards as conditions become warmer. Clearing the coniferous forest leads to invasion by broadleaves such as aspen, birch and poplar, but these short-lived species are gradually replaced by conifers as the succession proceeds towards climax forest again. There is little topographical variation in parts of this zone, so vast areas are composed primarily of one or two tree species. The principal trees in the northern hemisphere are larches, pines, spruces and silver firs, all of which are able to grow on poor soils and with low summer temperatures.

Forests similar to the taiga occur on the higher parts of mountains further south. Here the day-length is shorter but the temperatures higher. At these higher altitudes, as in northern regions, desiccation reduces growth, the trees growing better where snow lies. At high altitudes and in northern latitudes, coniferous trees are narrow in shape compared with those growing at lower altitudes. The abundant aquatic life typical of taiga is absent because of drier conditions and better drainage, but the better soil conditions encourage plant life. Soils under the conifers here are often brown earths as they originate mainly from basic rocks, and the higher temperatures and moderate rainfall promote decomposition.

All the forest zones described above are of necessity generalizations, one merging into another, with no hard and fast boundaries occurring unless climatic conditions change abruptly. Within zones, tree species typical of other zones will be found as a minor component where the conditions favour them. Species found on mountains, for example, may show a sharp contrast between north- and south-facing slopes.

FORESTS OF BRITAIN AND EUROPE

Most of Europe lies in the last three zones described: the Temperate Broadleaved Deciduous Forest, the Mixed Coniferous and Broadleaved Forest and the Northern Coniferous Forest. The British Isles falls naturally into these zones (although it lacks the full range of species) but the maritime climate reduces the mean annual summer temperature and raises the mean winter temperature, allowing a longer but slower growing period than on the Continent. There are only 35 native British tree species (Mitchell 1974) and there is no doubt that but for the formation of the English Channel after the last ice age, there would be a wider range and British forests would be similar to those in Europe, as they had been in the last interglacial phase. However, Europe itself is poorer in tree species than Asia and America. On these latter continents, vegetation was able to move south freely as the ice advanced, there being no mountains running east to west to form a barrier, as the Alps do in

Plate 1.2 Alpine fir (*Abies lasiocarpa*) forming the tree line at 2000m on Mt. Hood, Oregon, USA. (Photo: E. H. M. Harris)

Plate 1.3 Unmanaged mixed coniferous and broadleaved forest in the Bialowieza International Biosphere Reserve, east Poland. The forest contains 28 tree species including *Picea abies, Pinus sylvestris, Quercus* species, *Tilia cordata, Carpinus betulus* and *Ulmus glabra. Larix, Taxus* and *Fagus* are absent. (Photo: E. H. M. Harris)

Europe. Thus a much richer flora and fauna were able to survive elsewhere; for example, there are over 400 tree species in China.

Although in Britain there have been climatic changes since the ice age, as well as extensive human exploitation and modification of the forest cover, climatically the conditions are now those typical of Mixed Coniferous and Broadleaved Forest (Zone 6) over most of the country. The major part of Scotland and the higher areas of England and Wales are in the Northern Coniferous Zone (Zone 7), whilst the more continental southern and eastern areas of England fall into the Northern Broadleaved Zone (Zone 5.2). To be successful, therefore, wildlife conservation in woodlands today should be considered in this context.

The human impact is second only to climatic influence in changing the landscape. There are no forests in Europe today that have escaped human intervention and very few in the rest of the world. Both the Mixed Forest Zone and the Temperate Broadleaved Zone not only allow good tree growth but also provide excellent land for farming. Extensive areas of former forest land are therefore now under agriculture and some has subsequently been built upon. The amount of land affected has fluctuated throughout the centuries, reflecting economic demands and population pressures. In times when much of the land was not farmed for various reasons, it reverted to woodland, coming back into cultivation when circumstances changed. In

New England on the eastern seaboard of the USA, mixed forest cleared by the early settlers has largely reverted to broadleaved woodland over the hundred years since the farming population moved west. Both in Britain and Europe, the forests were intensively managed from early times and their structure and composition were modified to favour the species that were most useful at different periods. For example, oak was the major species required in Britain until the twentieth century, so it was selected to form the dominant species in our older woodlands. In Hungary today, beech is selected in preference to other forest components as its timber is the most important product (Harris and Harris 1988b). In contrast, Norway spruce is favoured in some forests in Austria and beech is therefore removed to obtain maximum timber production (Harris 1985).

A few places exist in Europe where mixed forest remains in a semi-natural state because human intervention has been less drastic. Bialowieza Forest on the Polish-Russian border is an example; this was because the forest was preserved for hunting; the New Forest in Britain also survived for this reason. Similar royal hunting forests, where human influence has been minimal, are found in China, Korea and the Russian Far East. Animal and plant species in these forests are in many cases the same species that occur in similar climatic conditions but in widely separated parts of the world. The habitats that they exploit under these more natural conditions can usefully be evaluated and the knowledge attained applied to the management of commercial woodlands in order to make these more suitable for a wide variety of species.

CHAPTER 2

British Forests and their History

Many millions of years ago, according to the recently developed theory of plate tectonics, Britain was part of a great land mass known as Pangea that eventually separated into the continents that exist today. The continents are still 'drifting' apart. The altered climate and physical conditions in the separate masses, compared with those that existed in Pangea, led to complex changes in the plant and animal life and this developed separately in each continent. Even small alterations in the prevailing climatic conditions over short periods affect species dramatically, favouring some and destroying others. Major climatic changes had catastrophic effects on animal and plant life. Side by side with major and minor climatic fluctuations, plants and animals are constantly struggling to reach an equilibrium within a habitat as it exists at a particular time.

The earliest trees appeared towards the end of the Paleozoic era about 250 million years ago and were the ancestors of the present-day conifers. The maidenhair tree *(Ginkgo biloba)* is the only close relative to a species from that period which has survived until the present day. In the Mesozoic (about 150 million years ago), the evolution of Gymnosperms, which includes all the conifers, proceeded rapidly. The major change at this time was the adaption to drier conditions: reproductive methods were modified so that the trees were no longer dependent on water for fertilization to take place as their ancestors had been. Towards the end of the Mesozoic the Angiosperms or 'flowering plants' made an appearance, eventually including all broadleaved trees.

From the beginning of the last major era 50 million years ago, the Cainozoic (which includes the present), there is evidence from the fossil record of a wide range of trees growing in Britain. The subsequent major climatic change known as the ice age, beginning 1 million years ago, had far reaching effects on the whole of Europe. Although usually spoken of as if it was one continuous phase, it was in fact a series of major climatic fluctuations of both warm and very cold periods. The number of periods and the extent and duration of the 'interglacials' in between is still a subject that is being intensively studied.

In Britain, the last two glaciations were the most extensive and about 17000 years ago the whole country was covered with ice, with the exception of the extreme south, some of the Scottish islands and possibly land areas to the west and some mountain tops. The land area of the west coast was more extensive then, as the sea level was considerably lower because of the large quantity of water locked up in the ice. The persistence of some elements of the British flora (Arctic-alpines and the west coast or Lusitanian flora) is thought to be connected to the occurrence of these ice-free areas in the west. Recent work on Scots pine *(Pinus sylvestris)* indicates two separate colonizations after glaciation, one from the east and one from the west of Scotland (Kinloch et al. 1987). There are indications that this also applies to other tree species which may have survived and recolonized from the west rather than from Europe. For example, pollen studies show oak and hazel appearing earlier in the west than the east (Huntley and Birks 1983). During the fluctuations in ice sheet cover, the vegetation ebbed and flowed as the climate changed. When conditions improved, about 14 000 years ago, the ice finally retreated and the bare soil exposed was colonized in the south by vegetation moving in from the European continent, which at that time was joined to Britain, both across the North Sea plain and from France.

Knowledge of the vegetation cover at that time is based on pollen analysis, and charcoal and plant remains (Zeuner 1952; Godwin 1975). Pollen grains of most species are distinct and can be identified from remains preserved in peat layers of bogs and lakes. Most trees are wind pollinated and have hard pollen grains that do not decompose but a few are insect pollinated and their pollen has not been preserved so well or in such large quantities. The latter include maples (sycamore), fruit trees and limes. Analysis of pollen deposits and other tree remains gives an indication of the conditions prevailing when the layer in which they are found was laid down (see figure 2.1). These deposits can be dated using various techniques, including carbon dating. From this kind of information, maps have been produced showing the tree pollens present at various periods (Huntley and Birks 1983). A high proportion of tree pollen indicates woodland and a high proportion of grass pollen suggests open conditions. These conclusions have limitations: pollen is only preserved in certain circumstances, so the results cannot be taken to apply to the whole country; some pollens do not preserve well, whilst others, such as hazel, do and so are recorded in large quantities. Thus a false impression may be given of the relative abundance of a particular species. Correction factors are applied to minimize this bias. Windborne pollen can be carried a long way, as much as 5° latitude (Woillard 1979), so the presence of pollen does not always indicate tree cover near by.

As the ice retreated, conditions became similar to those of the tundra in northern Russia and Canada today, with a sparse vegetation of lichens and mosses. Winters were cold and extensive wet areas formed in the summer when the frost melted. With the warming of the climate, trees and shrubs began to colonize the bare ground. These were pioneer species such as birch, aspen and pine with windborne

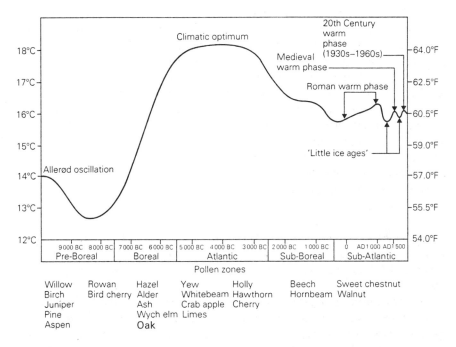

Figure 2.1 Summer temperatures in central England since the ice age Native trees and two Roman introductions are shown against the pollen zones in which they first appear. Their appearance is related to the temperatures, especially in summer, prevailing at the time. Some species never colonized northern Britain.

After a significant temperature rise in the Boreal followed by a 3000 year maximum in the Atlantic period, there has been a partial reduction and several fluctuations; the latter still continue. Since the 1960s, the average summer temperature in Britain has fallen; the increased levels of CO_2 responsible for the world rise in temperature are not yet showing a local effect. Mean April temperatures declined in Britain from the 1960s to the 1980s but October temperatures rose, indicating a later growing season.
(Climatic data from Barry and Chorley 1976 and Parker 1988)

seed, capable of surviving on the poorest soils and under harsh conditions. In many cases they are able to grow on poor soils or even on unweathered material, because bacteria help them to fix nitrogen from the air or they have fungi to improve the uptake of phosphorus from the soil (Kormanik 1979). These pioneer species gradually add humus to the soil and eventually create woodland cover in which less tolerant trees can survive. Thus as the climate warmed, species such as oak and hazel were able to move in and continued to do so until Britain was cut off from the European continent by the flooding of the North Sea plain and the opening of the English Channel. This took place between 6000 and 8000 years ago, or as it is more usually expressed, 6000 to 8000 before present (BP).

Until then the British climate had been continental in type, with cold winters and warm summers. Once Britain became an island surrounded by sea and under the influence of the Gulf Stream, warmer and wetter conditions prevailed, favouring the spread of broadleaves. The removal of the land bridges halted not only the further invasion of trees, but also other plants and animals unable to cross the increasing water barriers. Ireland was separated early: only two-thirds of British plant species are found there, and there are also fewer animals native to Ireland. At the time of separation, the lowland areas of Britain were covered in forest consisting of species typical of the present-day mixed and northern broadleaved forest on the European mainland; but spruce, silver fir and larch were missing, although they had been present in Britain in the previous interglacial period.

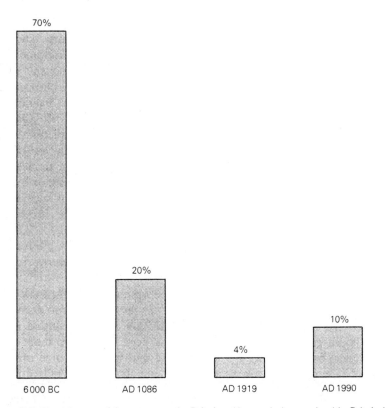

Figure 2.2 The history of forest cover in Britain Human beings arrived in Britain in about 9000 BC; fire was first used to clear forest by 8000 BC. The Domesday book of 1086 provides the first records of land use in Britain. In 1919 the Forestry Commission was set up, after the severe fellings of the First World War.

Although humans were present in Britain during the last interglacial, this was only as hunters of animals and gatherers of fruits and nuts. After the final retreat of the ice they returned and Mesolithic hunter-gatherers gradually increased, living in the forest, especially near water (Zvelebil 1982). There is evidence of the use of axes to fell trees as early as 9000 BP. Human presence in the uplands at this time is now well documented and grazing animals were probably pastured above the tree line (Darvill 1986). Fire may also have been used early as a tool to clear the forest, and the pine in particular would be easier to burn. The later Neolithic charcoal remains contain a wide range of tree species and it has been suggested that the increasing dominance of oak, during the period from 9000 to 5000 BP, may have been the result of selective clearance, as other trees are smaller and softer to cut. Even today, the clearance of mixed forest in China has resulted in a predominance of oak *(Quercus mongolica)*, especially on warm, south-facing slopes (Harris and Harris 1987). The dramatic rise in hazel pollen at the end of the postglacial climatic phase known as the Boreal may also have been due to clearance by humans. Certainly, the clearing of forest cover by cutting or by low intensity surface fires enables broadleaved trees and shrubs such as oak and hazel to sprout again quickly and to dominate other vegetation (Wang 1961; Rowe and Scotter 1973). As hazel nuts were a valuable food source for Mesolithic people the growth of hazel may have been encouraged intentionally in this way, as hazel flowers and fruits a few years after coppicing. Hazel coppicing is known to have occurred in 5000 BP (Rackham 1974).

Four thousand five hundred years ago, a significant change took place. People began to settle in one place as the population grew and unrestricted movement became more difficult. As they settled down, crops were grown and domestic animals kept. More and more forest areas were brought into agricultural use and from this time onwards human beings have had an increasing impact on the landscape (see table 2. 1).

With the advent of a more settled way of life in Neolithic times and the gradual development of agriculture side by side with continuing hunting and gathering, more trees would have been felled to grow crops. This is thought to be one of the reasons for the dramatic decline of lime in the pollen record at this time as it grows on the best soils for agriculture and it is very easily cut.The process continued for a long time (Baker et al. 1978) and still goes on in Europe today (Harris and Harris 1988a). Forest areas would also have been opened up for grazing animals as well as cultivation. Fire was probably used to clear the trees, or they may have been ring barked and the animals grazed beneath. Grazing and further burning would then have prevented regeneration. The foliage of some tree species was important for fodder and as animal numbers increased, overcutting for forage probably destroyed many trees. The decline of elm in the Neolithic has been attributed to this cause but attacks of Dutch elm disease have also been suggested (Moore 1984b). Pollarding and coppicing for forage would have prevented

Table 2.1 Forest succession in Britain since the ice age

Age	Period	Climate	Trees present	Man
	Pre-glacial	Warm	Ginkgo, Sciadopitys, redwood, monkey puzzle, swamp cypress, tulip tree, walnut, plane, laurel, fig, Indian bean, southern beech	
	Pre-glacial	Colder	Norway spruce, Scots pine, yew, hazel, birch, hornbeam, beech, alder, oak, elm, rowan, field maple, hawthorn	
	Glacial	Very cold	No record	
	Interglacial	Cold to warm	Norway spruce, silver fir	
20 000 BC	Glacial	Very cold	No record. Trees thought to have gone	River drift man. Neanderthal man.
	Interglacial	Cold to warm	No record	Aurignacian man.
15 000 BC	Glacial	Very cold	All trees thought to have gone	
10 000 BC	I Early Dryas	Cold	Dwarf willow and birch on tundra. Juniper	
9000 BC	II Allerød	Warmer but cool	Birch, aspen, juniper, Scots pine	Palaeolithic man (food gatherers and hunters).
	III Later Dryas	Cold	Dwarf willow, birch on tundra. Dwarf birch.	

Date		Zone	Climate	Vegetation	Human activity
8000 BC	IV	Pre-boreal	Continental climate	Willows, juniper, rowan, birch, pine, bird cherry, ash	Fire used as tool to clear forests.
	V	Boreal	Warm continental climate	Extensive pine and hazel. Elm, oak, small-leafed lime, alder appear	Mesolithic man (hunters, fishers).
7000 BC	VI	Boreal	Warm continental climate	Extensive pine and hazel. Alder and lime increase. Oak, elm (wych)	Man fells trees with axes and begins to herd animals.
				In Neolithic charcoal: yew, crab apple, poplar, whitebeam, buckthorn	Neolithic man herding some domestic animals and cultivating some crops.
5500 BC	VII	Atlantic	Coastal climate with mild winters and high rainfall	Oak, elm and lime increase. Hawthorn, holly, cherry	Bronze age.
2500 BC	VIII	Sub-boreal	Climate more continental	Oak/ash forests. Elm and lime decrease. Beech and hornbeam appear about 2000 BC	Extensive cultivation for agriculture. Forest clearance
				Charcoals include: hawthorn, service tree, elm, plum, cherry, blackthorn	Iron Age.
				Probable introduction of: walnut, white poplar, sweet chestnut, peach, apricot, quince, fig, medlar.	Romans.
400 BC	IX	Sub-Atlantic	Cool climate with cooler summers	Beech/hornbeam forest. Less oak.	

elm from flowering and this too could have led to a decline of elm in the pollen record.

As farming and civilization increased, more trees were used for fuel, not only for domestic use but also to provide charcoal for smelting minerals. Bronze, made from copper and tin, had by then become important for making tools and ornaments. At about 3000 BP the heavy plough took over from the hoe as a tool for cultivation and larger fields became possible. The introduction of iron tools made forest clearance easier and the number of settlements increased, especially in the lowlands as the population rose. Dramatic rises in the proportion of herb pollen in relation to tree pollen indicate the extent of the change during this period.

By Roman times, Britain was still well wooded in the north and west but elsewhere, especially in the south, there had been a considerable reduction in woodland cover. Mining for iron, copper and lead by the Romans then extended as far afield as Cornwall, Yorkshire, Wales and the Lake District, and wherever there was mining trees were felled to provide the charcoal needed for smelting. With the extensive removal of tree cover came soil deterioration; the continual removal of soil nutrients as meat, milk and crops broke the nutrient cycle that occurs under forest cover. In addition, the removal of trees accelerated nutrient leaching from the soil, as water passed through it more rapidly and carried nutrients away into lakes and rivers. Soil deterioration from this cause was more rapid in the west and north because of the higher rainfall and steeper terrain. In flatter areas, the removal of tree cover led to a rise in the water table and an increase in waterlogged conditions.

It is clear from the Domesday survey that by 1086, 80 per cent of Britain's forest cover had been removed and this gives an indication of the pressures on woodland. Trees had been exploited for a long time to provide building material, firewood, tools, fibres, fodder and charcoal. Domestic animals found food in the forest and their grazing and trampling prevented regeneration. Pasture woodland was important as the widely spaced trees fruited heavily. Pigs, both wild and domestic, ate acorns and other mast, reducing the seed available for regeneration. Regulation of random clearing and grazing became a priority and by Norman times each village had its own area of woodland to supply its needs. The removal of produce was allocated and controlled, and the number of animals per person limited. The popular conception that woodland was more natural then than it is today is probably unrealistic. Demand exceeded supply and rights existed for the removal of dead wood for fuel and small branch wood for firing ovens. Ferns were also important and were collected for bedding. Thus a more accurate picture is of a clear forest floor with all debris collected. Full use of all forest materials can still be seen in other parts of the world.

All sizes of timber were in demand. Oak standards grown over coppice were needed for farm buildings, houses and ships. Only royal forests survived comparatively intact as they were set aside for hunting and operated under strict laws; such areas were a mixture of woods and open ground. Rigorous forest laws also existed in Scotland as woodland was already sparse in the southern part of that country; abbeys, such as Melrose, cultivated large areas of land and pastured large flocks of sheep. In Wales some Cistercian abbeys were already practising forest management by the thirteenth century (Linnaird 1982).

In 1457 laws were passed in England to encourage tree planting and a subsequent law of 1483 allowed the enclosure of woodland against grazing animals for seven years. This gave the coppice a chance to grow, as the grazing of cows and pannage for swine was an important use of forest. Coppice provided small sized material for many purposes; the main species employed were hazel and oak, with sweet chestnut and hornbeam in the south of England. Trees were also cut back at head height above the reach of grazing animals, a practice known as pollarding, which provided fuel and animal fodder. By the sixteenth century increasing amounts of wood were being used for charcoal, especially for iron smelting, brick and glass making and the manufacture of gunpowder. Much of the coppiced oak in the west and north of Britain was used for charcoal and tan-bark. The scrub oak in those areas today is the remnant of woodland subjected to centuries of intensive use. Laws were passed in Elizabethan times in an attempt to restrict the cutting of trees but they do not seem to have been very effective. As a result of the shortage of wood attention turned to Scotland where some reserves were left although timber was already being

Plate 2.1   A wood in the English Midlands. All material is utilized, including dead wood, leaving bare ground, producing conditions similar to those of medieval woodland. (Photo: E. H. M. Harris)

imported. Pine and spruce had been imported from the Baltic as early as the twelfth century to England and the thirteenth century to Scotland. The exploitation of Scottish forests continued until the middle of the nineteenth century and large areas of felled woodland were used for sheep grazing. Red deer, deprived of their natural woodland habitat, learnt to survive on the open hill. Much of the natural Scots pine was floated down rivers to coastal ports to be used for ships' masts.

In the seventeenth century it was finally realized that the destruction of woodland could not continue. John Evelyn wrote his famous book *Silva, or a Discourse of Forest Trees* in 1664; this was the first attempt to collect together information on trees and forestry practice. It is evident from his book that a considerable amount was already known about woodland management by this time and that many present-day forestry principles were in use. The importance for shipbuilding of free grown oak with large branches had become a major priority by then for both the Navy and commercial vessels.

Pressure for woodland clearance was, however, relentless. The wool trade had long been vital to Britain's export trade and clearance for sheep grazing continued. It was the prosperity generated by the wool trade that eventually provided the resources to finance and implement the Industrial Revolution and, in turn, that led to a movement for planting trees which gained momentum in succeeding decades. Much of this planting was in parks and round country houses, providing visual amenity and shelter. Changes in agricultural practice, initiated by the enclosure of land, enabled many landowners to establish plantations, some on former woodland sites and others on poor agricultural land. The Duke of Atholl established large areas of larch between 1764 and 1826 in Perthshire.

Suddenly, however, the demand for certain types of wood fell: the use of coal instead of wood for fuel reduced the demand for firewood, and the building of iron ships reduced the demand for free-grown oak. The availability of cheap timber imports from abroad, especially in the late nineteenth century, discouraged planting by landowners who invested in trade and commerce instead. By the end of the nineteenth century it was finally appreciated that timber resources had fallen to a dangerously low level and the Office of Woods, Forests and Land Revenues was formed. This government body began afforestation at Hafod Fawr in North Wales in 1899 and represented the first official attempt to replenish national reserves. In Scotland similar plantations were established at Inverliever in 1909. The First World War saw millions of acres of woodland felled, much of it planted by the far-sighted landowners of the eighteenth century; the timber was used to supply pit props for the coal mines necessary for industry and to keep the Royal Navy at sea. After the war, forest cover had declined to 4 per cent, the lowest in Europe and among the lowest world-wide. Thus the record of forest clearance in Britain is one of the most devastating in the world and today's forestry policy should be viewed against this background. Full accounts of forest history are given by Anderson (1967), James (1981) and Linnard (1982).

The Forestry Commission was formed in 1919 to establish a home-grown timber resource with the aim of eventually supplying about a third of domestic needs. This target is about half-way to being achieved and forest cover has now reached 9 per cent, which is still far less than it was in 1086 and much less than in most other developed countries. The Commission is responsible for encouraging an expansion of productive forests in Britain and does this in two ways. Firstly it is the national 'forest authority' and stimulates new planting and replanting of mature woods by means of grants. It also controls the clearance of woodland by issuing felling licences for thinning and final felling; the latter include a requirement to replant. This control and grant support has to be in accordance with a *plan of operations* (a plan of work) designed for the woodland concerned and approved by the Forestry Commission. Secondly, the Commission has established a large area of state forest (1 million ha) which it manages itself; much of this is on poor land, mainly marginal hill land, and it is therefore inevitable that the Commission has used a high proportion of introduced conifer species, principally Sitka spruce, for this purpose.

In order to support its own afforestation objectives and to achieve a high standard of forest management for privately owned woodlands, the Forestry Commission runs a research and development programme for its own purposes and for the private sector. This research is of a high standard and has contributed significantly to the successful establishment of forestry in Britain, which now has one of the best forest industries in existence although it is still very small by world standards. As a result of this research, forest management has developed quickly; the national forest resource is beginning to stimulate valuable consumer industries in regions where employment is needed, because Britain has a suitable climate for growing trees, both for paper pulp and structural timber. This is of considerable significance to world conservation as it will, to some extent, compensate for the drastic felling of both coniferous and broadleaved virgin forest taking place elsewhere in the world. Britain, as a major importer, bears some responsibility for the devastation that is taking place; the use of home-grown timber, thereby reducing imports, will enable the pressures on overseas forests to be reduced. It is paradoxical to criticize deforestation in other countries (when the developed nations have been guilty of the same offence in the past, contributing in no small measure to the devastated forests now existing abroad), whilst at the same time condemning rather than welcoming the efforts now being made in Britain to redress the balance. Now a major concern, the importance of 'sustainability' in all forms of forest management, both in Britain and abroad, is beginning to shape future forest policies (see chapter 9).

TYPES OF WOODLAND COVER IN BRITAIN

There are two main types of woodland cover in Britain today: semi-natural woodland and plantations.

'Semi-natural' woodland

Although virgin forest untouched by human influence no longer exists in Britain, a few areas are left that have always carried trees (Rackham 1981). These have been cut by man and grazed by domestic animals for centuries, but such woods still retain ecological conditions that can only be reproduced in plantations by the passage of time and so are deserving of preservation. Planting trees, even native ones, will not immediately produce the conditions required by some of the more fragile species that survive in such ecosystems; many of these are dependent on woodland continuity and some require the presence of old trees. There is a need for detailed study of the factors that will maintain the equilibrium and the management needed to perpetuate this kind of woodland. A considerable proportion of the most important sites are on Sites of Special Scientific Interest (SSSIs) or in the ownership of county trusts or other conservation organizations, and this is as it should be.

Semi-natural woodland is classified either as *primary* woodland, which is thought to be growing on sites that have always been under trees, or as *secondary* woodland. The latter is on land thought to have been under agricultural use at one time but which has reverted, developing many of the characteristics of natural woodland (Peterken 1981).

Plantations

The planting of trees, as opposed to a reliance on coppicing or natural regeneration, became common in the seventeenth century, although plantations of oak are thought to have been established by the Cistercians as long ago as the thirteenth century (Linnaird 1982). The word 'plantation' is often taken to mean coniferous woodland but plantations can also be of broadleaves or of mixtures of broadleaves and conifers.

Many species now grown for timber production have been introduced from other countries and have proved to grow well in the British climate. These exotics, as they are called, have a part to play in conservation because they can provide wildlife habitats that are just as suitable for animals, birds and many other forms of wildlife as woodlands composed only of native species. The practising forester has a valuable tool that can be used in various ways to maximize productive forestry and provision for wildlife. This is our main theme and we will discuss it more fully in later chapters, where we show that these new ecosystems can provide conditions in which a wide range of species can exist.

CHAPTER 3

The History and Principles of Conservation in Britain

THE DEVELOPMENT OF BRITISH CONSERVATION PRACTICE

The history of the conservation of woodlands is inseparable from the history of conservation as a whole because woodlands provide a habitat for many forms of wildlife. The realization that woodlands were being lost came early but the study of their natural history did not become popular until the nineteenth century, although individual naturalists, such as Gilbert White, were studying the natural world many years earlier. The rise in interest was the result of a number of social changes, the most significant being the increased leisure time available, especially among the middle classes. Clergymen in particular formed a large proportion of the eminent naturalists who emerged during this period.

Many societies devoted to studying all branches of natural history, or concerned with the protection of animals, came into existence in the wake of this rise in popularity. The inclusion of the words 'field club' in the title of many of them reflected the emphasis on outdoor activities and expeditions into the countryside. In the latter half of the nineteenth century, the development of the railways meant increased access to the countryside for a growing number of people. Such outdoor activities required little funding or organization and kept down the cost of joining societies concerned with the study of nature, thus attracting a wide membership. Many of these societies still retain the words 'field club' in their title today. They cater for a wide range of interests and play a prominent part in conservation (Allen 1978).

Initially, the interest in wildlife was not in how the animals lived but in the collection, identification and recording of specimens. Vast numbers of all kinds of animals and plants were collected, not only by people interested in natural history but also by those who indulged in the Victorian enthusiasm for collections of all kinds. Whilst much knowledge was gained and the foundations of present day natural history were laid, this emphasis led to drastic reductions in the numbers of some species. In addition, birds and animals were shot for sport or because they interfered with it. The lists of predators and other species destroyed by some of the big estates at this time make depressing reading. The countryside took many years to recover from such depredations and some species were lost altogether.

When it was realized towards the turn of the century that damage was being done, attention was turned to protection or preservation and a number of bodies were formed to implement these aims (Lowe 1983). Out of this movement came the Commons, Open Spaces and Footpaths Preservation Society in 1865, the Royal Society for the Protection of Birds (RSPB) in 1889 and the National Trust in 1895. These were organizations concerned primarily with protection but the first attempts at active conservation were also emerging. The setting aside of reserves for the protection and conservation of animals and birds had begun some time earlier at the instigation of individual enlightened landowners, but the first cooperative venture was the acquisition as a reserve of Breydon Water in the Norfolk Broads in 1888, foreshadowing future trends (Allen 1978; Lowe 1983; NCC 1984). The formation of the Society for the Promotion of Nature Reserves, now the Royal Society for Nature Conservation, followed in 1912.

The trend away from the collection of specimens towards living plants and animals studied in their natural surroundings continued. Cameras, binoculars and note books took the place of guns and collecting equipment, but unfortunately the passion for the physical collection of specimens has not yet altogether died. In the first half of the twentieth century, Huxley's studies on behaviour in animals, Eliot Howard's on bird territories, Tansley's classification of British vegetation types and Elton's animal ecology and population studies were major contributions to this new approach. People began to appreciate that living things were not static or fixed but changing and adapting all the time in response to their environment. In 1913 the British Ecological Society was formed to cater for interest in the developing science of ecology, so important to the understanding of relationships between plants and animals and their habitats. A new form of collecting began to emerge in the shape of extensive surveys and distribution records showing the status of herbaceous plants, trees, animals and insects, and in turn providing information on rarity and change in population densities. Such surveys, to which the amateur naturalist makes important contributions by collecting data in the field, continue today.

A survey with important implications for conservation was the Land Utilisation Survey of 1930 carried out by Dudley Stamp, which highlighted the rapid loss of agricultural land, and led to land classification and the concept of planning resources for the future, including the formation of nature reserves. The process was temporarily halted by the Second World War but a second land use survey has since been completed. Land use changes are now monitored by the Land Use Survey Group of the Institute of Terrestrial Ecology; 32 basic land classes are recognised.

During the war, however, a Nature Reserves Investigation Committee was established under the Society for the Promotion of Nature Reserves with the object of looking at the provision of, and need for, nature conservation and nature reserves. Its recommendations, and others from the British Ecological Society, were

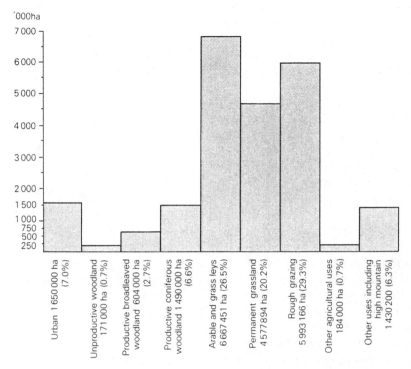

Figure 3.1 **Land use in Great Britain** The area of Great Britain (excluding water) is 22.7 million ha. Of the totals shown, 8.5 per cent is under protection for nature conservation.
(Data from NCC 1984; Forestry Commission 1987; *Farmers Weekly* 1987)

considered by committees under the chairmanship of Sir Julian Huxley for England and Wales and James Ritchie for Scotland. After the war, these deliberations led to the formation of the Nature Conservancy (NC) in 1949, with national funds available for the first time for conservation. The NC was originally intended to be closely linked with the setting up of National Parks, which would include areas of ecological interest, but later the Parks became the responsibility of the separate countryside commissions (NCC 1984). The first reserve to be established by the NC was Beinn Eighe in the Scottish Highlands containing a large area of semi-natural Scots pine woodland. With encouragement from the NC, local county trusts soon multiplied and these purchased reserves of their own. County trusts now cover almost all the UK and since 1958 have been coordinated by the Society for the Promotion of Nature Reserves. In 1981, that Society became the Royal Society for Nature Conservation (RSNC). Other bodies such as the RSPB also increased their

Plate 3.1 Natural Scots pine (*Pinus sylvestris*) remnant of the northern coniferous forest, heavily exploited in the past. Glen Affric, north Scotland. (Photo: E. H. M. Harris)

membership dramatically and began to purchase reserves. By 1987 there were 233 National Nature Reserves. Of the total area designated, only a quarter was owned by the Nature Conservancy Council (NCC); the greater part was in private hands or with the national trusts and county trusts, or with conservation bodies such as the RSPB and Woodland Trust, who together administer over 2000 woodland reserves.

In 1949 the NC was also empowered to designate SSSIs (see page xxx) in order to preserve sites that held rare species or special geological formations. In the early years, the location of these sites and the reasons for designating them were not always divulged to owners of the land concerned and as a result many were damaged. After 1968, the NC was given powers to make management agreements with landowners for the maintenance of such sites and, although many successful agreements have been concluded, a large proportion still remained vulnerable to damage and deterioration. The Nature Conservation Review, published in 1977, lists areas of national biological importance and the reason for the importance of each. It also contains details of all SSSIs that had been designated. Since the passing of the Wildlife and Countryside Act of 1981, all sites designated as SSSIs and still of conservation value are being notified to landowners. These include some new sites that have been designated since the passing of the Act, whilst others have been eliminated.

As well as promoting conservation by the establishment of reserves and designating SSSIs, the NC initiated research into the study of ecology so that the results can be applied to the management of animal and plant communities. By 1973, eight research stations existed and work was also being carried out under grants to universities and other bodies. The NC also took over the records of the first plant distribution scheme carried out in the 1950s by the Botanical Society of the British Isles, which laid the foundation of the Biological Records Centre at Monk's Wood Experimental Station. This Centre now holds distribution records for a wide variety of wildlife species.

From the earlier work of establishing nature reserves, the NC progressed to the wider concept of conservation, realizing that reserves on their own were only part of improving and conserving the environment. Increased public awareness was encouraged through the creation of the Council for Nature in 1958 by bringing together all wildlife and natural history organizations. One of the Council's schemes was the creation of the Conservation Corps to provide young people with practical experience of nature conservation through voluntary work. This proved very popular and eventually became the British Trust for Conservation Volunteers in 1970. The Council's promotion of National Nature Weeks and Countryside Conferences also stimulated public interest in conservation. The Council for Nature has now been absorbed by the Environmental Council (formerly CoEnCo) which is involved in wider concerns of the environment, including recreation and pollution.

Throughout this period, films and talks on natural history were promoted and television programmes brought the subject to a mass audience. An increasing number of books on natural history contributed to the process, among the most significant being the volumes of the Collins 'New Naturalist' series.

In the early 1970s the NC was replaced by two new bodies: the Nature Conservancy Council and the Institute of Terrestrial Ecology. The Nature Conservancy Council (NCC) became autonomous; it was responsible for recommending conservation policy to the Secretary of State for the Environment and was funded through that Department. The Institute of Terrestrial Ecology (ITE), formed from the research section of the NC, was set up under the Natural Environment Research Council (NERC). The new NCC also set up its own research team to look at various conservation problems.

At the same time as interest in conservation grew, not only in Britain but also abroad, there was a growing awareness that the conservation of a species in many cases could not be confined by political boundaries. UNESCO (United Nations Educational, Scientific and Cultural Organization) initiated the 'Man and the Biosphere' programme in 1970 to conserve a network of ecologically important areas, of which there are at present 243 throughout the world in 65 different countries. The first was designated in 1976 and the aim was the establishment of large areas containing both natural and man-made ecosystems. The 'core' of such reserves was to be strictly maintained in its 'natural state', free from human disruption, and was to be as diverse as possible, with the intention of preserving animal and plant communities within natural systems and of safeguarding their genetic diversity. Surrounding the undisturbed 'core' are 'buffer' zones in which fertilizers, pesticides and herbicides are not used; there is a further unrestricted zone outside this.

The reason for the designation of these reserves was the provision of facilities for long-term ecosystem monitoring, and information on both natural and human-induced changes. The unrestricted area outside the buffer zone provides information on managing areas for different purposes (forestry, agriculture) in ways that still sustain the ecosystem (Von Droste and Gregg 1985; Von Droste 1987).

There are several important international agreements on conservation. In 1973 the 'Washington Convention' or Convention on International Trade in Endangered Species of Wild Fauna and Flora (CITES) restricted trade in threatened and endangered species. The 'Bonn Convention' signed in Bonn in 1979 on the Conservation of Migratory Species of Wild Animals is concerned with the protection of species which migrate through different countries. The 'Berne Convention', the Convention on the Conservation of European Wildlife and Natural Habitats, is concerned with the protection of 400 species considered to be endangered and the habitats in which they live. This Convention was ratified in 1982 and took effect in all EC countries in September of that year (NCC 1984). The 'Ramsar Convention' or the Convention on Wetlands of International Importance, especially as Waterfowl Habitats, was initiated at Ramsar, Iran, in 1971 to protect the wetlands for migratory birds worldwide; by 1982, 237 Ramsar sites had been designated (NCC 1983a).

The International Union for the Conservation of Nature and Natural Resources (IUCN) sought to extend conservation in much wider terms by its World Conservation Strategy, launched in 1980 and subsequently formally adopted by many countries, including Britain. Its main aims are summed up as 'living resource conservation for sustainable economic development' and it urges all countries to he members of the Bonn, Berne and Ramsar Conventions, as well as the World Heritage Convention which seeks to preserve natural and cultural areas of international importance (IUCN 1980).

As a result of the Berne Convention it became necessary to alter existing legislation in Britain and previous Acts concerned with wildlife protection were superseded by the Wildlife and Countryside Act of 1981. The NCC became responsible for licensing the taking of birds, animals and plants from the wild and for administering the Endangered Species Act 1976. The Wildlife and Countryside Act also required the notification of all SSSIs, both former and new designations, to the landowner. Powers were given to the NCC under the Act to protect SSSIs and to negotiate compensation agreements for profit forgone. Management plans are currently being prepared for all SSSIs and are designed to keep them in a suitable state to support the species for which they were designated. Failures to manage in the past led to losses, such as that of the large blue butterfly (Thomas 1980). In future, sites will he monitored to detect changes.

NCC policies included a wider look at conservation with the possibility of reintroducing declining or lost species and the re-creation of lost or declining habitats. This is a step forward from the narrow concept that has prevailed to date, succinctly expressed by Morton Boyd (1987) in the following words, 'nature conservation is a new culture of the 20th century dedicated to the survival of native species in their habitats'.

In 1991 the NCC was divided into four bodies which came together in the Joint Nature Conservation Committee. England is administered by English Nature, Wales by the Countryside Council for Wales (incorporating the Welsh part of the Countryside Commission), Scotland by Scottish Natural Heritage and Northern Ireland by the Environment Service (part of the Department of the Environment, Northern Ireland).

The dynamic nature of SSSIs and the need for management is now fully recognised. Cooperation with land owners is being encouraged under Wildlife Enhancement Schemes. These provide some of the managing costs and there is liaison through the agriculture departments with the Stewardship Scheme (in England and Wales) and Environmentally Sensitive Areas.

The concept of dividing the country into Natural Areas, rather than units based on local government boundaries, is being developed and may in future be a basis for selection of SSSIs. Special Areas of Conservation (SACs), which will incorporate the most important SSSIs, and Special Protection Areas (SPAs) are also being designated under the European Habitats and Species Directive, which was adopted by the member States of the European Union (formerly the European Community) in 1992. This Directive gives legal backing to the resolutions of the Berne Convention and the Rio Summit. SACs and SPAs will form the Natura 2000 Series, designated to protect unique and special wildlife sites throughout Europe.

The importance of assisting scarce species is also progressing under the Species Recovery Programme. This aims to 'achieve long term, self sustained survival in the wild of species of plants and animals currently under the threat of extinction'. Among the species reintroduced in Britain under the Programme are red kite, dormouse, large blue butterfly, natterjack toad and Plymouth pear.

CONSERVATION PRINCIPLES IN BRITAIN

As we pointed out in the previous section, conservation in Britain is directed towards maintaining and improving semi-natural habitats and their associated species in as natural a state as possible, because there are no remaining areas unmodified by human activity. Conservation sites can be envisaged as a series of reserves reflecting the best examples of ecosystems and variations within them, representative of different regions. The selection of sites that merit protection is based on a number of criteria. The order of importance varies and depends on the context in which they are applied. They include diversity of both habitat and species within a conservation area; species rarity and population size; the degree of 'naturalness'; the size and habitat of the site; amenity and aesthetic considerations; the 'representativeness' or 'typicalness' of the site; and its uniqueness and fragility (Ratcliffe 1977; Usher 1986). These criteria were developed mainly by botanists and only some can be applied to other fields (Disney 1986; Fuller and Langslow 1986).

Diversity of habitat and species

It is not difficult to classify sites into broad ecological divisions such as heath, grassland and woodland, but further division into discrete and definable communities is often a matter of personal choice. The classification of woodland types is approached in different ways; numerous subdivisions can be made, usually based on dominant tree and plant species whose presence and abundance indicate factors such as soil, water availability, temperature, aspect, land use changes and past management. Habitat diversity is especially important for species using more than one type.

If the species diversity of an area is high, numbers of an individual species may be low. For some species successful breeding depends on substantial total numbers of individuals and it is then more important to select a uniform habitat sustaining high numbers of that species. An example of a species benefiting from a particular woodland habitat in Britain is the dormouse. Coppicing once constituted a habitat of perpetual thicket in various stages that provided food, cover and a good supply of nuts from which to accumulate winter fat for hibernation. Accounts written in the early 1900s report this animal occurring in large numbers and although climatic changes may have exerted some effect, the reduction of coppice habitat has contributed to a significant change in the status of the dormouse since coppicing ceased to be important as a form of woodland management.

Estimation of species diversity is often difficult and various methods have been developed to provide a consistent basis from which to work. Mapping all species within an area is the usual method employed and transects and random quadrats (small square areas) are traditional in botanical work. Transects and spot checks at various times of the year are used for recording birds, and trapping is a further method for estimating species diversity and the population size of insects and mammals. Mathematical indices, such as the Shannon-Weaver Index, have been developed to measure diversity. Further research on this topic is important, as it is one of the main criteria on which the selection of sites must first be made.

Species rarity and population size

Rarity appeals to most people and conservation is most often directed towards preserving rare species but the definition of rarity is not simple and raises a number of questions. For example, when protection of a rare species is successful, at what stage does the species become common enough to allow the removal of restrictions? What is the basis for deciding that the numbers of a common species have reached a critical level for survival? When does a common species reach pest status and require control? What attitude should be taken towards harmless but exotic species present in small numbers: do they merit protection? If a rare, exotic species causes damage, what is an appropriate conservation strategy? An example of this last problem is the edible dormouse *(Glis glis)*, which we discuss in chapter 11.

Rarity may be related to a number of factors, such as a small population size over a wide area, a large population in a restricted area, or local isolated populations of species that are common in other parts of their range. Rare species can be relict (left behind) species, or new immigrants arriving naturally or by deliberate or accidental introduction.

Population size is involved in the concept of rarity: the question of what is a viable population for a particular species is the subject of much debate and often attracts unfounded speculation. A considerable amount of research needs to be addressed to these questions to enable conservation strategies to be based on scientific criteria.

The degree of naturalness

'Naturalness' is usually high on the list of priorities for conservation when areas for protection are being designated. In countries where 'wilderness' still exists, the delineation of appropriate areas is not too difficult. In Britain and the rest of Europe, settled and modified continuously since the ice age, estimating the degree of naturalness presents major problems; judgements are often subjective and based merely on appearance. For example, many beech woods and native Scots pine sites

Plate 3.2 Unmanaged broadleaved woodland on an ancient woodland site in the Chilterns, southern England. (Photo: E. H. M. Harris)

prove on investigation to have been planted and managed in the past, although they now exhibit a natural appearance. The concept of 'ancient woodland' seeks to attribute naturalness to age and continuity of woodland cover. Heather moorland has been modified and managed but it is accepted as a semi-natural ecosystem, whilst chalk quarries provide sites of importance for a number of rare species. Thus the degree of naturalness of a site should not be the overriding criterion.

There is a widespread belief that naturalness is necessary for the preservation of species. However, in many cases where sites have been left untouched, the dynamic nature of plant and animal communities has produced a succession, which in turn has caused changes, resulting in the loss of the species and diversity for which the sites were originally designated. This is now recognized and management plans are becoming an integral part of conservation administration. Management is particularly important on sites whose subclimax state depends on intervention.

The flaw in the extreme interpretation of the concept of naturalness is that it takes no account of adaptability and the natural genetic changes that occur in species and which allow them to exploit new habitats. An excessive adherence to the naturalness principle rules out the reintroductions of lost species, the introduction of exotics, and the re-creation of habitats. Any change initiated by 'artificial' activities is considered to affect naturalness adversely. This extreme approach restricts the conservation opportunities that arise in managed woodland and focuses too much attention on natural habitats rather than on species preservation, which becomes a lesser objective. Many species, however, have benefited from new habitats, such as gardens, abandoned industrial workings and flooded clay pits; the last can provide attractive wildfowl refuges.

Area and size

The size of an area to be conserved needs to be related to habitat type and also to the requirements of the 'target' species (the main species to be preserved on that site). Predatory animals and birds high in the food chain require larger areas to sustain them than do prey species, and the size of an area needed by the prey will depend upon their body size and the food-producing quality of the range that they occupy. Small areas with contrasting habitats on their boundaries may contain a higher species diversity than a large uniform area, though in general as the size of an area increases more species are able to coexist within it because of variation in habitat types. Studies on appropriate area size for conservation are necessarily long term, in view of the fluctuating population levels of both animals and plants and the dynamic nature of vegetation changes.

How much of a particular habitat should be reserved and its arrangement into large or small areas depend on the species to be encouraged or already present. A single area cannot sustain all species that need protection, so it is essential to state the objectives to be sustained. Positive management benefiting one species may be the opposite of that required to ensure the survival of another equally important species. It is often stated that the conservation value of an area will depend upon its degree of isolation, because isolated sites mean that immigration is limited and the possibility of extinction greater. This concept and some of the other theories put forward for the design and size of reserves have arisen from attempts to apply the equilibrium theory of island biogeography to them; the validity of such applications has been questioned by Simberloff (1986).

Amenity and aesthetic considerations

Appearance and amenity value are bound up with the feeling for naturalness and reflect personal views of a landscape at a particular time. Change, especially rapid change, is disliked but it is taking place continuously. Amenity and aesthetic considerations therefore mean we should look beyond preservation of the status quo and plan ahead for the future, bearing in mind that features that are not acceptable now may become so.

Representativeness and typicalness

Conservation sites are usually selected to represent particular ecosystems and their subdivisions. Within a specified area, such as an English county, one site may be more 'typical' than another and thus of higher value. Conversely, similar sites within a neighbouring county may be less 'typical' and there the representative selection may differ. The assessment of a typical site cannot therefore be rigidly determined and the influence of other criteria affects selection; very often in practice the assessment of typicalness again comes down to individual judgement.

Uniqueness and fragility

Uniqueness is a quality that is part of all sites, as any one site is unique in terms of its location, physical factors, history and species communities. Usually, sites that contain a number of important features are those selected. Fragile sites are those whose importance is threatened by disturbance that would upset the value of the habitat; for example, draining a wet meadow will result in floral changes.

Summary

The above criteria should all be taken into account when sites are selected for conservation. Certain questions remain however. Do sites need protection where the species involved are at the edge of their range but are common in neighbouring regions? How justified is it to spend protection resources on those marginal species whose continued presence is dependent on uncontrollable factors, such as climate? How far should subdivisions within ecosystems be preserved; for example, how relevant is it to describe and designate a number of woodland types on a similar area when these may reflect past management rather than inherent community differences?

Habitat protection is certainly needed and this is the primary role of nature conservation organizations, without whose efforts many important sites would have been and will be lost. However, a large number of individual species readily exploit new habitats, and they are by no means all common species. If this adaptability helps to build up their numbers, these colonizations and the new communities that result should not be disregarded as a conservation resource. House martins make use of the artificial cliffs provided by houses, and bats make use of the artificial caves offered by domestic lofts. The range of species conserved in parks and gardens has not been assessed but must be considerable, and old peat diggings, quarries and artificial bodies of water all provide valuable habitats for the conservation of a wide range of plant and animal species. This is also true of productive planted woodlands, both coniferous and broadleaved, and their contribution in providing suitable habitats for both common and rare species should not be dismissed. This kind of conservation practice outside semi-natural reserves is sometimes referred to as 'extensive conservation', and is the theme of this book. In later chapters we compare habitats and associated species in natural forest with the habitats provided, or that can be provided, in managed forests.

CHAPTER 4

The Principles of Forest Management

FOREST SUCCESSION

Forest tree species can be broadly divided into three major groups. The first group consists of trees that colonize open ground, often on inhospitable and poor soils. The trees in this group are light-demanders and are referred to as *pioneer* species (table 4.1); pine and larch are examples amongst the conifers and birch and aspen amongst the broadleaves. Pioneer species produce abundant, light seed which is dispersed by wind. This enables colonization of bare ground to occur rapidly or, similarly, recolonization of cleared ground. Forests formed of pioneer species are often single-species forests: birch forest in Britain is an example. They are gradually replaced by trees of the second group, unless natural succession is arrested by some external cause.

In contrast, the second group require the protection of other trees or vegetation for their seedlings to survive. These trees are semi-shade-bearers, appearing later in natural plant successions, and are called *secondary* species. Examples are spruce, oak and maple. They are able to tolerate a certain amount of shade, especially when their seedlings are small, but need light for their development early in life. Finally, the third group comprises species able to survive and regenerate beneath a closed tree canopy. Amongst the conifers silver fir, western hemlock and western red cedar are examples, whilst beech and hornbeam are examples of broadleaves. Trees in this group are referred to as climax species (table. 4 1).

Once bare ground has been colonized, succession takes place (Finegan 1984) until dense forest, referred to as the *climax forest*, is formed, but the process takes several hundreds of years to complete, for example, taking between 400 to 600 years in the North American taiga (Wright and Heinselman 1973). Thus the usual course of events is the covering of open ground by light-demanding pioneer species that are gradually replaced by more-shade-bearing secondary species. Finally, climax forests of shade-bearing species develop and persist until the cycle is started again by some natural disturbance opening the tree canopy, such as windblow or fire.

Table 4.1   Characteristics of pioneer and secondary tree species

Pioneer trees (light-demanders)	Climax species (shade-bearers)
Seed light, windborne	Seed heavy (but dispersal can be rapid)
Seedlings need full light to develop	Seedlings develop in shade
Germination is inhibited by red light	Germination is not inhibited by red light
Trees are relatively short-lived	Trees are long-lived
Trees grow quickly	Trees grow more slowly
Seeds are produced early in life	Seeds are produced later in life
Trees seed frequently (usually annually in temperate latitudes)	Good seed years erratic
Photosynthesis requires high light intensity	Photosynthesis takes place in low light intensity
Low timber volume producers	High timber volume producers
Colonize disturbed ground	Regenerate under an existing canopy, either pioneers or the same species

However, succession can be arrested in its development by climatic or other factors, and subclimax vegetation is then maintained, usually as a result of human activities. In Britain, the burning of heather, to benefit sheep or grouse, perpetuates heath and moorland where there would otherwise be trees. Present-day afforestation can be seen as continuing the natural process of succession to some degree.

In a forest consisting of climax, shade-bearing species, the composition can only be altered by some change. The Chiltern beechwoods are an example of this as they are so dark that the wider range of trees that would naturally occur cannot regenerate. Minimum disturbance and change may be produced by the breaking off of small or large branches, the result of wind, disease or old age. On a slightly larger scale, whole trees fall. If the gap thus created in the canopy is large enough, seeds will blow into the bare ground and grow, probably mixed in with seedlings which were already present but required more light to develop. What follows depends on the competition for light and for nutrients. In the larger gaps, light-demanders may persist, but if the gaps created in the canopy are small, shade-bearers will dominate any light-demanders that have started to grow. Forests thus regenerating within themselves show 'cyclic' rather than successional replacement (Whitmore 1985).

Quicker and more drastic changes take place when larger gaps are created, and can involve groups of trees or, in some cases, very large areas. These major disturbances take place periodically in a natural forest, and are part of the natural process of continual change (Pickell and White 1985) that creates widely varying habitats for a wide variety of life forms. Such changes in forest structure are caused by a number of factors. Fire is important for maintaining a diverse structure in taiga (Washburn 1973): one fire in Siberia in 1915 destroyed 1.8 million km^2 of forest (Seitz 1986). Flooding regularly causes changes in tropical forest in South America; volcanic eruptions, like that at Mount St. Helens in 1983, lay large areas bare. Earthquakes (Garwood et al.1979), landslides, erosion, avalanches and ice storms are other factors. Hurricanes and cyclones occur in the tropics (Whitmore 1985), and gales in temperate climates, such as those occurring in Britain in 1953, 1968, 1976, 1987 and 1990, frequently restart the successional process. Overgrazing and physical damage by animals is common: elephants in one area of Africa destroyed their own habitat through overpopulation. Insect pests, such as the spruce bud worm in Canada, devastate large areas of mature forest, whilst drought, such as the summer drought in Britain in 1976, kills many trees.

As the successional process restarts, on either a small or large scale, light-demanders colonize the open ground. A volcanic eruption 400 years ago on Chang Bai Shan, on the border of China and Korea, destroyed the forest on the Korean side of the mountain and that side is still covered only with pure larch, a light-demander that recolonized the affected slopes (Poore, pers. comm.), whilst the undamaged Chinese side has mixed climax forest, undisturbed for a thousand years.

Much taiga forest never reaches a true climax condition, consisting mainly of pioneer larch, pine and birch, with spruce as a secondary species. This is because the harsh climatic conditions prevent the development of more sensitive species. There are implications in this for the range of species that can be grown in upland Britain (see plate 1.1).

Human beings, even in the period before settled agriculture started, have accelerated and altered natural processes, removing forest cover completely and often preventing regrowth by agricultural practices. Large areas formerly under forest now consist of agricultural land, whilst other areas are seriously degraded as a result of severe erosion problems. The reafforestation of degraded land can only be undertaken by the use of hardy light-demanding pioneer tree species. Vast areas exist in all parts of the world that have been cultivated or grazed for thousands of years and which can only be recovered by repeating, through planting, natural succession.

FOREST MANAGEMENT SYSTEMS

People have exploited forests for millennia and are still doing so. Their motivation was certainly to obtain timber and other wood products but at least equally important was the need to clear land for agriculture, a process that increased in pace as standards of living rose. By the Middle Ages, in what are now called the developed countries, it was appreciated that clearance must be controlled and forests managed for the natural resources they provided, rather than merely exploited. This meant ensuring regular regeneration of new crops to replace those that were harvested. Management took many forms and still does so today. In its simplest form, forest management merely ensured that the natural processes of regeneration were allowed

48

Plate 4.1 Windblown pine plantation. Wind is a normal hazard diversifying both plantations and natural woodlands. Chilterns, southern England. (Photo: E. H. M. Harris)

to occur; but natural processes are slow and many hundreds of years may elapse between the regeneration of one natural stand and the next on the same site. People wanted to harvest the timber yield more quickly and thus to hasten the natural process, something that can be done without loss of total yield.

Indeed a much enhanced yield can be obtained without exhausting the soil, particularly from coniferous and mixed forests as their soils in the northern hemisphere are more resilient than the fragile soils of tropical forests. Trees in the latter depend for their nutrients on the surface litter layer which cannot be recovered if it dries out following felling.

However, even artificially hastened woodland processes are slow and any over exploitation takes time to heal. A woodland cut too heavily will take many years to recover its full potential: once the 'bank balance' has been overdrawn, it takes a very long time to build up the capital again. Avoiding this danger of over exploitation by using the forest's natural ability to regenerate itself is the cornerstone of forest management. Forest management, practised at first in Europe, has reached a high level of development on this basis and many sophisticated methods have been used. They are all based on the fundamental principle of *sustained yield,* a principle which also provides an excellent foundation for wildlife conservation.

Plate 4.2 Western hemlock (*Tsuga heterophylla*), 44 years old being regenerated by the group selection system on greensand in Wiltshire, southern England. Several of the north western American conifers are now regenerating freely in Britain. (Photo: E. H. M. Harris)

Sustained yield, as the fundamental concept of forest management, aims to produce the maximum possible yield of wood products from a given site indefinitely. The maximum yield is the most that can be harvested whilst leaving enough of the growing stock to regenerate the forest. This principle can be applied equally to a fairly small wood or a very large forest; in the first only individual trees are taken out every few years, whilst in the second large areas are cut and regenerated. The important thing in each case is that enough is regenerated to replace the timber removed.

The growth of individual trees, small woods and large forests, natural or man-made, all follow similar growth patterns; the amount each grows in any one year depends upon the capital (in the form of living material) available at the end of the previous year. A young tree or young wood therefore does not grow a lot in one year, but as it gets older it produces more because that year's growth is added to the capital available to produce the growth for the next year (see figure 4.1). So in the early life of a tree (and this applies to young woods and forests too) each year produces more growth in the current year than in the previous year. There comes a time when the tree matures and the rate of growth slows down; this varies from about 40 years with birch to about 90 years with oak, for instance. Woodlands with a variety of age classes (trees of different ages) do not slow down in growth if there are

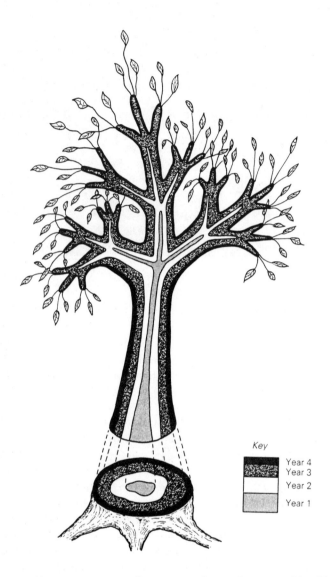

Figure 4.1 How a tree grows Every year a sleeve of new wood is added under the bark and over the wood of the previous year; this occurs in both the stem and the branches. At the same time, new woody growth extends the length of twigs. There is a ring of wood for each year of the tree's age at the bottom of the stem, but higher up there are fewer rings. In a branch there is a ring for each year since that branch first developed.

Key

Year 4
Year 3
Year 2
Year 1

sufficient young trees to replace the older dying trees. A mature woodland will produce surplus growth beyond that needed to replace the dying trees and this can be harvested as usable timber.

Woodland growth, or the growth of individual trees, is described as *increment.* The growth in a single year is called *annual increment,* usually measured in cubic metres per hectare. To manage a forest on the basis of sustained yield necessitates an estimate of increment, and then a decision on how much of this needs to be added to the *growing stock* (the capital of the forest) for future growth and how much can be allocated to a harvestable yield of timber. In a mature and fully stocked forest almost all the increment can be harvested without over exploiting the growing stock. In a very young forest none should be harvested as all the new growth is needed to build up the capital for future growth. As forests grow, some of the increment can be harvested as an intermediate yield by taking out the poorer trees as thinnings and thereby allowing the rest of the increment to be added to the better trees as capital for future growth. A useful rule of thumb is to allow 50 per cent for capital growth and 50 per cent as an intermediate yield. This is conservative in British conditions for conifers, as they grow well and fairly fast. The Forestry Commission management tables, which provide guidance on the amount that can be cut at various stages in the growth of a forest crop, are based on the removal of 70 per cent of the increment as thinnings.

The principle of sustained yield (see figure 4.2) based on the removal of a proportion of the increment has a wide variety of applications and can be used in many different ways. It can be applied to every type of forest or woodland and can meet a range of objectives beyond mere timber production. Sustained yield is the only proper basis for managing woodlands in perpetuity: if they are left unmanaged and nothing is removed, they will at first become very dense while the growing stock builds up; they will then reach the maximum standing volume that a particular site can maintain. If undisturbed, the older trees will die and fall down to be replaced by young seedlings. However, as we have already pointed out, this does not always happen under natural conditions, particularly in the windy British climate where whole woods or large parts of them are blown down before they are mature, providing a wider variety of habitats than mature woodland alone.

Each wood or forest should have its own method of management to achieve sustained yield, but there are many classic methods that can be drawn upon for ideas. They fall broadly into four groups: selection forest, clear cutting systems, shelterwood systems and coppice systems. In the first two, regeneration can be by natural seedlings, by planting or by a combination of these two methods so as to establish a new crop, or as it is called, the next *rotation.*

Selection forest

Selection forest is sometimes thought of as being closest to natural conditions, as it aims to maintain mature forest in perpetuity throughout the whole forest area, with

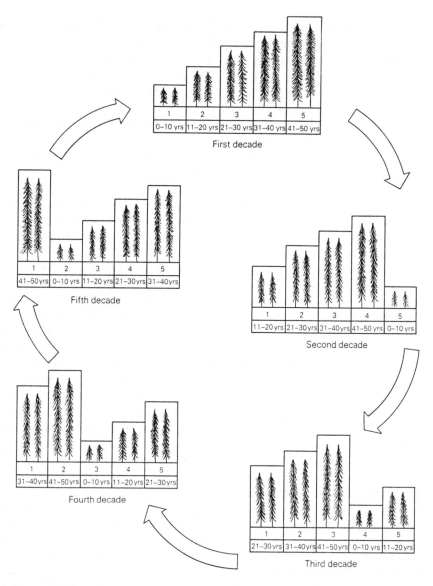

Figure 4.2 Sustained yield Mature forests are managed to provide a sustained yield by harvesting and regenerating a proportion of the oldest areas in relation to the rotation age. The figure shows five 10-year age-classes on a 50-year rotation. Block 1 is the youngest age-class in the first decade; it eventually becomes the oldest age-class in the fifth decade, during which it is felled. Each block for felling need not be a contiguous unit, but may be represented by smaller areas of the same age scattered throughout the forest. This provides a great diversity of habitats. The principles can be applied equally well to natural regeneration, regeneration by planting, and a combination of the two.

young trees coming up continuously below the canopy of mature trees. The *allowable cut* is difficult to calculate as the increment is difficult to estimate but the harvest is taken mainly from amongst the mature trees, with some removal of smaller but saleable trees too. The system is only practical for shade-bearing species as any young seedlings have to develop under the partial shade of the mature trees. In Britain this has limitations, as there is only one shade-bearing broadleaved timber tree, the beech, used under British conditions. Selection forest must therefore depend on the use of at least some shade-bearing conifers, such as western red cedar or western hemlock, in Britain, where it also suffers limitations imposed by the windiness of the climate causing periodic breaks in the canopy.

A modification of the true selection system is the *group selection system,* typically used for beech in the Chilterns. Here small holes are opened in the canopy by felling a group of mature trees, thereby allowing just enough light to the forest floor to encourage the development of natural seedlings but not enough to stimulate a lot of competing weed growth. Beech produces heavy crops of seed only periodically, in the so-called 'mast years' when a long summer in the previous year has built up food reserves. Each time a mast year occurs the groups are expanded. This is a long, slow process but probably much closer to the way forests regenerate naturally. Group regeneration is also very suitable for artificial regeneration by planting as the young trees benefit from the shelter of surrounding woodland, and if there are landscape reasons for avoiding clear felling the woodland character is not lost. In any group system the size of the groups is critical to success. Beech is a strong shade-bearer and quite small groups of 0.1 ha or so are sufficient for the seedlings to start and dark enough to discourage a heavy growth of competing weeds. For light-demanding trees such as oak and especially ash and sycamore, groups of at least 0.25 ha are needed. Sophisticated management systems involve planting light-demanders in the centre of groups and shade-bearers around the margins partly under the canopy of the old trees. This is sometimes seen occurring in naturally regenerated forests too; for example in mixed forest of spruce, silver fir and broadleaves on the European continent, gaps caused by windblow or disease can be seen with the shade-bearing conifers around the margins whilst sycamore and the other light-demanding species seed into the centre. In a forest composed of oriental spruce *(P. orientalis),* European silver fir (*A. alba*) and various broadleaves, near Bakuriani in Georgia, the spruce bark beetle *Dendroctonus micans* kills the spruce in groups at about 120 years of age. Successional natural regeneration occurs in these roughly circular groups, by light-demanding broadleaves (maple, ash, oak, dogwood, hawthorn and spindle) in the centre, surrounded by hornbeam and beech, with spruce and silver fir occurring around the margins under the partial shade of the forest canopy (Harris and Harris 1988c).

Much more commonly in the natural forest the young trees come up in pure groups: ash for example, where there is a lot of light, oak where there is less and spruce in the darker areas where the gaps are small. This is particularly evident in the Bialowieza Forest in Poland where ash, lime, Norway maple and Norway spruce regenerate as small, pure species groups, according to their light and soil moisture requirements (Harris and Harris 1988a). This gives a useful guide to artificial regeneration by the group system because single-species groups, or two species at the most, are usually the most successful. A mixture of species can then be achieved by planting several species and many groups throughout the regeneration area, but with each group composed of one species.

Clear cutting

Clear cutting systems are more common in Britain than selection systems. There are many management advantages in clear cutting and it need not be on a large scale. As we shall see later, clear cutting is seldom incompatible with wildlife conservation, and indeed is often advantageous, providing habitats for 'open ground' species. Whilst with selection systems there are trees of all ages, or a range of ages, in every part of the forest, clear cutting systems apply to even-aged woodlands; that is, trees of one age in a particular part of the forest and of other ages in other parts. Such a forest when mature (and if managed for sustained yield), will have even-aged stands of trees representing all ages up to the age when final felling occurs: the *rotation age*. The yield will come both from the mature stands when they are felled and from younger stands when they are thinned.

Regeneration can be by natural seeding or by planting. Natural seeding is very suitable for pine because it is a light-demanding species which seeds profusely and grows best in light soils where there is little competing weed growth. A typical method is to clear fell strips of mature woodland at right angles to the prevailing wind so that seed from the narrower strips of mature woodland left standing blows into the cleared areas. When these are established the remaining strips of *mother trees* are felled; by then there are sufficient seedlings for regeneration in these strips as they have been left narrow enough to let the light in under the mother trees. However, planting is the usual way of regenerating clear cutting systems, and has the advantage that the next crop can be established quickly and with certainty. It also means that the most desirable species, or mixture of species, can be obtained and the most suitable seed provenance used. If timber production is the prime objective, a fast-growing high wood-quality provenance can be planted; if the aim is to perpetuate native woodland, seed of local origin can be used.

In the clear cutting system, the area regenerated corresponds to the allowable cut; that is, the amount of increment that can be harvested. This will be equivalent to the annual increment if the forest is fully stocked (that is, *normal;* see figure 4.3) but less if it is understocked and the growing stock needs to be built up. If it is overstocked the amount will be more than the annual increment as some of the capital can be 'cashed in'. Incremental growth is normally measured in cubic metres per hectare per annum; complex management systems (particularly in France and Germany) regulate the allowable cut in cubic metres, but this requires expensive and

55

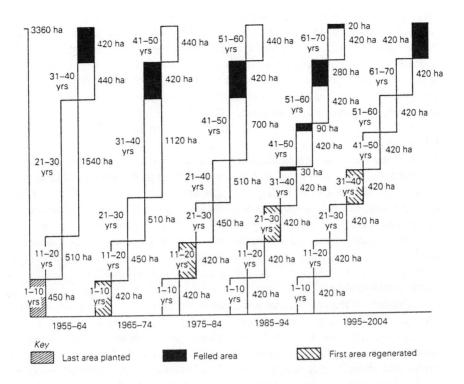

Key

| Last area planted | Felled area | First area regenerated |

Figure 4.3 Creating a 'normal' forest The histogram represents a 3360 ha Scots pine forest on a uniform sandy site. Planting started in 1924 and was followed by large planting programmes. After a wartime break between 1939 and 1945, planting was completed in 1954. The 'rotation of maximum volume production' for this low-yielding species on this poor site is 80 years.

The large pre-war plantings have left an uneven age-class distribution and a very uniform forest structure. The figure shows how a 'normal forest' can be created by the end of the first 80-year rotation, providing an even age-class distribution (each 10-year class occupying one eighth of the forest) and a range of habitats.

Premature felling is required, starting in 1955 with 420 ha from the 31 to 40 year age-class even though it is only halfway to rotation age. A further 30 ha of the same age-class, this time from the final planting, is felled in 1994 as part of a series of adjustments made towards the end of the rotation. Premature felling will substantially reduce the yield from the first rotation as any crop harvested before 80 years will not have reached the age for maximum volume production. The sacrifice is necessary to achieve sustained yield from year 80 onwards, and to maintain a permanent diversity of habitats and thus excellent conservation conditions.

The black areas in the figure are felled and replaced in the decade commencing with the year indicated, and then become the youngest age-class of 1 to 10 years in the following decade. The 10-year age-classes need not be in 420 ha blocks, but can be distributed in any number of smaller areas to a total of 420 ha. It is probably convenient to regenerate 42 ha every year in several separate coupes, but 84 ha every second year would be equally suitable.

frequent measurement of individual trees. Regulation by area is usually sufficient and this is straightforward under a clear cutting system. To take a simple example, if a 100 ha forest is being managed on a 100-year rotation, 1 ha is felled and regenerated each year. If it is more convenient, and better prices can be obtained for larger parcels of timber, the regeneration can be periodic; for example, 5 ha every five years. In practice, modifications will always be required. In all probability growth will vary throughout the forest, with the large faster-growing trees occurring on the lower ground. In such a case a sustained yield can be obtained by felling smaller areas on the lower ground and larger areas higher up, which can have conservation benefits in widening the variety of habitats.

From a conservation point of view it is important to appreciate that there is no need to fell the whole year's regeneration area (called a *coupe*) in one place. Whatever the area to be regenerated is, it can be divided up into desirable sizes, shapes and locations throughout the forest. The clear cutting system is not very different from the group system when coupes are small or are divided into small areas.

Shelterwood systems

On the European mainland beech and other broadleaved trees are usually managed under a *shelterwood* system; only applicable to natural regeneration and seldom adopted in Britain. As a crop of trees matures it is thinned heavily in order to develop large seed-bearing crowns and to allow enough light to reach the forest floor for seedlings to develop there. When sufficient young trees are established to form the next crop the mother trees are removed, but it is sometimes necessary to supplement the seedlings with planted stock. Some advantages of the shelterwood system are the maintenance of woodland cover and, towards the end of the rotation, the presence of two layers, the upper canopy and the young crop below, so that for a short time there is some diversity in the structure of the woodland. There is no bare-ground phase, however, the young crop being well established before felling. As Hart (1995) points out, 'Although the (shelterwood) system results in even-aged stands, it is a much more environmentally acceptable method than clear cutting during the regeneration stage.'

In Britain, where objections are raised to the complete clearance of small woods and attention to the appearance and landscaping and of woodlands is important, a modification of the shelterwood system is often useful. Thus, in derelict and unmanaged broadleaved woods that are to be replanted, there are usually a few well shaped and comparatively large trees scattered through the woodland. Some of these can be left standing as only a few big trees remaining minimize the appearance of devastation and leave the area still looking like a wood. The crowns of these trees do of course cast shade on the young trees planted beneath and in due course will need to be felled for the benefit of the new crop but by then, the young trees will be giving the site the appearance of woodland again.

Coppicing

Coppice systems are only applicable to broadleaved trees as most conifers do not sprout easily from cut stumps. Such systems were extensively used in Britain and mainland Europe right up to the early decades of the twentieth century but in Britain are now mainly confined to the growing of small-sized sweet chestnut in Kent and Sussex. However, research is being conducted into the use of coppice systems to provide renewable energy. Coppicing consists of cutting semi-mature trees at ground level and allowing the stumps to grow again. Many broadleaved species do so vigorously because an extensive root system already exists in the ground. The roots are not killed by cutting off the stem; the nutrients and water that would have gone into the stem are available instead for the rapid growth of new shoots, which arise from dormant buds under the bark just above ground level. (Most conifers do not coppice freely because they have fewer dormant buds.) The cut stump from which the new sprouts arise is called a *stool,* and it produces new shoots for several coppice rotations. Eventually the root system loses its vigour and needs to be replaced. Traditionally this was done by bending one coppice shoot down to ground level and getting it to 'layer' into the soil. When it had taken root it was used as a new stool in the next rotation.

Coppice systems were well suited to the requirements of a wide range of important industries in the past. They were easy and reliable to regenerate, and, most importantly, small-sized material was produced which could be cut with the hand tools available and carried out of the wood by hand or by horse. Fuel wood, charcoal, hedging stakes, thatching spars, gate material, hurdles (including wattle for wattle and daub walls), tool handles and tan-bark (from oak) are a few of the most common examples. A wide variety of species were used, particularly oak, ash (for tool handles), sycamore (for domestic utensils) and hazel (for hurdles). Rotations varied: about 8 years for hazel, from 12 to 15 years for sweet chestnut and ash, and up to 40 years with oak for tan-bark.

Much of the scrub oak in western Britain (in Devon and Cornwall, in Wales, in the Lake District and west Scotland) is the remnant of important tan-bark and charcoal industries, but is on poor sites unsuitable for the growth of oak timber. On the better lowland sites oak 'standards' were sometimes grown widely spaced as individual trees between the coppice stools. This system was called *coppice with standards* and produced large joinery and structural timbers at the same time as the small-sized coppice material, which was usually hazel. The standards were cut on a rotation equal to several coppice rotations (up to 150 years), and a few taken at each coppice cutting so that they were of varying ages and sizes.

Oak grown in this way had large branches and fairly short boles (stems), formerly required by both shipbuilding and house construction which used naturally shaped or 'grown' timbers cut from the large limbs, as well as straight planks from the bole. Coppice systems, both simple coppice and coppice with standards, have little application today because they were designed to produce small-sized material and 'grown timbers' for which there is now only limited use. Large and long boles for planks and veneer butts (the large sections of timber from which veneers are manufactured) are required now, and at present large quantities of suitable timber are imported from tropical and other natural forests overseas. The growth of more of these valuable trees at home, both broadleaves and conifers, would create a surplus of small material both as thinnings and as branchwood when the tree is felled at maturity, and hence ample supplies of small wood. The demand for fuel wood, popular to a limited extent again as firewood and charcoal for barbecues, can be more than satisfied by this surplus low-grade material and branchwood that is inevitably produced when high-quality sawmilling and veneer timber is grown.

Coppice systems therefore have no significance for future timber production and are unlikely to be revived on any large scale in the future. Their conservation value has often been acclaimed because the regular and frequent cutting allows direct light to reach a cleared forest floor for a few seasons every 10 or 15 years. The increase in light brings forth a few seasons of abundant herbaceous growth on what was formerly a shaded forest floor with few plants under the dense coppice, so that a variety of previously dormant small plants appear and flower.

This is followed by an increase in insect, small animal and bird life for a few years until the new growth in the next coppice rotation closes in again. Coppice also has the benefit of continuously providing favourable conditions on nearby sites, as when one year's 'cant' (cut area) grows up another nearby is cut. Animals, birds and insects can move easily to the most suitable areas, whilst most flowering plants lie dormant as seeds or vegetative organs until the next cutting, although some with very light seed blow into the newly cleared ground from elsewhere and germinate. However, all these species are abundant elsewhere because they are 'tough and adaptable species' and none are confined to coppice woodland (Hambler and Speight 1995).

Coppice management is practised in many woodland nature reserves to obtain these specially favourable conditions and often to provide some by products, such as firewood, as well. However, coppice as a habitat is limited in the range of bird and insect species it will support because it is in effect perpetually in the thicket stage (see below), and therefore does not offer the range of habitats found in managed *normal* forest with its full range of age classes. In particular, coppice lacks lichens, for which ancient woodlands in Britain's moist, west coast Atlantic climate are of international importance, and saproxylic (dead wood eating) insects dependent upon dead wood (Hambler and Speight 1995). Nor can it support a very wide range or abundance of species because both are related to the nutrient status of a site. Nurtrient loss occurs under all forms of coppice management due to the frequent removals of woody material, particularly under short rotation intensive coppice systems using the highly productive clones of willow and poplar that are now being used for biomass production. By contrast, under high forest systems nutrients are continually being built up, unless 'whole tree harvesting' is practised (Range and Nys 1996).

THE STAGES OF FOREST DEVELOPMENT

As a planted forest matures it goes through various stages (figure 4.4). The first after the trees are planted (or regenerated) is the *establishment* stage. Establishment can be on ground that was not under a tree crop before, when it is called *afforestation,* or on ground where a tree crop has been felled, when it is called *restocking.* When the branches of the young trees begin to grow together and the new crop starts to dominate the site, the woodland is described as being in the *thicket stage.* Up until then the young trees have probably needed *weeding* so that they are not smothered by the ground vegetation. Once they are in the thicket stage they can usually look after themselves, but if unwanted young trees, particularly birch, invade the crop it has to be *cleaned* to keep the crop trees growing well. As the trees grow up they start to compete with each other and the lower branches get shaded out and die. The crop is then in the *pole stage* and the lower branches may need to be removed; this is called *brashing.* Soon the crop will need its first *thinning,* which is the removal of some trees in order to favour the better trees (figure 4.5). The latter are already growing faster than the rest and thinning merely encourages them to develop by removing their competitors: a process that would occur naturally but much more slowly. The thinnings give an intermediate yield of harvestable poles.

Thinning takes place every few years to encourage the growth of the best trees and to select the species required in the final crop (if it is a mixture) to grow to maturity. As the woodland matures and conditions become more open again, the trees are encouraged, by further thinning, to develop large crowns. First thinnings are often *low thinnings* which space the trees out evenly over the ground; later thinnings should be *crown thinnings* (see figure 4.6). Here, the final crop trees are selected and nearby trees whose crowns are competing for light are removed. The

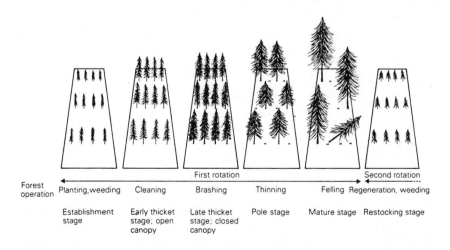

Figure 4.4 Forest stages

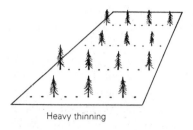
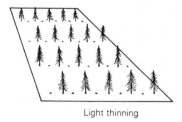

Heavy thinning   Light thinning

Figure 4.5 Heavy and light thinning   Thinning is carried out to allow the better trees to grow large without restriction from their neighbours. Light thinning removes a few of the smaller trees; heavy thinning removes many more and larger trees. The trees left after a heavy thinning grow fatter than those left after a light thinning because the total amount of wood added in the seasons following the thinning is the same for a given unit area. Thinning has no significant effect on the height growth of trees.

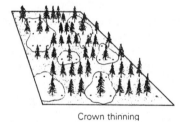

Low thinning   Crown thinning

Figure 4.6 Low thinning and crown thinning   Thinning can be carried out in many ways, but most are a variation of low and crown thinning methods. Low thinning removes the smaller trees and allow the dominant trees to grow on. Crown thinning is not concerned with smaller trees that are no longer competing with the dominants, and instead removes some of the larger trees competing with the best dominants, thus isolating the final crop trees. Light can then reach the whole crown of the final crop trees and some also reaches the forest floor.

From the conservation point of view there are advantages in a crown thinning. It allows different light intensities to reach the forest floor, and smaller trees can be left to die to produce a resource for fungi, insects and birds.

timber quality of the final crop trees can be improved by the removal of branches; this is called *high pruning*. Thinning allows more light to reach the forest floor and the dark conditions that occurred during the late thicket stage and early pole stage are replaced by semi-shade. Low-thinnings produce a uniform shade but crown thinnings, as well as favouring the better trees, produce patches of different light intensities on the forest floor. After thinning, the canopy gradually closes again and the light reaching the ground is correspondingly reduced until the next thinning. The pole stage is followed by the *mature* stage when the tree crop is ready for

Plate 4.3 A young stand of Scots pine (*Pinus sylvestris*), recently 'crown thinned' to allow light to reach the best trees. The trees that will form the final crop are being high pruned, adding value to the timber. Both operations allow light in, stimulating the growth of ground vegetation as well as the trees. (Photo: E. H. M. Harris)

harvesting. Felling takes place in *coupes* and felled areas are known as *clear cuts*. These different stages of a forest or woodland, with varying light conditions, provide various kinds of habitats suitable for a wide range of animals and plants. If different parts of the forest are at different stages of growth, a mosaic of different wildlife habitats is created. Managed forest with a range of age classes is therefore an ideal medium for conservation, and clear cutting often provides the most diverse habitats. The size of felled coupes is varied depending upon the species of tree being regenerated and the type of wildlife it is hoped to encourage, as different sizes of open areas attract different species. We must emphasize that, contrary to popular concept, the amount felled at any one time in a 'normal' forest is only a fraction of the whole and this is seldom in a single coupe.

FERTILIZERS AND THE CONTROL OF WEEDS AND PESTS
Few chemicals are used in forestry and none is employed on a regular basis. This contrasts strongly with their extensive and frequent use in agriculture. Chemical use in forestry falls into three categories: as fertilizers, herbicide weed killers and insect pesticides.

Fertilizers
Chemicals in the form of fertilizer are usually applied on poor upland sites; but they are not needed on old woodland sites or fertile lowland and seldom during the replanting of upland forest. Even on upland sites there is usually sufficient nitrogen for tree growth, although it is often not in a form available to the shallow root systems of recently planted trees. An application of phosphate at the time of planting mobilizes the nitrogen already present that is needed for growth and makes it available to the young trees. If their growth is slow, and this is typical of spruce planted on heather sites, a second application of phosphate a few years later will usually mobilize sufficient soil nitrogen to stimulate satisfactory growth. Phosphate is usually applied in the form of superphosphate from the air over the whole site at the rate of 150 kg/ha. This stimulates the growth not only of the trees but of the vegetation as well, so the latter becomes more luxuriant but not to the extent that it suppresses the young trees. It is only on the poorest sites that any further application of fertilizer is needed, as once the young trees start to grow their root activity mobilizes the whole nutrient cycle.

Fertilizers are usually only needed when secondary species such as spruce are being established in the first rotation on impoverished sites on which the nutrient status is low; for instance, as a result of previous heavy sheep grazing. Pioneer species such as pine and larch do not need fertilizer, and indeed they can be used in mixture with spruce to ameliorate the site for the latter instead of applying fertilizer. On the poorest soils in the north of Scotland, 'self-thinning' mixtures are planted that consist of lodgepole pine *(Pinus contorta)* and Sitka spruce *(Picea sitchensis)*. The pine roots aerate the soil and mobilize the existing soil nutrients to the benefit of the spruce, which would not grow initially without the presence of the pine. Later the spruce will grow faster than the pine, suppress it and eventually kill it. This is a useful technique in remote areas where there is no sale for small first thinnings. The method reproduces the natural forest of northern latitudes where pine and spruce grow together in this way, and is an example of one of the alternative options available through an imaginative approach to forest management.

Once a new tree crop has been established (that is, it has come to dominate the ground vegetation) and right through to the end of the rotation, there is no need for the addition of fertilizer as sufficient nutrients are mobilized in the soil for tree growth. Indeed, quite soon the falling needles and leaves add more nutrients to the soil than are extracted by the growing trees. When thinning takes place, only the stems of the trees are normally removed whilst the bark, twigs and leaves are left to rot down. In ascending order they consist of higher proportions of living material than the stem (some of which is dead heart wood) and thereby add significantly to the build-up of nutrients. At the time of felling, the same applies if only the stem (trunk) is removed, as is the usual practice. However, there has been some development of 'whole-tree' harvesting recently, in which the entire tree is removed and most of it utilized. On low-grade sites this may reduce the nutrient status, but in most British conditions the litter and humus that have accumulated during the fairly long rotations that the climate makes necessary are quite sufficient to sustain the subsequent crop.

In Georgia in the USA, pine is grown on short rotations of 30 years on derelict (degraded) cotton land and whole-tree harvesting is practised. Foresters have expressed concern about the potential loss of nutrients but so far, in two and three rotations, there has been no noticeable reduction in tree growth. However, recent work at the Hubbard Brook Laboratory in New Hampshire indicates that stem harvesting does not deplete soil nutrients, but that whole-tree harvesting is critical on certain nutrient poor sites (Harris and Harris 1990).

Herbicides

Herbicides are the most common chemicals met with in British forestry, and the objective is to kill or control competing plant growth. Chemical formulations that do not affect animal and human life are mainly used and new less-toxic compounds are being developed all the time. They are applied to herbaceous and shrub growth or regrowth that is competing for light and soil moisture with a newly planted tree crop. If the weeds get above the young trees and cast heavy shade over them, tree growth will be suppressed and some young trees may die, particularly if they are light-demanding species such as larch. Even more important is root competition from the established weeds, which are more efficient at extracting moisture from the soil than are the undeveloped root systems of the recently planted trees. If this competition is removed by hand or mechanical cutting, the roots of the weeds are stimulated to grow more actively and throw up more aerial parts, so competition for moisture is increased although the competition for light will have been temporarily eliminated; this is the reason for the use of chemical weed killers in forestry.

Paraquat and glyphosate are the main chemicals used. Paraquat is employed to kill young seedlings and clear the ground. It is a contact chemical and only kills those parts of the plant that it actually touches. Glyphosate is used on herbaceous weed growth and the green foliage of woody growth competing with planted stock and its action is systemic; that is, it is absorbed into the leaf and stem tissues of the weed growth. Both these chemicals act upon the chlorophyll in the green parts of the plants treated. Both are inactivated on contact with the soil so natural seedlings will continue to germinate on the ground where the chemicals have been used. Indeed, paraquat can be used immediately before planting as it has no residual effect, though it does seem to encourage the growth of moss. The use of these chemicals can diversify the ground flora as they allow the germination of dormant seeds previously prevented from germinating by the existing weed growth.

In farm woodlands, where either grassland is being converted to woodland or old woods are being replanted, two particularly useful chemicals are 'Kerb' granules (propyzamide) and 'Timbrel' (2,4,5-T). The first is used to control grass competition after planting in the winter or before planting. It is stable at low temperatures and acts slowly without affecting the tree roots out of the growing season; then as temperatures rise it is dispersed safely. The second is applied to cut stumps when broadleaved woodland is replanted to prevent the vigorous coppice regrowth that would otherwise compete with the young trees for some years and be expensive to clear.

It is interesting that Ravenscroft (1989) found that the use of glyphosate against brambles and another chemical asulam, for bracken control in East Anglia, assisted colonization by breeding nightjars. This was because the chemical treatment kept the ground sufficiently clear of bramble and bracken to maintain areas suitably bare for nesting for five years after replanting. As a result, nightjar numbers are now higher within the plantations than on the surrounding heathland.

Weedkillers are only used at planting time and in the first few years after planting, and then never more than once a year and often only in one year or alternate years. They may be used to clear patches before planting or for a year or two after planting. There is no need to spray the whole area, at most a square metre around each young tree; thus no more than half, and usually much less, of the site receives chemical. Even so, diverse ground vegetation is enabled to flourish when blanket spraying against dense bramble has been practised before planting.

Pesticides

Insect pest control is a complex subject and only a brief outline can be given here. Pesticides are not widely used in British forestry. Britain's variable climate and comparatively small forest areas combine to prevent the build-up of insect epidemics, although these are a major threat to large forest areas in countries where the climate is more uniform. In addition, the extensive use of exotic species in Britain provides a long initial period of freedom from attack before either the natural pests of the introduced tree arrives or local insect pests transfer to the introduced species. Sitka spruce enjoys a remarkable freedom from insect attack and disease and demonstrates this well. However, the spruce bark beetle *(Dendroctonus micans)* from the European mainland arrived in Britain via imported timber and has damaged spruce on the Welsh borders. There has recently been an example of a transfer from a native host to an exotic species: pine beauty moth *(Panolis flammea)*, endemic on native Scots pine and a pest elsewhere in Europe, transferred to introduced lodgepole pine in 1976 causing extensive damage, but this occurred where lodgepole had been planted on very poor sites and was therefore already under stress. At first, biological control was attempted using the bacterium *Bacillus thuringiensis,* but this was not successful. Spraying with fenitrothion, an organophosphate, was necessary to avoid extensive losses. This chemical has been used in a limited way since, but research is under way to develop less toxic methods of control. Fenitrothion is very toxic to insects and potentially toxic to some other animals but does not persist. Work in Canada has shown that half is broken down to harmless substances in 2 days and 95 per cent in 10 days. Only a small amount is taken up by the surface layer of conifer foliage and remains there for at least a year or until the leaf dies naturally. Fenitrothion cannot be detected in the soil at all 2 months after application, and in the acid ponds typical of pine-growing regions it cannot be detected after a month.

When severe insect outbreaks do occur, chemical pesticides are the only method of rapid control to prevent an epidemic getting out of hand. Pine looper, the caterpillar of the bordered white moth *(Bupalis pinaris)*, found on Scots *(Pinus sylvestris)* and Corsican *(Pinus nigra* var. *maritima)* pine on sandy sites in Eastern Britain has periodically been controlled in this way. Continuous monitoring is carried out on susceptible sites and chemicals are only used when populations are threatening to reach epidemic levels. However, chemicals only provide temporary control. Long-term control of the pine looper and other insect pests is more effective and more environmentally acceptable if a natural predator can be introduced.

Biological control of this kind has been successfully used on larch sawfly *(Pristiphora erichsonii)* in the Lake District and is the accepted means of controlling (but not eliminating) the spruce bark beetle mentioned above. The predator is the parasitic wasp *Rhizophagus grandis,* which has been successfully introduced into the bark beetle population in Georgia (former USSR) and elsewhere in Europe including, more recently, Britain. Biological control can also be effected by introducing diseases specific to a particular pest; to a limited extent viruses have been used in this way. Pheromones (natural chemical attractants) have also been used to lure bark beetles into traps. Another example of biological control widely used in British forestry is the treatment of conifer stumps after felling to prevent the spread of the conifer heart rot 'fomes' *(Heterobasidion annosum)*. Urea sprayed on the cut surface of the stump shortly after felling encourages the development of fungi spores other than fomes, and these inhibit the development of fomes spores when they arrive. In pine areas the fungus *Peniophora gigantea* is introduced directly into the cut stump for the same purpose. This treatment has controlled one of the most devastating diseases of conifer crops.

In the long term biological control is the only satisfactory method of insect pest and disease control as it is permanent and unlikely to have side effects. However, as it cannot be applied to a new pest species until careful testing has been carried out there will always be a place for chemical control as a short-term measure in emergencies.

In comparison with agriculture, forest crops are grown on a much longer time scale so control of insect pests and diseases is not as frequent or as concentrated. Often control may only be necessary during a small part of the crop's life. However, we can expect an increase of insect pest outbreaks in the future although they are comparatively rare at present. Pest outbreaks are not confined to planted forests and occur regularly in natural forest where they provide an important pathway for succession, so from a conservation point of view they can be beneficial by providing diversity. In British conditions, conservation benefits follow storm damage: structure is diversified and unharvestable timber can be left to be colonized by insects and, eventually, by hole-nesting birds.

CONSERVATION AND PRODUCTIVE FORESTRY

In this book we discuss and suggest ways of combining wildlife conservation with productive forestry, as we believe that only in this way can conservation be extensively practised throughout the countryside. Forest management systems suited to present-day timber requirements must therefore be considered for their conservation value, and perhaps modified so that conservation and productive forestry are married together. This can be done much more easily than is often appreciated.

Selection forestry in its purest sense, although still practised widely on the European mainland, does not meet the criteria of either a high level of conservation or of wood production, although it is operated in a few places with this aim in mind. True selection forest is limited to a few shade-bearing species and is permanently dark, thereby providing only a limited range of wildlife habitats. Harvesting is difficult and expensive because individual trees have to be selected, felled and removed from dense forest, and the yield comes in a range of sizes making marketing difficult. Proper calculation of the allowable cut is not easy but is important, because it is just as wasteful to undercut as to overcut. Selection systems do not necessarily contain large, old trees, any more than any other type of forest, as the trees are harvested when they are mature. The selection forest also contains little open space and light, except at its edges or near roads and rides, as it has an unbroken canopy. Indeed, in Hart's (1995) comprehensive account of alternative silvicultural systems, a recurring theme in examples of the selection systems he reports upon is the dominance of the ground flora by one species, usually great wood-rush, Yorkshire fog or bramble. This inevitably limits the range of wildlife species compared with the abundance that occurs following clear felling. However, the diverse structure provided by trees of varying sizes is attractive to some species. In the US national parks fire control has led to forests progressing towards climax species and it is now considered that this has reduced the range of wildlife habitats (Rowe and Scotter 1973).

Clear felling has recently been introduced into many US State forests in order to encourage the regeneration of light-demanding species and to make areas more attractive to wildlife, particularly deer. Deer are predated by wolves and bears, and it is interesting that wolves have deserted an area of wilderness set aside for their preservation and have followed their prey into managed forest (Harris and Harris 1990). A similar use of managed forest by bison occurs in Poland. The large and ancient Bialowieza forest on the Polish-Russian border is maintained as wilderness, and bison have been re-established there; but they prefer clear felled coupes for grazing and the mature stands for shelter and protection in the adjacent managed forest (Harris and Harris 1987).

Group selection is much more compatible with both conservation and production as a fairly wide range of habitats is achieved. Thus direct sunlight reaches part

Plate 4.4 The European bison (*Bison bonasus*), lives wild in Bialowieza Forest, Poland but uses the managed forest surrounding the reserve rather than the unmanaged forest. Here a family group has come out of the seclusion of the mature forest to feed on the abundant vegetation in a clear cut, regeneration coupe. (Photo: E. H. M. Harris)

of the forest floor, helping plants and animals as well as the young trees, and shelter is provided. Of particular importance is the flexibility of the system: group size, shape and distribution can be varied to satisfy a whole range of needs. In existing small woodlands some form of group selection is usually the most appropriate way of meeting the joint objectives of conservation and wood production.

In large, generally conifer, forest areas, particularly if they have originated from planting and are therefore of fairly uniform age, as are many upland forests at present, a clear cutting system is generally the most appropriate. This cannot be put into operation until the time has come to regenerate the forest, but it is not necessary to wait until the end of the first rotation. *Premature felling*, at say 60 years on an 80-year rotation, on part of the area is possible, although some potential yield is thereby forfeited. Similarly, some parts of the forest can be stood over beyond rotation age in order to even out the age-class distribution in the second rotation, but again there will be a reduction in the yield harvested (see figure 4.3). The object should be to achieve a more balanced distribution of age-classes in the second rotation than in the first, aiming ultimately (perhaps over several rotations) at an equal area of every age-class represented. Such a system is enormously versatile and provides both a wide range of wildlife habitats and a steady out-turn of wood products easily harvested but in a different location each year. Thinnings can be married in with felling, so that costly harvesting equipment is fully utilized as work is concentrated in one or a few areas. Each year, new open habitats are formed, suited to light-demanding plants and open-country birds, whilst other areas of the forest remain untouched for several years and provide contrasting habitats suitable for different species. Many woodland animals need both the dense forest areas for shelter and seclusion as well as the open areas for feeding, and move regularly from one to the other. Deer in particular have built up to very large numbers in Britain in the forests that have been established this century and, now that these forests are being regenerated by the clear cutting system, conditions for them are ideal, so much so that red and roe deer are a pest in many places. Clear cut regeneration areas provide conservation opportunities similar to those under coppice systems, and the shorter the rotation the nearer they come to the regular cycle of renewal typical of coppice. Such short rotations may be a necessity in areas of high windthrow risk on high-elevation sites. Clear cutting does not necessarily mean regeneration by planting. In some large Forestry Commission forests planted between the two World Wars, Sitka spruce is regenerating so freely in the felling coupes that no planting is needed; examples are Clocaenog Forest in North Wales and Fernworthy Forest on Dartmoor.

It is reasonable to assume, though no direct comparisons have yet been made, that a mature upland forest in the second rotation with annual cutting and regeneration, and thus a full range of age-classes, will have both a greater variety of species and greater species numbers than the first-rotation forest because there is a wider range of habitats. The full extent of these developments are not yet apparent in Britain but the early Forestry Commission forests, which are now some way into their second rotation, are already becoming of increasing wildlife interest and conservation value. That this can be combined with the most productive way of managing a large forest provides an enormous opportunity for conservation; the potential needs to be more widely appreciated and to be fully exploited in forest management plans. (For detailed accounts of plantation management and silvicultural systems, see Savill and Evans 1986; and Matthews 1989.)

CHAPTER 5

Managed Broadleaved Woodlands and their Conservation Value

As broadleaved trees provide a variety of important habitats, their conservation value is considerable. However, they are expensive to grow and in British conditions can only compete on commercial terms with conifers on the best soils and on warm, sheltered sites. This is because Britain is at the northern extremity of natural broadleaved woodland and most species prefer a more continental climate. In order to maximize wildlife conservation in today's productive broadleaved woodlands, it is necessary to consider British broadleaved woodland in its world context and to examine both its origins and how it has been managed in the past. (We are not concerned here with unmanaged semi-natural woodlands.)

The postglacial tree cover of Britain is mainly broadleaved, with the exception of the Scottish Highlands and some high land elsewhere, but as we explain more fully in chapter 6, the small number of species present is the result of the comparatively short period between the retreat of the ice and the breaking of the land bridge with continental Europe. This short period of only 4000 years did not allow many tree species to reach Britain before land connections with the mainland were severed. Only two of these were conifers, yew *(Taxus baccata)* and Scots pine, the latter colonizing early in the dry, cold period that followed the ice. Colonization by broadleaves reached its peak during the Atlantic period (Godwin 1975; Rackham 1981; Peterken 1981). Much of lowland Britain then consisted mainly of mixed broadleaved forest, typical of the northern broadleaved forest that stretches across the northern hemisphere today. This was, to quote Godwin (1975) 'a polyclimax rather than a monoclimax' forest; that is, it consisted of several equally dominant species. The transitional mixed forest zone of conifers and broadleaves, present elsewhere in the northern hemisphere, was absent because several European conifers did not reach Britain. Typical mixed and northern broadleaved or hardwood forest contains a wide variety of tree and shrub species, the genera being common to all continents. For example, Scots pine occurs right across Eurasia, Korean pine *(Pinus koreana)* occurs in China, whilst white pine *(P. strobus)* is typical of this forest type in the eastern USA and the western white pine *(P. monticola)* in the western USA. Indeed, Korean pine and white pine are closely related and cross-pollination produces fertile seed (Wang 1961). Similarly, the genus *Fraxinus* (ash) occurs as *F. excelsior* (the European ash), as *F. mandshurica* in north-east China, and as *F. americana* in the northern USA.

Oak, both sessile oak *(Quercus petraea)* and pedunculate oak *(Q. robur)*, were important elements in the British forest of the Atlantic period, but were not as dominant as is often assumed. Many woodlands neglected since the Second World War now exhibit a wide range of species with a composition typical of northern broadleaved forest, whereas previously many of them were almost pure oak because this species had been selected for (Peterken 1981). There is evidence suggesting that birch and pine may have hung on during the ice age in west Scotland, an then spread out again from there to merge with a reinvasion from the European continent (Brown pers. comm.; Moore 1987). It has also been suggested that oak may have done likewise (M. E. D. Poore, pers. comm.); if so, this is likely to have been *Q. petraea,* which is more common in the west, whilst *Q. robur* may have reinvaded from mainland Europe during the Atlantic period. In any event, it would have been sessile oak that occurred on the drier, shallower, more acid soils with pine, birch and hazel. Pure stands of oak may have existed on southern slopes as they do in other parts of the world, especially after felling which enables oak to dominate. There is often a contrast between northern and southern slopes, with pines forming a more important constituent on the former (Wang 1961, Harris and Harris 1987). In the USA, felling the coniferous element in mixed forest allows the broadleaves to sprout and initially to take over the secondary forest. Conifers eventually grow up and once more become dominant unless cut for timber (Harris and Harris 1990).

As we have pointed out more fully in chapter 2, recent work on oak suggests that it first invaded Britain after the ice age from the west, spreading north through what is now the Irish Sea, and was already abundant in Cornwall 9500 years ago. Whether this was sessile oak is not known but it seems likely, whilst pedunculate oak may have arrived from the Continent of Europe about 1000 years later, before the North Sea developed. At the time of maximum warmth (the climatic optimum), 7000 to 5000 years ago, oak had reached most of Scotland and was growing to larger sizes than we can expect today in Britain, evidence of which is preserved in peat bogs. It has been suggested that global warming could produce large oak trees again but this is mere speculation.

BROADLEAVED TREE SPECIES IN BRITAIN

With the advent of the warm and wetter Atlantic period the broadleaved forest pushed northwards but the more warmth loving species, such as lime *(Tilia* spp.), were mainly confined to England and Wales. Pedunculate oak then probably colonized the deep, moist lowland soils that it so much prefers, whilst sessile oak was

Plate 5.1 Oak, (*Quercus petraea*) at the beginning of the second growing season after planting. Dorset, southern England. Tree shelters encourage rapid early growth of sensitive broadleaved trees and allow 'spot' weeding with chemicals, encouraging a variety of natural vegetation. (Photo: E. H. M. Harris)

Table 5.1 The world distribution of British trees and shrubs

Circumpolar	Eurasia	Northern Europe and western Asia	Southern Europe and western Asia
Juniper	Scots pine	Yew	Black poplar[a? b]
	Aspen	Downy birch	Grey poplar[a]
	Bay willow	Silver birch	White poplar[a]
	Goat willow	Crack willow	Smooth elm[a?]
	Bird cherry	White willow	Field elm[a]
	Rowan	Alder	Cherry plum[a]
		Wych elm	Whitebeam[b c]
		Wild cherry	True service[d]
		Hazel	Wild service[b]
		Sessile oak	Field maple[b]
		Pedunculate oak	Sycamore[a?]
		Ash	Beech[a? b]
		Holly	Hornbeam[b]
		Hawthorn	Small lime[b]
		Midland hawthorn	Large lime[a? b]
		Crab apple	Box[b]
			Horse chestnut[a]
			Sweet chestnut[a]
			Pear[a]
Reticulate willow	Osier		
	Dwarf birch	Blackthorn	
	Sea buckthorn[b]	Elder	
	Alder buckthorn[b]	Spindle	
	Guelder rose	Wayfaring tree[b]	
		Buckthorn[b]	
		Dogwood[b]	

[a] Introductions and possible introductions
[b] Occurs naturally in England and Wales only; the natural distribution is now masked by planting.
[c] Whitebeam (*S. aria*) is found in England and Wales, and in Galway in Ireland; a number of subspecies and crosses have been described
[d] Now extinct as a wild tree
The trees that have been in Britain for a long time all come from northern Europe and Eurasia. The more recent arrivals and those confined naturally to England and Wales are all from southern Europe.

pushed up the hillsides to between 150 and 300 metres in Scotland and no doubt in other places as well (Anderson 1967). The remains of oak in bogs from this period illustrate optimum growing conditions then for broadleaves, including oak which belongs to 'a vast genus whose main territory is warm temperate' (Jones 1974). As well as the warm conditions and relatively unleached soils, examination of ice cores has shown that there was a rising level of carbon dioxide (CO_2) in the atmosphere

(Delmas et al. 1980) and CO_2 concentration enhances tree growth (Kauppi 1987; La Marche et al. 1984). The large trees found in bogs may therefore be a reflection of a number of factors, such as temperature, rich soils and ample CO_2. Since then, cooler conditions and leached soils have restricted broadleaved tree growth in Britain, and the production of good hardwood timber is now limited mainly to the south of the country. The 'primeval' forest (Rackham 1981, Peterken 1981) there-fore consisted of a wide range of broadleaved species, their proportions varying with climatic fluctuations as shown by the analysis of pollen deposits and other evidence. The proportion of oak has almost certainly been increased by human activities from early times, compared with that in the 'primeval' forest. Human influence on the landscape is now known to have had an impact much earlier than was first thought (Zvelebil 1982 and 1994; Darvill 1986). At first it would have been more difficult to clear oak than the smaller, softer birch, alder, cherry, ash, lime and pine. The presence of so many shrubs typical of the northern temperate forest in Neolithic charcoals also reflects the ease of cutting (Turrill 1948). Oak was there-fore left at first, but soon became an important tree species as it has a wide range of uses, making it particularly valuable. As many of the early uses depended upon coppicing rather than felling, it would have been perpetuated easily. Fuel wood and charcoal would have been more manageable as small coppice poles than as large trees, and coppicing seems to have been one of the earliest forms of management (Godwin 1975; Rackham 1977). Up until the early decades of the twentieth century, oak bark was the main source of tannin for curing leather, and coppice growing on long rotations of about 40 years provided greater quantities than indi-vidual trees. The bark of large trees felled for other purposes was also used.

With improved felling facilities, such as iron tools, large-sized oak timber was important from the iron age onwards for the construction of houses and barns and for shipbuilding, and was usually grown as oak standards over coppice. The coppice was usually hazel (important for fencing, wattle and daub walls, and thatching spars) and the widespread use of this forest management system favoured oak over other broadleaved species. Oak was also 'open grown' in wood pasture to produce the branching timbers needed for house building and shipbuilding. Animals were grazed under the trees, which were often pollarded to provide fodder (Peterken 1981). The wide spacing of the trees encouraged fruiting so that a good crop of acorns was produced, as pannage for pigs was very important in Britain and throughout Europe from early times (Ellenberg 1988). A few examples of wood pasture remain (the New Forest is one example), and together with open grown trees in parks now form an important habitat for epiphytes and for insects requiring old wood (Harding and Rose 1986). The coppicing of upland oak for mineral smelting and bark for tanning has now ceased but some derelict western woods surviving from this important process hold important communities of Atlantic epiphytes (Ratcliffe 1977). The value of oak as a wildlife resource has been documented in many publications, including *The British Oak* (Morris and Perring 1974) and is important but has too often been overstated in popular literature; in fact, 'conifers can support several times the abundance of invertebrates on oak' (Hambler and Speight 1995).

Other major broadleaved species found as dominants with oak in the northern broadleaved forest are elm *(Ulmus)* and lime *(Tilia)*. Both were important species for animal fodder and the decline of elm and lime in the late Atlantic (6000 BP) is now usually attributed to human influence. In the case of elm, Dutch elm disease has been implicated (Moore 1984b), but a rise in cereal pollen predated the decline (Maloney 1984). However it has been pointed out (Lines, pers. comm.) that elm was coppiced at least until 1965 in Yugoslavia for animal fodder and that this, due to the short life span of a coppice crop, may have prevented it from flowering and thus leaving a pollen record. The more fertile southern soils occupied by lime make good agricultural land, and the clearance of lime woods for this purpose still continues in some parts of Europe (Harris and Harris 1988a). Britain's native elm is wych elm *(Ulmus glabra)*, and Richens (1983) considers that it invaded from Germany over the area that forms the present North Sea, because the insects associated with it are not found on it in France. The recent drastic decline resulting from Dutch elm disease has largely eliminated all species of elm from many parts of Britain and until the disease dies out or resistant strains of elm appear it cannot be considered in commercial or conservation terms. Few species of animals or plants are in fact associated exclusively with elm, although it was a major food source for the large tortoiseshell *(Nymphalis polychloros)* and the white letter hairstreak *(Strymonidsia w-album)* butterflies and a few other insects.

Small leaved lime *(Tilia cordata)* and, to a far lesser extent, large leaved lime *(Tilia platyphyllos)* formed a major component of the Atlantic forest in southern Britain and in some places comprised from 60 to 70 per cent of the canopy (Pigott 1988b). Lime rather than oak is known to have been the dominant tree species in Denmark and probably in many parts of southern Britain on fertile valley soils (Godwin 1975; Pigott 1988b). It was a dominant tree in Epping forest into Saxon times but then appears to have been cleared, favouring the growth of oak and hornbeam (Baker et al. 1978). The soft timber of lime was easily worked and the bast fibres (from the bark) were important for making rope. The flowers provided honey, and wooden beehives were already being used in Neolithic times; lime flowers are still an important honey source in parts of mainland Europe (Harris and Harris 1988a). Large leaved lime prefers drier soils and a much warmer climate than now exists in Britain, where it is therefore restricted to warm limestones and is a relict species (Pigott 1988a), but in such situations has a rich ground flora growing beneath it. Small leaved lime also prefers a more continental climate than now exists in Britain, so it is limited today to the south and east of the country although it extends much further north, as far as 63° on the European mainland (Pigott 1988a). The soft timber of lime means that it has very limited commercial use today, but it is the traditional wood for carving and is connected in particular with the work of the English carver Grinling Gibbons (1648-1720). Some insects are associated solely with lime (Winter 1983) and as it supports a good aphid population it is a food source for insect-eating birds: the sweet 'honey dew' secreted by the aphids attracts other insects to feed on this sticky substance. The few lime woodlands that occur in the eastern counties of England hold good bird populations. As lime grows on base-rich (that is, non-acid) soils, an abundant flora is found under mature stands. There is some evidence of soil improvement, such as occurs with birch (see chapter 6); for example, Piggott (1988b) found that on plateau clay with flint soil in Hampshire the pH under beech declined to 3.8, whereas under lime it increased to pH 5.6. This may be related to the high content of bases in the leaf litter, as both lime and maple have been shown to raise the pH value in planted spruce woods (Ellenberg 1988; Piggot 1988a).

Ash *(Fraxinus excelsior)* is an important and valuable timber tree when it grows on deep, fertile loams; although it regenerates freely on lighter soils, it is of less commercial value there as it does not grow fast enough to produce high-quality white timber. It is more warmth-loving than wych elm but hardier than lime, beech *(Fagus sylvatica)* and hornbeam *(Carpinus betulus)*. Ash was a favourite timber in the past for tool handles and cart wheels. High-quality ash still finds a good market for sports goods and internal joinery. For the former it must be grown fast. Like lime, however, its range has contracted in Britain since Atlantic times, although ash woods remain on the warmer limestones in the north and are of high conservation value because of the basic soil conditions and the light foliage sustaining a rich ground flora. The production of fast-grown timber is mainly limited to fertile soils in the south of Britain although it regenerates freely in cleared woodland on a much wider range of sites. It is tempting to utilize such seedlings but unless these are on moist, deep loam soil, quality ash timber will not be produced. The insect fauna associated with ash is not large because of its light foliage, although some species are specific to it.

Birch *(Betula* spp.) is an important timber tree in many parts of Europe and research is being carried out to improve the quality of birch grown in upland Britain; field trials are now underway and if successful it may become more economic to grow this species. Benefits in upland forests would include soil improvement and species diversity.

The uses of broadleaved timber today, particularly of oak, are much more limited than in the past. There is little call for small-dimension material other than as pulpwood and chipboard, for which it is less suitable than coniferous timber, although there are still some traditional local markets; birch for brush backs (mainly in the south east), sweet chestnut (in Kent and Sussex for split fencing material), sycamore (for turnery poles), beech and cherry (in the Chilterns for the furniture trade) are examples.

However, as large-dimension timber, only high quality material is required. It is used internally for decorative purposes and for the best furniture, where not only is

Table 5.2 'Native' British broadleaved trees compared with broadleaved trees of the eastern USA and China that will grow in Britain

Genus	Number of species in: Britain	USA	China
Birch (*Betula*)	2	11	28
Willow (*Salix*)	4	15	30
Poplar (*Populus*)	1 (2?)	10 + hybrids	50 + hybrids
Rowan and Whitebeam (*Sorbus*)	4	2	50
Alder (*Alnus*)	1	8	8
Cherry, fruit trees (*Prunus*)	2	30	140
Hazel (*Corylus*)	1	3	7
Elm (*Ulmus*)	1 (2?)	6	23
Maple (*Acer*)	1	13	150
Ash (*Fraxinus*)	1	18	20
Oak (*Quercus*)	2	91	76
Lime (*Tilia*)	1 (2?)	4	12
Hornbeam (*Carpinus*)	1	1	30
Beech (*Fagus*)	1 (?)	1	5
Holly (*Ilex*)	1	14	118
Box (*Buxus*)	1		11
Crab apple (*Malus*)	1	6	20
Hawthorn (*Crataegus*)	2	149	17
Walnut (*Juglans*)		6	4
Sweet chestnut (*Castanea*)		5	4
Horse chestnut (*Aesculus*)		7	8
Plane (*Platanus*)		1	1
Mulberry (*Morus*)		2	9
Wing nut (*Pterocarya*)			7
Hickory (*Carya*)		12 + hybrids	4
Hackberry (*Celtis*)		5	20
Hop hornbeam (*Ostrya*)		3	4
Zelkova (*Zelkova*)			3
Honey locust (*Gleditsia*)		2	8
Magnolia (*Magnolia*)		8	30
Tulip tree (*Liriodendron*)		1	1
Sweet gum (*Liquidambar*)		1	2
Black gum (*Nyssa*)		3	6
Sassafras (*Sassafras*)		1	2
Kentucky coffee tree (*Gymnocladus*)		1	1
Sumac (*Rhus*)		17	24
Mescalbean, pagoda tree (*Sophora*)		2	23
Persimmon (*Diospyrus*)		2	56
Bean tree (*Catalpa*)		2	5
Wafer ash (*Ptelea*)		1	
Osage orange (*Maclura*)		1	
False acacia (*Robinia*)		4	
Euodia (*Euodia*)			25
Amur cork tree (*Phellodendron*)			2
Tree of heaven (*Ailanthus*)			2

Castor oil tree (*Kalopanax*)	1
Maackia (*Maackia*)	6
Foxglove tree (*Paulownia*)	7
Golden rain tree (*Koelreuteria*)	4
Dove tree (*Davidia*)	1

Source: Information from: USDA 1949; Clapham et al, 1952; Wang 1961; Mitchell 1974; Preston 1977; How 1984; Sylva Sinica 1985, 1986; Schauer 1986; Harris and Harris 1987.

Britain has very few species of tree because it became an island soon after the ice age; most of them migrated from warmer, southern Europe. China has the greatest number of species because it is part of a large land mass and species survived by moving southwards, migrating north again as the ice melted.

the strength important but also the fine finish obtainable with hardwoods that shows up the attractive grain and colour. These two characteristics of broadleaved timber, giving a smooth finish and attractive grain, must be exploited to the full if broadleaved trees are to be grown commercially and at the same time provide their special conservation resource. To obtain the right kind of material, straight, knot-free stems are needed in large sizes. Broadleaved timber is often also used for veneering, a process in which a thin layer of high-quality wood is glued on to less-valuable material. In this way the fine figure can be displayed and a large amount of valuable veneer can be obtained from a single first-quality stem. Veneering requires knot-free logs that are at least 90 centimetres in diameter but they need not be very long.

Rapidly grown, fat trees are therefore needed to produce suitable timber and the number of sites in Britain where such quality hardwoods can be grown is limited for most broadleaved species. However, the heavy thinning that is needed to produce good trees is of great benefit to conservation as more light reaches the forest floor, with all the benefits that follow from this. High-quality broadleaved stands that can support the necessary intensive management with useful intermediate yields are an important conservation resource. It is therefore encouraging that, as standards of living rise, there is an increasing demand for high-quality broadleaved timbers. Unfortunately, they are becoming difficult to obtain; most are imported (at a cost to Britain, at the time of writing, of over £2 billion annually) but the natural forests from which most of these imports come are themselves getting scarce. The developing countries that own them naturally wish to add value to their own economies by producing manufactured goods rather than exporting timber in the round. This means that there is an increasing place for growing broadleaved timber in Britain but only if it is of the best quality, which means that it must be grown on the best land available to forestry. It must be of good quality because only the high-quality market is available, unlike that for coniferous timber which has a wider range of uses. It must be grown on the best land because large sizes are needed and broadleaves on the whole are much slower to mature than conifers, many of them being on the edge of their range in Britain; much better growth is obtained in more continental climates. There is also increasing competition from high-quality temperate hardwood timber coming into Europe from the eastern United States. This is from natural second growth forests on abandoned farmland which are at present growing about twice as fast as they are being harvested (Harris and Harris 1990), so only the very best British hardwood timber can expect to find a ready market.

Profits are much lower from broadleaved than from coniferous timber, and indeed a profit can only be made from broadleaves when they are grown on comparatively short rotations. The faster-growing species, particularly sycamore, cherry and ash on fertile sites, will become increasingly important and already command good prices if they are straight, clean (free from knots) and fat (that is, over 60 centimetres in diameter). These three species can be grown to maturity in half the time taken for oak (60 rather than 120 years), but ash needs to be on deep, rich soil in a warm situation, whilst sycamore is much less demanding yet can command the highest price. There may be a future for southern beech (*Nothofagus* spp.) as it will grow even faster than conifers on frost-free sheltered sites and the wood is very similar to European beech. The species that grow best in Britain (*N. procera* and *N. obliqua*) both come from a wide latitude range in South America, but only provenances matching the British climate are likely to be successful in Britain. Several species of *Nothofagus* have already acquired a wide range of insect species in Britain (Welch 1987). Cherry supports few insects but the fruits serve as food for a number of birds and animals; as it prefers more basic soils, it encourages a wide range of ground flora as it matures. A neglected cherry orchard in Hertfordshire was observed to contain a similar ground flora to nearby oak woodland (personal observation). It is not known why cherry is not damaged by grey squirrels but this may be because the bark strips horizontally rather than vertically or that it has a bitter taste. In the Chilterns it is unfortunately becoming subject to attack by the edible dormouse (*Glis glis*) and is also damaged by muntjac (*Muntiacus reevesi*) if not protected.

The remaining British broadleaved timber species, hornbeam, beech, sycamore (*Acer pseudoplatanus*) and sweet chestnut *(Castanea sativa)*, cannot claim to be definitely native under the criteria usually applied of having arrived without human assistance, but if the time they have been present in Britain is taken into account, all would qualify. Both beech and hornbeam pollen only occurred after human settlement in postglacial Britain and after the English Channel was formed. Hornbeam is associated with beech, ash and sycamore on the European mainland. Although an important coppice species for fuel and charcoal in the past, it has no place now as a timber species. Hornbeam does not have nuts that would have been valuable as food, but the seed is windborne and the species is more likely to have invaded unaided than beech.

Beech was late in leaving its mountain refuges in both eastern and south-west Europe after the glaciation and did not reach northern Europe until about 4000 BP

Plate 5.2 Well managed 50 year old beech (*Fagus sylvatica*), established with conifer nurses in a line mixture from which larch (*Larix decidua*) has been removed. The sparse ground vegetation includes violet helleborine (*Epipactis sessilifolia*). Chilterns, southern England. (Photo: E. H. M. Harris)

(Zeuner 1952). It is possible, in view of its importance in early times as a food source for both people and animals, that like sweet chestnut it may have been introduced. Beech is on the north-west edge of its range in Britain and requires a continental climate for good growth. In the mountains of continental Europe it occurs at higher altitudes than oak and grows much more vigorously than in Britain; it is not confined to limestone there and grows on a range of mountain soils. According to Ellenberg and Klotzi (1972), its occurrence on a wide range of soils means that it has no characteristic plant indicator species (that is, vegetation thought to be typical of beech wood). These only become evident under marginal conditions, such as the sites we associate it with in Britain. Although beech is usually regarded as natural in southern Britain and is associated particularly with the South Downs, the Chilterns and the Cotswolds, there are records from the seventeenth and eighteenth centuries that suggest that much of the beech on the South Downs and the Chilterns may have been planted, so its presence there and the attractive landscape created may not be as natural as was once supposed. On the Chilterns, beech has replaced the traditional hazel coppice and oak standards and was planted originally as firewood for the ever-increasing demands of expanding London (Rodens 1968). The furniture trade of High Wycombe and the surrounding district developed later, as a result of the availability of a timber that turns easily for chair legs, and this encouraged foresters to favour beech, so that by the middle of the nineteenth century it was traditional to plant beech, usually in mixture with pine or larch, in the Chilterns. Beech still commands a good price if it is clean and white but the timber is easily discoloured, especially if grown on soils containing iron; it is not durable out of doors and does not season easily. It has probably been grown rather too widely in Britain as a timber species; although it will grow almost anywhere, it is only on the best sites that it reaches a large enough size in a reasonable time to be economic. Beech woodland does not sustain many insect species, and a few of these are specific. However, it is extremely important for orchids and sustains some of the rarest plant species occurring in Britain. This is because the dense canopy of beech casts a heavy shade suppressing almost all vegetation, and the orchids consequently have little to compete with; they also like the lime-rich sites that beech is usually grown on. Suitable conditions for orchids can also occur under evergreen conifers, particularly under pine on calcareous soils (Summerhayes 1951); the same orchid species are found under a much wider variety of tree species in mainland Europe (see chapter 10).

As sycamore can produce the most valuable broadleaved timber grown in Britain, but has a poor reputation for conservation, special consideration needs to be given to it here. Sycamore is not normally regarded as a native tree but there is no evidence for its introduction. It comes from the mountains of central and western Europe, where it grows with beech, spruce, ash and lime, and is a tree not to be dismissed from either a forestry or a conservation point of view in Britain, whether it is native or not. The popular view is that it was introduced by the Romans but there

Plate 5.3 A heavily thinned stand of naturally regenerated sycamore (*Acer pseudo-platanus*) with abundant and varied ground flora. Northumberland, northern England. (Photo: J. A. Harris)

is no evidence at all to support this; why the Romans should introduce a non-food species is open to question, as other trees that the Romans are known to have introduced provide edible fruits, such as walnut, mulberry and medlar. The discontinuity in the distribution of sycamore in western Europe suggests that it may not be natural in Britain: in France it does not quite reach the English Channel but it does come as far north as Paris.

Many writers have referred to sycamore being introduced from France in the Middle Ages but none quote their reasons. There is a fourteenth-century carving of what is clearly a sycamore leaf in St Frideswide's shrine in Christ Church Cathedral at Oxford so the carver was clearly familiar with the tree. Sycamore is mentioned by Chaucer in the *Canterbury Tales,* written in about 1380, and as he was once a forester himself he would have been familiar with it. The earliest known written record of sycamore being grown in England is in Turner's *Herbal* of 1551. Loudon, writing in 1838, thought the tree had been in Britain a lot longer than that. Firmer though negative evidence is the lack of identified sycamore pollen in ancient deposits in Britain, but it is not easily distinguished from that of field maple. Sycamore is a maple and its older name was the great maple, which causes some confusion in the literature. Pollen would be scarce in any case as sycamore occurs on the drier, more basic soils so is less likely to be preserved. As sycamore is pollinated by insects, pollen amounts are small compared with the large quantities produced by wind-pollinated trees; nor is the pollen grain hard like windborne pollen. The wood of sycamore too is so like that of field maple that it cannot easily be distinguished. Maple was used in many old wooden artefacts, furniture and musical instruments, but the species cannot be identified by examining such items. The native field maples growing today are too small to be significant timber trees, though larger specimens may have occurred in the past. With no positive evidence concerning the arrival of sycamore, it would be foolish to dismiss this useful species simply on the grounds of its supposed non-native status.

Native or not, sycamore grows remarkably well all over Britain, from the sheltered and fertile south, where it will thrive on heavy clay, to the exposed and rugged north on shallow non-acidic soils. It does well on dry limestone hills in Yorkshire and on salt-sprayed coasts all around the country. In the hills of Wales and Scotland it has been used for generations to shelter farmhouses and it grows higher in the hills than any other large broadleaved tree. Even in these exposed places, sycamore grows straight and fat enough to provide valuable timber in about 70 years, where other broadleaved trees would be valueless scrub. The most northerly wood in mainland Britain is a sycamore plantation on the Queen Mother's Castle of Mey Estate on the exposed coast of Caithness and there is sycamore growing well on Orkney. As well as being tolerant of salt spray, it is one of the trees least affected by atmospheric pollution and has been the most successful tree used for planting in the industrial areas of northern England.

Sycamore improves the soil because its fallen leaves contain bases and rot down more quickly than most species, so that autumn leaves are seldom evident in the following spring. Thus the nutrients made in the leaves during the summer are added to the soil the following year. The tree is one of the most consistent seeders, producing prolific seed annually, unlike oak and beech which only bear seed periodically. Sycamore seedlings are therefore common every year, providing opportunities to regenerate woods naturally with a valuable timber species. For reasons of prejudice alone, the value of sycamore to wildlife is often underestimated or dismissed. The abundant seeds provide winter food for birds and animals. The hanging greenish-yellow flowers in May yield quantities of nectar attractive to bees, and 43 other insect species have been recorded associated with the tree. Indeed, in exposed situations, sycamore gives shelter to insects where otherwise there would be none. Whilst sycamore may not be host to a wide range of species, the total biomass it supports, particularly of aphids, is probably more than any other tree. The value of sycamore trees near rivers and their value to fish life has been studied in Wales and the Welsh Marches by Mason and MacDonald (Mason and Macdonald 1982; Mason et al. 1984). They showed that, of the trees studied, sycamore contributed the highest insect biomass to the river. Aphids feeding on sycamore produce 'honey-dew', and the insects attracted to feed on it also add to food falling from the tree into the river for fish (Royal Forestry Society 1987).

Sycamore bark is rich in mosses, liverworts and lichens: nearly 250 species have been recorded (Stern 1989). Harding and Rose (1986), recorded 194 lichens, and state that open-grown sycamore is the most important tree for lichens and bryophytes. Woodcock are abundant in sycamore stands because of the high earthworm populations (Royal Forestry Society 1985). The ground flora itself is rich under mature, well-thinned sycamore and resembles that found under the same species on the European continent, where it often grows in association with ash. The richness of the flora is a reflection of the type of soil that this tree prefers (see chapter 10).

The grey squirrel is the greatest enemy of sycamore, preferring it to many other broadleaves and conifers, both of which it attacks. Young trees are girdled at ground level and the stems in the crowns of older trees are damaged in a similar manner. Grey squirrel control is essential if sycamore and other broadleaves are to be grown.

Sycamore is a broadleaved tree that provides high-quality timber more quickly than any other on a wide range of sites. It can therefore compete in commercial terms with fast-growing conifers, yielding 4 to 5 per cent in money terms which is good by forestry standards, and it is one of the few profitable broadleaved species. High prices can be obtained for large saw logs, particularly if they show a 'ripple' or 'fiddle-back' figure, when a single stem may sell for several hundreds of pounds. This 'fiddle-back' is indicated by a ripple in the wood just under the bark; it is thought to be caused by trees leaning on uneven ground, thereby setting up stresses in the wood as it forms. Whatever the cause, these valuable trees only show their worth when they are sawn or veneered and the undulating fiddle-back pattern appears. This produces the fascinating contrast of light and dark seen on the back of violins and other string instruments. The sides, neck and bridge of string instruments are also made of 'maple', which is in fact often sycamore. Indeed, sycamore logs are exported from Britain to other parts of Europe and are then re-imported sawn as 'maple' for musical instrument making. Sycamore is a very popular furniture timber because it is strong, easily worked and comes up to a smooth finish with attractive grain. Sycamore indeed is a much-neglected tree, able to provide valuable timber quickly and to make significant contributions to conservation (Harris 1986).

BROADLEAVES AND CONSERVATION

The conservation value of broadleaves is generally accepted, but only high-quality broadleaved timber has an assured future and expanding market. Any attempt to grow broadleaved timber on exposed sites and shallow soils will be uneconomic and will produce mediocre woodlands; these may not then be managed and will therefore become of limited conservation value. Woods which now have high conservation value were usually managed for their produce in the past.

Until recently, forestry could not compete with farming for the best land and so was mainly confined to the poorer land. Now the success of post-war farming is producing increasing farm surpluses, and new opportunities for increasing the area of productive broadleaved woodland are being created. British farmers can no longer work in isolation but will be increasingly affected by the farming policy of the European Union. Significant reductions in the surpluses can only come about by taking some of the more productive land out of farming. Planting broadleaves on this better land in the lowlands will mean that woods with a useful future can be established more extensively than when only marginal land was available for afforestation. The slow return on traditional species such as oak will still deter farmers from planting them, but they have instead the opportunity of using the faster-growing broadleaves. Because farmers are dependent on their land for their livelihood not all the available land will be planted with broadleaved trees and some earlier return and income will be sought.

In 1988 a start was made on the new Farm Woodlands Scheme for a three-year experimental period up to 1991, in which annual payments are made to farmers who plant previously productive farmland with trees. There is a premium for broadleaves. The annual payments are for 40 years for oak and beech, 30 years for other broadleaves comprising more than half the crop and 20 years for crops which are mixtures or pure conifers. In 1986 the Forestry Commission set up the Broadleaved Grant Scheme with much higher rates than those available for conifers so as to encourage pure broadleaved planting. This was in response to public demand for more broadleaves for landscape and conservation reasons, and foresters using the scheme must make provision for these objectives as well as timber production. *Pure* broadleaved planting can only be a success, however, if the sensitive nature of broadleaves and the good sites that they need are taken fully into account. So far, increased planting of woodlands on farms has been of only limited success and the accomplishment of the broadleaved policy still depends upon how well broadleaved plantations are yet to be looked after in their much longer thicket stage than that of conifers or mixtures. If they are not kept clear of competing woody growth and free from grey squirrel attack, only scrub will result.

High-quality broadleaved timber has not been grown in Britain extensively in the past, so there is a lack of good parent trees to select seed from and very little work has been done to improve this state of affairs. It is not known, for instance, whether the valuable 'ripple' in some sycamore trees results wholly from site conditions or whether it is in part an inherited character. As the ripple increases the value of the wood by about three times, it is an important characteristic. Little too is known about the various degrades of oak. Very often when oak timber is felled it is 'shaken'; that is, it has cracks running out from the centre of the log in the shape of a star, or cracks running part way round some annual rings. Called *star shakes* and *ring shakes* respectively, they cause the sawn timber to fall apart and the fear of this reduces the price of standing trees. This may be caused by the site, as trees on dry sandy soils seem particularly prone, but recent work in France and Germany indicates that the predisposition to shake may be genetically controlled. Shake is also a problem in sweet chestnut timber. Another problem with oak is the production of *epicormics* on the stem below the crown; these are twigs arising from previously dormant buds in the bark which devalue an otherwise clean stem and make it useless for veneer. Although they tend to occur when a closely grown crop is opened out, perhaps because there is suddenly more light and nutrients available than the existing small crown can use, the fact that they are not such a problem on oaks in mainland Europe may mean Britain has poor stock and genetical improvement is needed. There certainly is a need to improve the growth rates of all broadleaved species if they are to compete with the much more profitable conifers and only in this way can the conservation value of broadleaves be fully exploited.

The limit to the growth of high-quality broadleaves in Britain is ultimately climate and the cool summers must mean that broadleaves, especially as a pure crop, will always have a limited place over much of the country. In areas of high summer temperature in central Europe, broadleaves are able to grow at high altitudes and summer temperature is even more important than an increase in mean annual temperature (Ellenberg 1988). A significant recent development overcoming climatic limitations during the first few years of the establishment of broadleaved crops is the *tree shelter*. This is a plastic tube about 1.2 m long that is placed over the young tree at the time of planting. For some species, particularly oak, the effect is to triple or even quadruple the height growth in the early years (Tuley 1985). Oak can he out of the top of the shelters in the second season and this early boost carries on for a few years, getting the young tree quickly above competing weed growth, though it pays to kill weeds round the tree for the first few years to reduce root competition. The shelter protects the tree from cooling and drying winds so that the young tree is growing in a favourable, humid environment with warmer conditions than are present outside. The lack of air movement may also raise the level of carbon dioxide, as when this is produced by respiration during active growth it is not blown away and thus enhances the photosynthetic rate. Carbon dioxide enrichment is used to increase growth rates in commercial glass houses and perhaps this occurs in tree shelters too. There is certainly evidence (Fearson and Weiss 1987) of better growth if the bottoms of the shelters are blocked. This probably prevents the carbon dioxide produced during respiration, which is heavier than air, from flowing out, and instead it accumulates at the bottom of the shelters.

The use of tree shelters has been found to have significant effects on oak with no reduction in girth; that is, the trees do not become weak and spindly. The effect on other broadleaved species is still not fully investigated but it seems that beech derives no extra benefit apart from the physical protection provided by the shelter, whilst cherry responds very well. Before the introduction of tree shelters in the early 1980s, the main problem in growing broadleaved crops was their slow and expensive establishment. This was one of the reasons why they were usually planted in mixture with conifers (see chapter 7), which provides an intermediate yield and gives protection to the young trees. Now there is an opportunity to plant pure broadleaved crops on the best sites, although if a slow-maturing species such as oak is used, it is wise to include other more quickly maturing species with it, such as cherry, to provide some early revenue. The latter will encourage the oak to go on growing upwards when it comes out of the top of the tree shelter and will then provide an early yield, at between 40 and 60 years. The oak can then be left to grow on and will benefit from the increased space provided to expand its crown.

Broadleaved woodlands have a number of characteristics that are important for conservation and that can be made use of in commercial managed forest. They allow more light to the forest floor in spring than conifers so that plants flowering at this time benefit from full light before canopy closure. Soils warm up more quickly in the spring and microbial activity is stimulated early. The palatable leaves of broadleaves host a different set of insect species to those of conifers and a wide variety of fruits, seeds and nuts provide food for a wide range of species. The leaves of most broadleaves, except beech, rot down more quickly than many conifers, those of lime and sycamore being rich in basic minerals (Ellenberg 1988), though larch and Douglas fir are important exceptions. The conservation value of broadleaves is greatest in the southern parts of the British Isles where the fauna, especially insects, dependent upon broadleaved trees is also concentrated, and it is in these regions that commercial broadleaves are best grown. None of this means, however, that broadleaved woodlands are necessarily superior in conservation terms to mixtures, which we will discuss in chapter 7. Mixtures are more than just a compromise as they have several advantages over pure crops. In pure crops, the species diversity of forms dependent on a particular tree species is low, although the number of individuals of a particular species may be high. In mixtures, the presence of different tree species increases species diversity.

CHAPTER 6

Conifers and Conservation

In this chapter we consider native species of conifers and, in particular, exotic introductions in order to discuss their roles in both conservation and forestry. Except on the very best soils, which are not usually available to forestry, conifer crops are the only means of producing a profit in British conditions, so the opportunities that they can provide for wildlife conservation need to be evaluated objectively.

The native fauna and flora of Britain today are poor, if measured in terms of species diversity, for two reasons. Firstly, as already described in chapter 2, the comparatively recent retreat of the ice 12 000 years ago, followed only 5000 years later by the separation of Britain from the rest of Europe when the Straits of Dover were formed, resulted in many plant and animal species not reaching Britain naturally. Secondly, and perhaps of even more significance, because Britain is on the edge of the northern broadleaved zone (see chapter 1), many broadleaved species such as lime, hornbeam and field maple only occur naturally in the south of England. The number of native trees, therefore, is small, usually regarded as amounting to only 35 species (Mitchell 1974). Of these, only three are evergreen and only two of these, Scots pine and yew, are conifers, leaving holly as the only native evergreen broadleaved tree. Scots pine is a versatile tree and is able to grow in a wide range of conditions. All other conifers grown in Britain are regarded as exotics (that is, introductions).

Much stress is laid on the conservation value of native species, especially in relation to trees. How useful is this concept? In the British context, native plants are usually defined as those thought to have arrived in Britain since the last ice age, though positive evidence is lacking for some species that are accepted on this criterion. Amongst herbaceous plants, many introduced agricultural weed species are now of high conservation value, for example corn cockle *(Agrostemma githago)*, now uncommon but only 50 years ago a common cornfield weed. However, trees are only accepted as native if they arrived here without human aid. Native bird species are defined rather differently. All breeding birds, winter visitors and rare visitors are accepted on to the British list; no distinction is made between early arrivals and more recent additions. Many introduced bird species have become established in quite recent times and reintroductions, such as the capercaillie and white tailed eagle, are all gladly added to the British list. Insects are treated in a similar manner. Additions to the poor British stock of native mammals are also welcomed, with the exception of those that have become pests, such as the grey squirrel. Even this species gives pleasure to many town dwellers, who often object to its control but paradoxically object to exotic conifers.

On the basis of the criteria adopted for birds and animals, the reintroduction of trees, such as spruce and fir, that grew here in the last interglacial, should not be out of place as they would have arrived here but for the formation of the English Channel. As a result of the truncated reinvasion by trees many niches were not filled, and shade-bearers in particular are scarce amongst the trees accepted as native. Strangely, the length of time a tree species has been growing or living under natural conditions in Britain is not usually regarded as a criterion for its being native, but this might be a more useful concept than applying an arbitrary date of several thousand years ago. Why the status of 'native' is so rigidly applied for trees is hard to understand, but perhaps it is because they represent 'wilderness' in a way that plants and animals do not. There is also the feeling that any animal or plant community developing in association with non-native exotic trees is somehow inferior and unnatural. This subjective view does not take into account the adaptability and genetic diversity present in plant and animal populations, and assumes a genetic 'uniqueness' of native populations. However, introduced species will evolve in time to suit the local conditions and thereby eventually become native.

To limit conservation value to native trees only is to oversimplify. Too often unequal comparisons are made. Age is often ignored, so that the value of an old oakwood is frequently compared with young plantations of exotic conifers, thus ignoring the fact that the mature coniferous forest can sustain a wide range of habitats because of its increased structural diversity. To many animals and plants it is the conditions provided by the woodland, not the actual species of tree forming it, that are the important factor. Such species do not distinguish between native and exotic trees if their own physiological requirements are met. This is clearly put by Hambler and Speight (1995): 'Many species use conifers and many of these can also use broadleaved trees. What matters to them is that a tree is a rigid structure with a lot of surface area, providing biomass and cover. 'Exotic' conifers, such as Norway spruce, support similar communities to the 'native' species. Most of Britain's woodland wildlife can also be found on the European continent, some of it in Norway spruce, so bringing spruce to Britain reunited woodland species with a habitat they are well able to use.'

In addition to natives and exotics, there are 'stateless citizens' amongst trees that fit no classification. Three natural hybrids have occurred in Britain with, in each case, exotic parents. London plane was the first and arose from seed resulting when oriental and western plane *(Platanus orientalis* and *P. occidentalis)* were cultivated together and cross-fertilized naturally. The second was hybrid larch *(Larix eurolepis),* arising from the pollination of European larch by Japanese larch. The third is the now widely used Leyland cypress (x *Cupressocyparis leylandii),* which has arisen naturally in several forms from two American species, Nootka cypress *(Chamaecyparis nootkatensis)* and Monterey cypress *(Cupressus macrocarpa)* when these two trees were grown together in Britain. It is now more familiar to most people than any other conifer, being found in most British gardens.

Few native broadleaved trees grow *well* in all parts of Britain, partly because their main range is on the European continent which has a markedly different continental climate to the cooler, Atlantic climate that prevails in Britain. It is not surprising, therefore, that exotics, especially conifers from similar climates elsewhere in the world, thrive in Britain and fill habitats for which there are no native species. Sitka spruce and several other north-west American conifers, especially those of the right provenance, are examples of trees that come from the Pacific seaboard and grow luxuriantly in Britain.

There have been three main stimulants in the past to augmenting the paucity of Britain's tree flora. Firstly, for economic reasons: nut and fruit trees were early introductions, examples include plum and apple in pre-Roman times, followed by walnut, apricot, peach and quince; the introduction of all these is usually attributed to the Romans. Secondly, for ornamental purposes, starting in the sixteenth century in the Elizabethan era, when times of peace initiated a period of building houses with decorative gardens. This was extended in later centuries when the formal gardens established in the eighteenth century extended beyond their boundaries into landscaped parks. Ornamental species were brought first from Europe, then with the discovery of the New World, from the eastern and later the western Americas. The numerous ornamental introductions from the Far East commenced in the nineteenth century, when this region opened up to Western trade, and continues today. From amongst these ornamental introductions, the third group of exotics arose, discovered almost by accident; these were the trees like Douglas fir *(Pseudotsuga menziesii)* that grew so well in gardens and parks that they became valued for their timber at least as much as for their attractive appearance. All of them are conifers and most of them come from the western coast of North America (from British Columbia to California), where the maritime climate is akin to Britain's. They have widened the spectrum of timber trees that can be grown economically in Britain to the extent that commercial forestry is now possible to a degree far beyond anything that could be achieved with native species alone. Britain is not alone in growing trees from other countries for timber. The world-wide use of eucalypts from Australia is but one example, and the widespread use of Monterey pine *(Pinus radiata)* from the USA is another.

Britain is on the western margin of the large Eurasian continental land mass, across which stretch vegetation zones defined by latitude and modified by altitude.

The taiga, boreal or northern coniferous forest zone is the largest single vegetation zone in the world, reaching from the Urals to the Pacific and extending into northern Europe before human interference. Taiga is primarily conifer forest, mainly pines, larches and firs, with some spruce and birch. The Highlands of Scotland, north of a line from Dunoon to Stonehaven, would have been taiga had Britain remained part of continental Europe and Tansley considered it the western extension of this zone (Godwin 1975). A small part of north-east Scotland, in Caithness and Sutherland, was probably forest tundra (that is, sparse woodland) in the distant past but even there extensive remains of Scots pine occur in the peat now covering the area. Stevens and Carlisle (1959), Burnett (1964) and Anderson (1967) considered that all Scotland was wooded up to the tree line before deforestation. Indeed, Chapman and Crawford (1985) reported regeneration of birch, aspen, rowan and hazel on 'treeless' Orkney when grazing pressure was removed. Taiga would also probably have occurred naturally in the higher parts of the Lake District and Snowdonia.

It has been stated that, if broadleaved trees had not reached Britain before the English Channel formed, the tree cover over most of the country would be Scots pine, whatever the rock formation (Godwin 1975). It is equally certain that but for the formation of the Channel, most of Britain's forests would have naturally been mixed forest, gradually merging southwards into northern broadleaved forest. The composition would have resembled the remaining fragments of such forest found abroad in transition zones; for example, Bialowiezia in Poland, Chang Bai Shan in China, and the Great Lakes region in America. These semi-natural and natural forests resemble forest that occurred in Britain during the last interglacial.

The Boreal or taiga forest over most of the northern hemisphere is in itself simple in composition, with a limited number of species, because of the effects of the ice age (Larsen 1980). The return of vegetation northwards followed a similar pattern on all continents, with the most northern forest being composed of only one or two dominant species (Wang 1961) and a wider range of species occurring as growth potential increases towards the south (Larsen 1980). The introduction of exotics to Britain has now started to create the 'mix' of genera that is present elsewhere in similar zones.

One of the characteristics of the mixed forest zone is that it contains species typical of the taiga and the northern broadleaved zones. Britain is in the position of being a meeting place of Boreal and temperate plant and animal species and this is well illustrated by examples elsewhere in this book. Trees, like other forms of life, have retreated southwards since the climatic optimum in the Atlantic period, so that much British wildlife is that which would be found naturally in taiga and mixed forest. The mainly pure coniferous and mixed plantations of Britain's commercial forests provide a new habitat for such species, one which many are already rapidly exploiting. The plant communities will not be the same as those that have developed in 'semi-natural' woodland but will be composed of a wide range of species. Even

Table 6.1 The main tree species in taiga forest in the Boreal region

Genus	Eurasia	North America
Pinus	Scots pine (*sylvestris*) Siberian stone pine (*sibirica*)	Jack pine (*banksiana*) Eastern white pine (*strobus*))[b] Lodgepole pine (*contorta*)[a]
Picea	Norway spruce (western Asia) (*abies*) Siberian spruce (*obovata*)	Red spruce (*rubens*)[b] Black spruce (*mariana*) White spruce (*glauca*) Sitka spruce (*sitchensis*)[a]
Larix	European larch (western Asia) (*decidua*) Dahurian larch (eastern Asia) (*dahurica*) Siberian larch (*sibirica*) Dahurian larch (eastern Asia) (*gmelinii*)	Tamarack (*laricina*)
Abies	Siberian fir (*sibirica*)	Balsam fir (*balsamea*) White fir (*amabilis*)[a] Grand fir (*grandis*)[a]
Thuja		Eastern white cedar (*occidentalis*)[b] Western red cedar (*plicata*)[a]
Chamaecyparis		Yellow cedar (*nootkatensis*)[a]
Tsuga		Eastern hemlock (east only) (*canadensis*)[b] Western hemlock (*heterophylla*)[a]
Pseudotsuga		Douglas fir (*menziesii*)[a]
Betula	Silver birch (*pendula*) Downy birch (*verrucosa*)	Yellow birch (*lutea*)[b] White birch (*papyrifera*) Grey birch (*populifera*)[b]
Populus	Aspen (*tremula*)	Trembling aspen (*tremuloides*) Balsam poplar (*balsamifera*) Black cottonwood (*trichocarpa*))[a]
Salix	Various species, mostly shrubs, including the circumboreal *reticulata* and *herbacea*	
Sorbus	Rowan (*aucuparia*)	Rowan (*americana* and *decora*)[b]
Prunus	Bird cherry (*padus*)	Bird cherry (*pennsylvanica*)
Alnus	Alder (*glutinosa*) western; (*mandshurica*) eastern Alder (*fruticosa*)	Red alder (*rubra*)[a] Sitka alder (*sinuata*)[a] Grey alder (*rugosa*)
Acer		Oregon maple (*macrophyllum*)[a]

[a] Species that grow on the west coast of north America at low elevations and constitute a maritime extension of the boreal forest. All grow well in the similar maritime climate of Britain.
[b] Species confined to eastern north America; they do not do so well in Britain.

Table 6.2 The main tree species in mountain coniferous forest in the temperate region

Genus	Europe	Caucasus	Eastern China	Western North America
Pinus	Scots pine (*sylvestris*) Arolla pine (*cembra*)		Scots pine Korean pine (*koraiensis*)	Ponderosa pine (*ponderosa*) Limber pine (*flexilis*) Jeffrey's pine (*jeffreyi*) Sugar pine (*lambertiana*) Western white pine (*monticola*)
Picea	Norway spruce (*abies*)	Oriental spruce (*orientalis*)	Jeddo spruce (*yezoensis*) Siberian spruce (*obovata*)	Englemann spruce (*englemannii*) Brewer spruce (*breweriana*)
Larix	European larch (*decidua*)		Dahurian larch (*gmelini*)	Western larch (*occidentalis*)
Abies	European silver fir (*alba*)	Caucasian silver fir (*nordmanniana*)	East Siberian silver fir (*nephrolepis*) Manchurian silver fir (*holophylla*)	Alpine fir (*lasiocarpa*) White fir (*concolor*)
Pseudotsuga				Blue Douglas fir (*menziesii* var. *glauca*)
Tsuga				Mountain hemlock (*mertensiana*)
Calocedrus				Incense cedar (*calocedrus*)
Chamaecyparis				Lawson cypress (*lawsoniana*)
Betula	Silver and downy birch (*pendula, verrucosa*)		Erman's birch (*ermannii*)	Mountain birch (*occidentalis*) White birch (*papyrifera*)
Populus	Aspen (*tremula*)	Aspen		Aspen (*tremuloides*)
Alnus	Grey alder (*incana*)			Mountain alder (*rhombifolia*)
Acer	Sycamore and Norway maple (*pseudoplatanus, platanoides*)			Mountain maple (*glabrum*)

This table illustrates the similarity between the genera and species in taiga and mountain forest. Forests in Japan have similar genera and species.

the semi-natural communities that now exist are themselves a reflection of past management and of land use changes.

Although rotations will be short in some exposed upland forests, management systems and objectives can be much more long term in coniferous and mixed forests grown in less vulnerable areas. Forests created today could be the semi-natural forests of tomorrow. The Braco pinewoods on Deeside, once thought to be natural remnants, are now known to have been planted in the eighteenth century, as were the supposed 'natural' beechwoods of the South Downs and Chilterns. The maximum potential of commercial forests will only be fully developed by sympathetic management, but even the contribution to wildlife of single-species conifer forests should by no means be disregarded because their short rotations create continuous diversity, similar to the habitats created by fire in the Boreal forest (Washburn 1973).

The Boreal or taiga forest is not restricted to a particular soil type, the main limitation over most of its range being climate. The principal species are pines, larches, firs, birch, aspen and willows. The taiga and the high-altitude coniferous forests are subject to climatic fluctuations and the tree line extends and recedes with climatic change. Indeed it has been suggested that some of the effects attributed to acid rain may be the result of this factor (Watt 1987). The northern part of the Boreal zone consists of more open forest than that further south and some authors apply the term 'taiga' to this zone alone (Viereck 1973). Other authors refer to all the forested area of the Boreal zone as taiga (Rowe and Scotter 1973; Larsen 1980; Flint et al. 1984) and the term is used in the latter sense in this book.

Disturbance in the taiga forest is usual and maintains it in a state of succession. Fire, wind, ageing and stress followed by disease outbreaks, all initiate regeneration (Washburn 1973; Sprugel and Bormann 1981). The role of fire is now considered to be of major importance in the taiga of the USA, where rigid fire control in national parks has led to progression towards a dominance of shade-bearing forest with a reduction in stands of pioneer species of varying ages. This in turn has led to a reduction in the range of wildlife habitats (Heinselman 1973). Fire also occurs naturally in the extensive taiga forests of Russia; the last major occurrence from natural causes, thought to be produced by a meteorite, was in Siberia in 1906 (Seitz 1986). Low-intensity fires or cutting in the taiga, both by humans and animals (such as beavers, which were once part of the British fauna), leads to aspen and birch regenerating by seed and sprouting from roots. These even-.aged broadleaved stands are eventually overtaken again by conifers. Birch has been shown to have an ameliorating effect on the soil (Dimbleby 1952; Miles 1981), with possible advantages to the conifers that follow.

Wind too is a major factor and is by no means limited to planted spruce in Britain. One example is 'wave regeneration' in silver fir forests, which progresses through even-aged stands. The old trees start to die and ground vegetation increases, whilst fir seedlings germinate as the canopy gets lighter. Wind then blows the senescent stand down and there is a flush of ground vegetation. The fir seedlings also respond to the extra light and begin to dominate the ground vegetation after a few years and then progressively shade it out. Between 35 and 40 years the stand begins to thin itself, the gaps in the canopy allowing ground vegetation to reinvade (Sprugel and Bormann 1981). Parallels with planted forest are evident, and all that forest management is doing is hastening natural processes. Harvesting by clear cutting has been shown to increase species variety in natural stands (Swindel 1983). Studies of nutrient cycles are also relevant in looking at natural and planted forests. Natural spruce forests show a loss of nutrients in the early stages of seedling establishment. As the young trees grow, nutrient loss decreases and a positive carbon budget builds up. At felling, the process is reversed again. Over the rotation the gains and losses balance out (Sprugel and Bormann 1981; Miles 1986; Mitchell and Kirby 1989). Discounting the effects of acid depositions, the complete loss of nutrients from the profile is small in most situations and is replaced by natural weathering as well as deposition from the atmosphere. Work in Europe has shown no reduction in soil fertility under planted coniferous forest (Ellenberg 1988). However, at the Hubbard Brook laboratory in New Hampshire, USA, it has been shown that whole-tree harvesting may deplete soil nutrients whilst stem-only removals do not (Harris and Harris 1990).

As pure stands of one species are normal in much of the taiga, so too are insect and disease outbreaks and large areas are attacked. It is thought that fire has some role in reducing populations of insects and disease (Washburn 1973). The taiga therefore consists mainly of even-aged stands of various species of varying ages in various successional stages. The shade-bearing species, such as the firs, form forest types that regenerate within stands of the same species in areas protected from disturbance. Succession from pines to fir takes about 400 years if it is not interrupted (Washburn 1973). The northern spruce forests at the tree line in Canada follow pine which grew originally on peat that had dried out. Similarly, the drier conditions in the Sub-Boreal (see figure 2.1) dried out peat in Scotland, allowing Scots pine to reinvade. Ploughing and draining for afforestation has the same effect. In Britain, plantations of pine-spruce mixtures (usually lodgepole pine and Sitka spruce) established on peat telescope this succession, the pioneer pine assisting the more useful spruce to grow in the inhospitable conditions which are thereby ameliorated. This is a natural process which afforestation merely reflects.

Susceptibility to wind, disease, insect pests and fire, and nutrient cycling and succession, follow similar patterns in both natural and plantation forests in northern latitudes. The simple species composition that prevails in the disturbed areas of natural forest is similar to planted 'monocultures', in that it often consists of discrete even-aged stands of differing ages, a state that will prevail once 'normal' forest conditions have been reached in Britain's reafforested uplands. Properly planned clear cutting in upland forests can be less drastic than disturbance by natural fire and can provide ground vegetation diversity sustaining a wide variety of species. Forests planted on open ground take time to acquire a woodland flora and at present in Britain the number of species occurring is limited (Hill 1987). This is not a permanent condition as much of the flora and fauna are composed of species that are adapted to constantly changing conditions. Already many species found naturally in taiga forest are finding conditions suitable in British plantation forests and careful management can encourage the presence of many more. This is a challenge that faces British foresters, particularly in the older forests where the greatest opportunities now occur, and is discussed more fully in chapter 10.

The upland coniferous forests, together with those in more lowland areas, are particularly important for wood production because there is little call today for small-dimension hardwood timber (that is, that obtained from broadleaved trees). So-called industrial wood, which is broken down into its constituent parts and formed into board or paper, comes mainly from softwoods (that is, the timber from coniferous trees). This is because the wood fibres of conifers are longer and provide the strength that is needed for this reconstituted material. Sitka spruce grown in Britain now provides some of the best paper pulp in the world; at the same time, like many other conifers it is a first-class structural timber, because the long fibres of conifers provide light but strong building materials. Strength, the most important characteristic of structural building timbers, is related to wood density, which in turn is related to the age of the tree when the wood is laid down. Thus the strongest and most valuable wood is on the outside of the lower part of the stem, whilst the core and upper parts of the stem are weaker. If good timber is to be produced, therefore, reasonably fast growth is needed, particularly when the trees are young, in order to grow good-quality mature wood outside the less valuable juvenile core (Zobel et al. 1987).

Rapid early growth therefore has both important economic and conservation benefits. As any form of woodland management is a long-term investment with heavy expenditure at the beginning and a long delay before saleable material or mature woodland is produced, the shortening of this delay is vital to economic success. Rapidly growing crops require early and regular thinning and this allows light to reach the forest floor much sooner than it would under conditions of natural thinning. This is one of the most significant ways in which production and conservation objectives can come together and be entirely compatible. Natural regeneration is often dense and self-thinning of natural stands takes place slowly. The result is that the number of trees per hectare is greatly in excess of the optimum achievable in managed forest. Yields from unmanaged natural stands are therefore much lower and the trees are smaller. Regular thinning, besides producing much more valuable final crop trees, is one of the most important ways of diversifying and thereby enhancing the habitat for wildlife in forests. Pine and larch allow light to reach the forest floor, and ground vegetation can be maintained throughout the rotation with regular thinning. Spruce casts a heavy shade and ground flora is lost in the early years but recovers slowly when thinning starts; thinning therefore needs to be undertaken early. The use of the fast-growing fine-branched provenances of Sitka spruce that are now available will reduce the degree of heavy shading. Mature spruce that has been managed exhibits a ground flora similar to that of neighbouring broadleaved woods, though the density may be less (Hill 1985); this is discussed more fully in chapter 10.

The colonization of plantation forests in the uplands is still little studied, but work carried out so far already indicates that they are being exploited by a wide range of species. Many studies have mistakenly compared these young plantations with old broadleaved forest, taking no account of age or succession. Oak and beech, in the thicket stage, can be as dense as a coniferous woodland in the summer and will remain in this stage for much longer due to the slower growth of native broadleaved trees. The number of bird species inhabiting the thicket stage of both conifers and broadleaves is similar on comparable sites.

Native tree species are often acclaimed for the wide range of insect species that are dependent on them and for the wide range of birds that they consequently attract. This often-quoted generalization needs more careful analysis and should not be confused with two other quite different factors: the all-important light-demanding characteristics of different tree species, and the age of trees and forests. Tree-feeding insects are usually specific to one or a few closely related tree species. Some trees have many insect species dependent upon them and these relationships have developed over long periods of evolutionary time. It is to be expected, therefore, that trees that have been growing in Britain for a long time will have many insects dependent upon them, but that exotics that have been introduced by seed, and thus without their related insects, will not. (There are, however, some native species with very few insects associated with them; holly and whitebeam are examples.) This was first pointed out by Southwood (1961) but this concept should not be taken in isolation. A more useful way of expressing it is not only in terms of how long a tree species has been in Britain, but equally importantly, whether it is closely related to other species already occurring in the flora (Strong 1979; see chapter 14 of this book). Several species of the southern hemisphere genus *Nothofagus* (southern beeches) have only recently been introduced to Britain but they are closely related to oak, sweet chestnut and beech. Welch (1981) has shown how rapidly native British insects have spread to *Nothofagus* growing in Britain. Exotic European conifers, which have been in Britain 200 to 300 years, are now subject to most of the insects that attack them in their native home; but the more recent North American conifer introductions are so far largely free of insect attack, which from a forestry management point of view is an advantage. As time passes, the number of insects transferring to introduced trees will increase. Indeed, the transfer in 1976 in Scotland of the pine beauty moth (*Panolis flammea*) from Scots pine to American lodgepole pine (*Pinus contorta*) (Watt 1986) is an example of pest problems that may become much more widespread.

Plate 6.1 Mature, well thinned Sitka spruce (*Picea sitchensis*) plantation with a ground flora containing species associated with oak woodland; these include wood sorrel (*Oxalis acetosella*) and yellow pimpernel (*Lysimachia nemorum*), both ancient woodland indicators. (Photo: J. A. Harris)

Many British birds too are associated with woodlands composed of native tree species, especially broadleaves, whereas the same birds in other parts of their range are also associated with evergreen coniferous species (see chapter 12). Options for extending conservation outside nature reserves are much reduced if the value of introduced trees as useful habitats is disregarded.

Imaginative forest management can provide a wide range of conditions attractive to woodland birds, which make up about 70 per cent of the land birds of Britain (see chapter 12). The same is true of other wildlife species, but birds are an excellent indicator of species richness because they are easily seen and popularly studied, whilst their requirements of light, food supply and living space are no different to those of other living things. Exotic tree species add to the diversity of woodland habitats and evergreens are particularly important in this respect because Britain has so few. An interesting study in Scotland showed that birds favoured woodland with an evergreen understorey (French et al. 1986). This is understandable because evergreens provide shelter in inclement weather and are particularly important in the winter as the level of animal populations is controlled largely by the number that can survive the winter in good enough condition to breed the following year. This important aspect needs to be more widely recognized: too often subjective judgements are made about the apparent lifelessness of evergreen woodlands just because it is not easy to see animals and birds in them. We tend to be out and about in the countryside more often in good weather and certainly more frequently in daylight, times when animals and birds are most likely to be in the deciduous part of their range or outside woodlands altogether. Judgements may also be influenced by our inbuilt fear of dense forest, almost certainly dating back to the time when we were in danger of being hunted ourselves. There has been little consistent scientific field work carried out on the value of our new evergreen forests as habitats; this is much needed but there is already little doubt that they have a valuable role to play. As Welch (1986) has pointed out in relation to insects, most conservationists concentrate on ancient woods and rare species and ignore all others, the old Scots pine woods in Scotland receiving particular attention. Many species found in these woods are also widespread in other coniferous forest types in other countries and occur increasingly in maturing plantations in Britain. The Forestry Commission and the Royal Society for the Protection of Birds have recently started a major study of bird habitats in an upland forest at Mynydd Du in Wales to see how the forest is used by birds, particularly in winter. This can be expected to yield some interesting and valuable data that may prove useful for the combination of conservation and forestry objectives in productive woodlands.

The maintenance of a breeding population of birds and animals is dependent on an adequate food supply all the year round. Most native trees, with the exceptions of birch, alder, rowan and hornbeam, do not flower and produce seed that can be exploited as food in the first two or three decades of their life, and only reach maximum production of seed several decades later. Many broadleaved trees only produce abundant crops of seed periodically; this is usually related to a long warm summer the year before allowing food reserves to be built up in addition to that needed for growth in the succeeding year. The extra food supplies are allocated to flower production but another good summer is needed in the following year for seed to ripen. Two such years were 1975 and 1976, and there was prolific beech mast (seed) in the latter so that many seedlings germinated in 1977. These 'mast years' are particularly typical of oak and beech, and provide a bonanza for seed-eating birds and animals which in turn are able to build up food reserves and breed well themselves. However, birds and animals need to maintain their populations in the lean years as they, unlike trees, breed annually. Food in the form of seeds is therefore needed every year and all the exotic conifers that have been introduced to Britain are particularly useful in this respect. They seed much earlier than most broadleaved trees (pine and larch at 10 or 15 years old and spruce soon afterwards) and also seed regularly once they have started to flower, providing regular supplies. This is significant for wildlife conservation because introduced conifers have a particularly important role to play in British forestry, both as high-volume producers and silviculturally (that is, in woodland management terms) as shade-bearers both in upland and lowland forests. Spruce, Douglas fir (known in the timber trade as Oregon pine) and larch are prime timber-producing trees when grown in Britain and there are many other less important species. World forests are dwindling fast and those countries that still have indigenous reserves are eager to put them to their own use and to add value by converting timber to wood products before they are sold; thus British forestry needs the wider species range that exotic conifers can offer if import bills and the depredation of the world's forests are to be sensibly reduced.

Exotic conifers are essential to British forestry for two reasons: the soils available and the silvicultural options they provide. The best forest soils of Britain have long gone to agriculture (and more recently to urban development) and though there may be some reversal as farm surpluses accumulate, forestry can never expect the best land. The soils of much of upland Britain are degraded by changes after deforestation of the natural tree cover, followed by centuries of intensive sheep grazing. In addition, many of the sites are very exposed; this is the principal limiting factor to tree growth in Britain. Of the native timber trees, only Scots pine can tolerate these conditions but it is limited to the drier sites. Several exotics, particularly Sitka spruce and lodgepole pine, will thrive and produce a good timber yield in all but the most inhospitable conditions. As woodland conditions improve these sites, a wider range of trees can then be grown but these are mainly exotics too, such as Douglas fir, western hemlock and western red cedar, all species that are imported in large quantities into Britain. On the sites where windthrow is a limiting factor, short rotation crops of early successional species may have to follow each other, as they do in the natural forest after windthrow and fire.

An aesthetic objection is sometimes raised to the growing of non-native trees but this is hard to justify in upland areas where the majority of afforestation takes place, as we explained in the previous paragraph. These bare sites, whether rundown sheep walks, heather moorland or wet peats, are not natural but are the result of past uses. Trees have been cleared from them in the past to sustain other monocultures: a grass monoculture for sheep or a heather monoculture for grouse. Previously grass and heather would have occurred mainly in woodland glades and windblown forest areas but are now maintained as dominants by grazing and burning.

Peat areas developed from a high water table following forest clearance caused by adverse climatic conditions or by human activity. It is sometimes suggested that upland areas, if they are to be returned to forest, should be planted with native species only. The justification for this in conservation terms is based on the narrow assumption that native species provide more suitable habitats for wildlife, but it has already been suggested that this is by no means true. The use of native species alone would mean that only Scots pine could be planted for wood production, and conservation forests of broadleaves alone would be very expensive to establish and maintain in upland areas unless they were limited to the boreal associates of birch, aspen and willow. Work is being carried out on the improvement of birch for economic wood production but commercial trials are only just starting (I. Brown, pers. comm.).

Plate 6.2 Managed forest, under a clear cutting system, provides a mosaic of age classes, benefiting a wide range of wildlife. British Columbia. (Photo: E. H. M. Harris)

There are strong reasons for extending the production of timber in Britain, particularly on the less good agricultural land, and many nature conservation objectives can be met at the same time when introduced species are used. Over much of Britain these two objectives can go forward together; indeed there is no other productive industry that can accommodate conservation so readily. Full advantage should be taken of the new habitats and developing communities that this will create, alongside the establishment of specialized nature reserves for semi-natural communities.

The second important forestry reason for using exotic species is silvicultural and arises from the lack of native shade-bearing trees. Beech is the only native shade-bearing timber tree grown in Britain and will only grow to timber size on the better sites, whilst there are several shadebearing exotic conifers that grow on a wide range of sites. Systems of management that do not involve clear felling depend upon at least some shade-bearers for regeneration. Such systems are often preferred for soil improvement, wildlife conservation and for landscape enhancement. An oak wood cannot easily be regenerated in small coupes with oak alone, but several evergreen shade-bearers will grow well under oak's partial shade. The oak itself will regenerate, or can be successfully planted, in larger regeneration coupes and on the forest edge.

Shade-bearers are extremely important for rehabilitating derelict woodlands by enrichment with more useful timber species, rather than clear felling them completely with the large-scale changes that then take place. Western red cedar *(Thuya plicata)* and western hemlock (*Tsuga heterophylla*), both from western North America, are particularly useful in this way and find a market in quite small sizes. Such additions to existing woodlands also add diversity to the woodland structure and thereby increase the range of wildlife habitats.

All these advantages of introduced conifers depend upon their fast growth and good yields when grown in Britain. Productivity is measured by *yield class*, which is a means of comparing all the species that are grown for wood production in Britain. Table 6.3 demonstrates this.

There has been some concern in recent years about acidification of soils and of water in streams and rivers from upland coniferous forests. Very complex issues and mechanisms are involved that are not yet fully understood. It is certainly an over-simplification to attribute acidification to conifers in all circumstances and a common error to disregard the acidification that can occur under broadleaves, especially beech and oak on certain sites. All woodland soils are acidic to some extent and the pH of the soil itself does not reduce the rate of decomposition significantly (at pH 3.7, the rate of decomposition is 85 per cent of that between pH 5.0 and 8.0 (Larsen 1980)). Other factors, such as temperature, are important and the breakdown of leaf litter is slow under dense shade-bearers. On poor soils, oak, for instance, can create quite acid conditions (Dimbleby and Gill 1957), whereas Douglas fir in the plots studied by Ovington (1953 and 1954) gave rise to mild or 'mull' soils. The acidity or otherwise of the underlying rock is of prime importance in determining soil type, and the freedom for water to move down the soil profile taking surface materials with it can contribute to acidification, particularly under conifers.

The more recently recognized but, again, little understood phenomenon of 'acid rain' in forest decline is not caused by conifers but does seem to have more effect on them. Their dense evergreen needles comb out solid particles from polluted air when wind blows through their thick crowns, and later this gets washed off into the soil. Soluble acids from industrial pollution accumulate in rain, cloud and snow and collect on all foliage to a varying degree, least of all on grassland but most heavily on dense conifer foliage, and eventually reach the soil. From both these sources, acids are collected and concentrated, particularly by conifers, and then find their way into streams and rivers; they may then release aluminium into streams, which is toxic to wildlife. Fish and other aquatic life that would normally tolerate quite low pH values are killed in less acidic streams by aluminium. At low concentrations, aluminium causes reduced growth rates. However, the amounts of acid deposition and composition vary in different parts of the country.

Keeping conifer planting back from water courses is now accepted practice, and is particularly important in areas where the underlying rocks are acidic and deposition is thereby concentrated. These acids are derived from the atmosphere and have a purely physical effect on the soil; they do not involve the humus layer and normal nutrient cycling. Liming stream and river banks has been shown to have a short-term

104

Table 6.3 Yields of introduced conifers compared with yields of native trees

Tree species	Native or exotic	Yield classes represented	Yield Class Range	Rotation of maximum volume production (years)	Yield at rotation of maximum volume production (m³)
Scots pine	N	14–4	10	66–99	924–361
Corsican pine	E	20–6	14	53–74	740–414
Lodgepole pine	E	14–4	10	52–88	728–355
Norway spruce	E	22–6	16	62–95	1362–569
Sitka spruce	E	24–6	18	46–68	1106–407
European larch	E	12–4	8	45–62	540–247
Japanese larch	E	14–4	10	41–60	575–240
Douglas fir	E	24–10	14	49–65	1175–649
Western hemlock	E	24–12	12	51–82	1222–982
Western red cedar	E	24–12	12	56–73	1343–852
Grand fir	E	30–14	16	51–58	1528–810
Oak	N	8–4	4	62–96	493–382
Beech	N	10–4	6	76–112	758–447
Sycamore/Ash	N	12–4	8	38–52	454–207

Source: Forestry Commission Management Tables.

1 The higher the yield class the greater the yield.
2 For any one species, yields culminate earlier in the higher yield class.
3 The rotation of maximum volume production is the age at which yield reaches a maximum before declining. Theoretically, if a series of crops of the same species are felled and replaced at this age on the same site, the maximum timber yield is produced from that site indefinitely.
4 Volume production must be considered with rotation age to be meaningful. If the same yield is obtained from two different species on separate sites, one on a shorter rotation than the other, the faster-growing species will be more productive because the second rotation can be started earlier on that site.
5 Some species grow satisfactorily on a wider range of sites than others, as their yields are comparatively higher even in the lower yield classes. For example, Sitka spruce exhibits 18 yield classes and oak only four. Even in its lowest classes, Sitka spruce yields a lot more and on a significantly shorter rotation.
6 The highest yield classes occur in the shade-bearing species.
7 Conifers yield a great deal more than broadleaves.
8 Exotics, particularly those from north-western North America, yield significantly more than British native species.
9 Yield is only an expression of wood production volume and takes no account of timber value. Value is usually related to species and size. Larch is a valuable species but yields are low. Oak is valuable in large sizes and this may justify rotations longer than the rotation of maximum volume production, because the unit value continues to increase after the growth yield has culminated.

buffering effect only and the latest recommendations are to lime the unforested headwaters where the lime is then released gradually. Sensitive areas that are most at risk are now being identified and guidelines drawn up (Royal Society 1988). Nevertheless a lot more needs to be learnt about these complex reactions.

Acidification resulting from pollution was occurring long before the advent of afforestation in Britian. A gradual increase in upland acidification has been underway in Scotland (Flower and Batterbee 1983) and Wales for a long time but not only in areas with forest plantations. The effects of the removal of the lime subsidy in upland areas have received little attention in this connection. Research has demonstrated that trees efficiently filter sulphates from the atmosphere and do so more readily than grassland, largely because of the increase in wind turbulence caused by trees compared with grassland. When sulphates enter mineral soils they normally combine with other minerals and thereby neutralize the acidity as they pass down the profile, though there are two important exceptions: firstly, when available neutralizing minerals are not present in the soil, secondly when rainwater runoff passes directly into water courses without going through the soil first. On mineral soils, therefore, it is necessary to keep conifer planting back from streams but at forestry sites established on peat that show little water movement (Pyatt et al. 1988) there is correspondingly little value in a buffer strip. The greatest concern about acidification is in Wales where the underlying rock consists mainly of Silurian shales, which are sedimentary rocks with a low mineral content and therefore with limited ability to neutralize acidity in rainwater runoff. In Scotland the geology is more complex, with a lot of igneous rocks; these usually have a higher mineral content than sedimentary rocks and the problems are therefore not so widespread.

A further problem affecting water quality is silt entering waters as a result of forest operations. This can affect aquatic life and damage spawning beds but can be reduced by sensitive management. Heavy rain after ploughing for new planting, particularly if it comes before drainage operations are completed, can cause large quantities of recently disturbed soil to be carried straight into water courses and even down into reservoirs. The damaging effects of this scouring and siltation can take several years to stabilize. The same can also happen when felling on steep slopes, especially if there is a lot of clay in the soil. Now the accepted practice is to stop drains before they enter water courses. Felling in groups or small coupes near water courses will also reduce this problem, whilst the maintenance of forest cover in these areas by the use of extended rotations and selective fellings are options to consider.

Other recommendations that have been put forward include leaving grass strips between conifer plantations, which themselves are to be kept well back (at least the height of the mature trees) from water courses and planting broadleaves within conifer plantations. The intention is that the broadleaves will have a buffering effect on acid rain runoff but this is by no means certain as broadleaved trees also collect atmospheric depositions during the summer, although not to the same extent as conifers. Grass strips will provide areas for deer to graze and will also allow the survival of light-requiring ground flora (Goldsmith 1983). However, wide strips of grass alone leave no stretches of shaded water and if a stream is shallow, deep pools where fish can find cool conditions will be few. Water courses containing fish need alternating light and shade so that fish can find a range of conditions. Shade in a stream is as important for salmon and trout as sunlight (Morrison 1988). Insects falling into water from overhanging trees are a major source of food for aquatic life (Mason and Macdonald 1982). Foliage from both coniferous and broadleaved trees also provides a food source (Ormerod et al. 1987). The optimum strategy is therefore an alternation of grass and clumps of trees, taking into account topography and aspect. Willows, birch and aspen provide a good range of insects, especially in upland forests, and provide the broadleaved genera normally present in the taiga. In lowland forests the range of broadleaves that can be used near water courses is wider, depending on the geography, but alder is a useful species that can be grown in most situations and the value of sycamore has already been discussed in chapter 5. *Forests and Water* (Forestry Commission 1988a) provides detailed information and guidelines.

The controversial issue of whether forests should be planted in the uplands, and if so where, needs to be approached objectively. There is little doubt that forestry in the past has failed to take conservation and landscape interests fully into account, but the historical reasons for this must not be forgotten. The earlier plantings were intended to cover the ground with trees and build up a growing stock after the devastation of two world wars. This limited approach is outdated and it is now normal practice to give consideration to both landscape and conservation.

The main objection to afforestation is that it displaces open-ground flora and fauna but any land use change will affect the balance of species; the value of some has to be weighed against the value of others. The so-called 'open'-ground species displaced by forestry contain a significant number of species that are normal inhabitants of taiga and high-altitude forest elsewhere in the world, and for which the open ground created by human activity in the uplands of Britain is not the main habitat.

There is truth in the argument that planted forests are not the same as natural forests, particularly to look at, but some natural even-aged stands are almost impossible to distinguish from plantations unless their history is known. One could equally well argue that gardens are not natural but their contribution to the survival of a wide range of species is unquestionable and cannot be ignored. Abrupt plantation edges are also criticized as unnatural but abrupt changes in conditions in natural forest occur on the edge of windblown and burnt areas, or where waterlogging increases rapidly with slope. Softening edges with scattered trees, such as other natural situations exhibit, has landscape benefits and also conservation benefits. A mixture of both edge types, depending on layout and topography, will provide a wide choice of habitats for wildlife. Leaving natural birch, either as blocks or scattered throughout the crop, will also widen the choice. Often habitats that look perfect to the human eye fail to fill expectations as far as use by wildlife is concerned. More work is needed to demonstrate how plantations can be managed to hold as wide a range of species as possible. We are only now beginning to see the opportunities provided by our still young upland conifer forests, but if the habitats they provide are looked at objectively in the context of their associated natural zone, it will be seen that they have considerable conservation potential. As Hambler and Speight (1995) point out, 'In Britain we must work hard to shed the image of conifer woodlands being poor for wildlife.'

CHAPTER 7

The Management of Mixed Forests for Timber and Conservation

This chapter is in three sections. First we look at mixed forest historically as the natural forest cover that might have occurred over much of Britain and therefore offers the greatest conservation potential as the transition zone between coniferous forest and northern broadleaved conditions. We then consider the forestry advantages and disadvantages of mixtures, and describe the ways in which they are established and managed. Finally, we discuss the value of mixtures for conservation. Throughout we emphasize that the best trees will be grown only if it is borne in mind that a mixture of conifers and broadleaves is the natural forest cover for the latitudes that most of Britain lies in.

NATURAL MIXED FOREST

We must first define 'mixed forest', as the term is often used loosely to describe either a forest of mixed conifer species or a forest of mixed broadleaved species. Correctly used, mixed forest describes woodland that contains a mixture of both conifers and broadleaved trees. In the few relatively unmanaged mixed woodlands that remain, evergreen and deciduous components occur in varying proportions, depending on soil type, aspect and altitude. In some places an intimate mixture of coniferous and broadleaved species can be found, whilst in others a particular species assumes dominance. The term 'mixed forest' is not being properly used if it merely describes a woodland of more than one species of tree; woodlands consisting of more than one species of conifer or of broadleaves thus need to be specifically designated 'mixed broadleaves' or 'mixed conifers'.

When considering mixed forest from the points of view of both wood production and conservation, it is useful to set it against the context of what would have developed in Britain without human intervention and if the separation from Europe had not taken place. Human influence is particularly important as it was at work almost immediately the ice retreated; indeed people had lived in Britain in small numbers during the last interglacial period. Had the separation not occurred, most of Britain other than the Highlands of Scotland and south-east England would have been in the mixed forest zone. This process could not develop because most of the European coniferous species are lacking from the postglacial British flora and, as a result, the forest cover was composed of northern broadleaves only. It is difficult to determine where the transition zone would have given way to northern broadleaved forest in Britain. The latter certainly extended further north in the Atlantic period, when it reached its maximum, but climatic conditions have deteriorated since then and the viability of broadleaved seeds decreases at high latitudes and altitudes in Britain. The transition zone can best be compared with the area where broadleaved trees become dominant over conifers as they do in altitudinally zoned forests. In Europe these descend as the coniferous mountain forest gives way to beech and then, on the lower slopes, to oak/hornbeam forest. The warm, dry conditions in which this last type of forest is found do not now occur in Britain. Perhaps the south and east of England should be regarded as being in the northern broadleaved forest zone, as the distribution of many native plant and animal species fit this pattern .

Since the 1960s there has been a marked swing to a more boreal climate with cooler summers and it is anticipated that this trend will continue unless masked by the rising level of CO_2 (Royal Society 1988). For this reason, and because much of Britain would have had mixed woodland such as that present in the last interglacial, mixed forest has an important role to play in wildlife conservation. The mixed forest zone, where the northern, mainly coniferous, species meet southern broadleaved species, is a rich area for wildlife wherever it occurs in the world. This suggests that if forests in Britain are designed to reproduce these conditions, they can be of great conservation value.

THE ADVANTAGES AND DISADVANTAGES OF MIXED FOREST

Looked at from the point of view of productive forestry, there are both advantages and disadvantages in mixed woodland; some of the advantages are difficult to quantify. Tree species differ in their requirements so it is often assumed, but difficult to prove, that mixed woodland makes more use of the available soil nutrients and perhaps of available light (Koiriukstis 1968). Trees root at different depths so mixed woodland may be less liable to windthrow and casual observation after gales seems to bear this out. The incorporation of forest *litter* (leaves and twigs) into the soil proceeds more rapidly when species are varied, particularly if leaves are mixed with conifer needles. Under beech or spruce alone, for instance, a mat of slowly decomposing litter accumulates which lasts for several years. More rapid decomposition is at least partly the result of more light reaching the forest floor when a variety of species occur, but also seems to be related to the different composition of leaves from different species. Evergreen conifer needles do decompose more slowly but total nutrient recycling is similar over the rotation (Miles 1986). In experiments, leaves, especially of lime and maple, spread over the needle litter of planted spruce raised the pH value considerably in six years (Ellenberg 1988).

Looked at from the point of view of wood production alone, there are many situations in which a single species, or monoculture, is more productive than a mixture. This is because one species will often perform much better than any other on a particular site, so 'dilution' with another species must reduce the yield. Examples of this are Sitka spruce in much of upland Britain and Corsican pine over most parts of south-east England. In both cases these two trees grow much faster and larger than any other species, so when they are grown pure they produce the highest yields possible. There has been a long-held view in Germany that where the natural mixed forest has been managed to favour the most useful species (in this case, spruce), after two or three rotations of pure spruce, yields fall, but more recent critical studies have failed to show decreases in yield between first- and second-rotation crops of Norway spruce (Miles 1986). The long-term effects of monocultures in British conditions still need to be monitored carefully, despite the attraction of the high yields that they can produce; these effects may vary widely with climate and soil type (Miles 1986; Ellenberg 1988). There is no danger on ground being afforested for the first time when a single pioneer species is often the easiest to establish; indeed, this is similar to what occurs in natural succession. An advantage of mixtures on some sites is that they can be used to telescope succession, which would take place over a longer period under natural conditions. It is common practice to establish more sensitive broadleaves with conifer nurses that are removed a third- to half-way through the rotation, completing in one rotation what would take longer to accomplish by natural succession.

The threat of disease and pests multiplying rapidly in monocultures is greater than in mixtures, and this threat is likely to increase as time elapses from the original introductions of the many exotic species used in Britain. This emphasizes the importance of mixtures in subsequent rotations. Where an epidemic occurs, it is likely that an insect pest or a disease attacking one species will not attack the others composing the mixture to the same extent, if at all. Damage by mammals is less selective. The great advantage of mixtures in this respect is that there are more management options available following severe damage if several species of tree are grown. Dutch elm disease would not have had such a drastic effect on the landscape of southern Britain if field elm had not been so dominant a hedgerow tree.

THE MANAGEMENT OF MIXTURES AND THEIR CONSERVATION VALUE

A wider range of management options is also provided by mixtures when there is uncertainty about future markets. Mixtures have the advantage that if markets change and what was expected to be a useful species when the plantation was established is no longer so saleable, then other species can be favoured as the crop matures and the new preferred species selected for during thinning. Much more important, though, is the uniformity and quantity that can be marketed, as buyers prefer one species in large quantities and this militates against mixtures. Indeed, the availability of large quantities of a uniform product tends to create its own market;

Sitka spruce, now in demand, is an example of this, whereas in the 1950s it was looked on with suspicion. A compromise that can be exercised in woodlands with varied aspects and soils is the planting of small blocks of single species to create a mixed wood of discrete blocks.

The management of mixtures as they mature is more difficult than the management of monocultures. Each species has different light requirements and to keep them all growing well requires both compatible species and skill in their management. To obtain the maximum conservation benefit by diversity, shade-bearers need to have been included in the original mixture to make use of all the available light. The species forming the main canopy cuts out 70 to 90 per cent of the light (Hill 1979), depending on the species, so shade-bearing trees, usually conifers, are needed to form an understorey to obtain full stocking. There are several western North American conifers suitable for this. If, however, these are allowed to become dominant, they will shade out the light-demanders. In Britain, most broadleaved trees are light-demanders, so the broadleaved components of a mixed wood can suffer because they grow less well on poor sites than do conifers. In some circumstances, conifers can be added to an existing crop to enrich it or as an understorey, whence the evergreen component adds diversity and can be of considerable conservation value. It can provide shelter (French et al. 1986), but it can also be used to control unwanted species, such as rhododendron and bramble. Mixtures composed solely of shade-bearers are the only ones that can be managed on a true single-tree selection system. Light-demanders require group regeneration and the groups need to be of sufficient size to allow such species to grow. True selection systems are similar to the later stages of natural succession when the forest regenerates within itself. The range of diverse habitats present in a natural succession is lost in the true selection system, together with the animal and plant species dependent on open and edge habitats.

Recently, the concept of *continuous cover forestry* has been introduced in Britain and also has some support on the Continent of Europe. It aims to simulate natural conditions, emphasizing maintenance of permanent woodland cover with the perceived benefits of better growth and no exposure of the soil, as well as avoidance of unsightly clear felling. There is, however, no firm evidence for enhanced growth under any form of selection forest, whilst cleared areas, provided these are not large, are no different from windblow, a natural hazard, and indeed a form of regeneration in boreal and temperate forests. The practice of continuous cover forestry will result in a very narrow range of habitats with low levels of light and limited benefits to wildlife, except for those species requiring dense forest conditions. It can only be successful as a form of management when applied to shade bearers, i.e. most evergreen conifers and beech. It is, however, sometimes the most apporapriate management system in very small woods.

Mixtures provide the widest opportunities for conservation and at the same time have some silvicultural advantages. Their most important silvicultural role is in establishing sensitive broadleaved crops; in many cases the coniferous element is

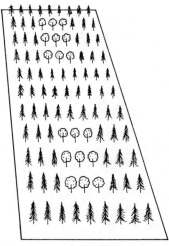

Three-line mixture   Group mixture

Figure 7.1 Types of mixture   Mixtures are crops of broadleaves and conifers growing together. Very often the conifers are grown on a short rotation to 'nurse' more sensitive broadleaves and to provide early yields of conifer poles that are saleable in small sizes. Two simple mixtures are shown here; both could comprise just two or more than two species. The line mixture contains fewer conifers and is easier to manage: complete rows of nursing conifers can be removed when they have served their purpose. Group mixtures present the danger of conifers shading out the broadleaves, and early thinnings are more difficult to extract.

not retained, the aim being a final crop of broadleaves. Several ways of establishing mixtures have therefore been developed and they fall broadly into two groups: line mixtures and group mixtures. *Line mixtures* (figure 7.1) are usually established as one, two or three lines of one species alternating with the same number of lines of another species. A typical line mixture consists of three or four species, each of them pure, but the objectives are seldom met where only single lines are planted or with more than four lines of one species. In the first the mixture is too intimate to achieve its purpose and in the second not intimate enough. The use of tree shelters which provide protection (see chapter 4) for the broadleaved component at wider spacing, may mean that single-line mixtures are more successful than in the past. The purpose of all these mixtures is to establish a good crop of a sensitive species, which is intended as the final crop, within the shelter provided by a less sensitive one called the *nurse*. Or it may be that the desired species is slow growing and would quickly become smothered by weed growth if this were not shaded out by the faster growing lines on either side. Line mixtures are a particularly good way of growing oak because young oak trees lack 'apical dominance' (the tendency to grow

Plate 7.1 A well grown stand of pole stage oak (*Quercus petraea*) nursed by Norway spruce (*Picea abies*), in a line mixture. Lincolnshire, east England. (Photo: E. H. M. Harris)

upwards), and easily become bushy. With some side shade from another, faster growing and particularly evergreen species, they grow upwards from the beginning.

The contrast in colour between lines of evergreen conifers nursing deciduous broadleaves looks unnatural on a hillside until the conifers are removed. This has led to the abandonment of line mixtures in favour of *group mixtures* (figure 7.1). Groups also provide the opportunity to plant a higher proportion of the nurse species as a matrix, in which to establish quite small groups of what is to form the basis of the final crop. Such groups mimic the single species groups that occur in natural mixed forest when it regenerates (see chapter 4). A much-used pattern is groups of nine plants of the final crop species making up not more than 20 per cent of the initial planting. The groups are evenly distributed with the intention that one tree in each group will eventually form the final crop. More complex mixtures can be planned with either mixed species in the groups or groups of two or more different species alternating within the matrix. Very complex group mixtures have been devised and they are usually planned with detailed thinning or extraction regimes in mind, but by the time these operations are due, markets may have changed or, more often, the trees will not have grown quite as expected.

An important commercial advantage of these nursing mixtures is that conifer thinnings are more saleable in small sizes than broadleaves, and the early returns provided by them offset the high cost of establishing slowly maturing broadleaved species. The dangers with conifer/broadleaved mixtures designed to provide a final broadleaved crop are that the nurses are often not taken out soon enough or that the nursing species grows too fast and dominates the final crop species. The latter can easily happen with single lines but seldom with three-line mixtures. The nurse species should be one that is saleable in small sizes so that there is an incentive to remove it before the future final crop trees are suppressed. Conifers are often good nurses for broadleaved trees because they grow faster initially, shading out competing vegetation, and are then saleable as small poles, but the time of their removal is crucial. The large areas of unthinned broadleaved/conifer mixtures in lowland woods have been a major factor in giving forestry a bad name and such unthinned woods have little conservation value. The need to take out the nurse species at the right time is much more critical in group mixtures than in lines, particularly if the groups are small, because the conifer nurses on the edge of the broadleaved groups grow faster and can quickly suppress them. The first thinning should be around the groups and subsequent thinning throughout the nursing matrix.

Damage from deer is claimed to be more severe where groups are used rather than lines (Royal Forestry Society 1984). Line mixtures also retain more of the ground flora than groups because there is a higher proportion of deciduous trees in the canopy. Line mixtures using narrow-crowned nurses, such as western red cedar, also mean more light reaches the forest floor allowing the retention of more ground vegetation, and they are less likely to suppress the broadleaved crop. The first thinning in a line mixture should be the removal of the two outside lines nearest the final crop

Plate 7.2 A young beech (*Fagus sylvatica*) group growing in a matrix of Norway spruce (*Picea abies*). The high proportion of conifer in group mixtures shades out more ground vegetation than line mixtures. (Photo: E. H. M. Harris)

species; this is usually much easier than extracting thinnings from around groups, and also allows more light to reach the ground vegetation before it is completely suppressed.

On the very best sites where a variety of species, both conifers and broadleaves, will grow equally well, intimate mixtures of several species on no set pattern are an attractive but risky option: risky, from a forestry point of view, because some of the species may fail, resulting in understocking and widely spaced trees with heavy branches. If successful, a wood of attractively varied appearance will quickly be established, together with a range of habitats of high conservation value and a variety of produce to market. The most extreme example of this in British forestry is the 'Bradford Plan' (see figure 7.2) practised at Tavistock Woodlands in Devon, where

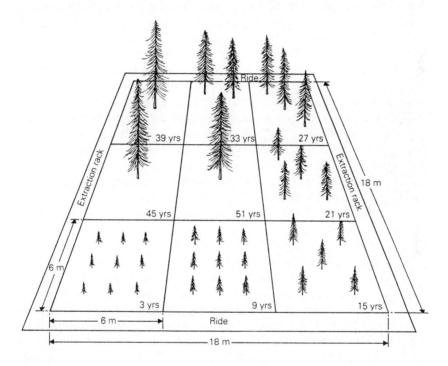

Figure 7.2 The Bradford Plan. The Bradford Plan, practised at Tavistock woodlands in Devon, is an example of a very complex system of management. It is based on very small plots each of sufficient size to contain one final crop tree by the end of the rotation. The system mainly, but not wholly, uses evergreen conifers, which grow fast and large in Devon. Each group of nine plots, representing nine age-classes separated by six-year intervals, is surrounded by extraction racks and rides.

Detailed and careful management is essential. The groups are managed on a six-year cycle with thinning, felling and planting all taking place within a particular group every six years. The intended result is a forest of varied structure and intimately mixed age-classes with no large open areas. Dark, dense and rather uniform woodland could be the final result, as the system will probably tend to favour shade-bearing species.

very detailed specifications on group size and species components are prescribed (Collin 1987). A more successful one is at Brocklesby in Lincolnshire (Royal Forestry Society 1984), where a proportion of two conifers to one broadleaved tree is planted with the aim of producing a final crop of broadleaves. Scots and Corsican pine, larch, beech and sycamore are all used in varying proportions.

Usually, simple mixtures are the most successful. Some useful examples are (1) three lines of oak and three lines of pine, larch or spruce; (2) two or three lines of beech and two, three or four lines of western red cedar; (3) squares of nine beech in a matrix of Norway spruce, with their centres about 15 metres apart; (4) rectangles of 12 or up to 20 oak in a matrix of Norway spruce; and (5) alternate squares of oak and cherry in a close planted matrix of Norway spruce. The species in the last example grow well together and on similar sites. There will be several types of material to harvest: the spruce as Christmas trees from about year 5 to year 15, and then as small poles and finally as timber to about year 50; the cherry as firewood and then timber from 20 to 60 years; and eventually the oak as the final crop. This requires careful management, but when most of the spruce has been removed, the cherry will cast a light enough shade to allow large crowns to develop on the clean-stemmed oak resulting from the early nursing by the spruce. Such a mixture provides a varied structure, particularly if a few spruce are left to the end of the rotation. Indeed, the conservation advantage of leaving a few large conifers in all mixtures is considerable, as they will provide all-year-round food and shelter as well as nesting sites for birds, especially birds of prey.

The mixtures described aim at dealing with two important aspects of British forestry. Firstly, a considerable amount of rehabilitation of old woodland is needed, but natural regeneration is often slow, with colonization initially by pioneer species such as birch, pine and larch, and a range of secondary species coming in much later. The latter are the more sensitive and usually the more valuable species, such as oak, beech and spruce. Mixtures are an attempt to telescope natural successions in order to re-establish lost woodland cover more quickly. Secondly, mixtures allow broadleaved trees, especially oak, to be grown where a pure oak or other broadleaved crop would produce no revenue for almost the whole of a human generation. Not all mixtures are beneficial, however, and experiments with spruce/oak mixtures at Gisburn in Yorkshire, Britain, showed no growth enhancement, whereas mixtures containing pine showed positive effects (Brown 1987). It seems that the beneficial effects usually occur when a pioneer species is associated with a secondary species, something we have already noted in chapter 6 in connection with lodgepole pine and Sitka spruce mixtures. Claims have been made for considerable conservation value to the ground flora of the very intimate Bradford Plan at Tavistock (Harris and Kent 1987), but it is questionable whether this will be sustained because the very small groups will eventually favour the shade-bearing western hemlock, already regenerating freely and likely eventually to dominate much of the woodland to the exclusion of other species.

The conservation value of mixtures is considerable. Structural diversity and a wide

Plate 7.3 Early thinning of this mixed plantation has produced fine and valuable stems of sycamore (*Acer pseudoplatanus*) and European larch (*Larix decidua*). The intensive management has encouraged abundant spring flowers. Yorkshire, north east England. (Photo: E. H. M. Harris)

120

range of species can be established easily on good sites. Although more difficult to manage, such woodlands accommodate a wide range of wildlife. In addition, any non-forest species that come in naturally, such as willow, crab apple, hazel and blackthorn, may be left, where they are not competing with the crop, and will extend the foods available to wildlife. If 'naturalness' is of conservation value, mixed woods of conifers and broadleaves most closely approximate what could have been the natural woodland cover over much of Britain, as the composition of the flora and fauna is typical of transition zones of mixed forest elsewhere in the world where northern and southern species intermix.

The 'broadleaved policy', a requirement of the Woodland Grant Scheme wherever deciduous woodland is being replanted (see chapter 5), insisting as it does on pure broadleaves from the start, mitigates against the sensible use of nursing mixtures for the establishment of woodland of high quality, both from the timber production and conservation points of view. After only ten years of this policy, the poor results are all too evident in very poor young oak and other species, growing without the benefit of evergreen side shade, wherever regular early maintenance has not been carried out. A return to more widespread use of mixtures, at least to establish broadleaved crops, even if all the conifers are removed in the first two or three thinnings, would be very much more successful. An example of an excellent broadleaved crop that was established in this way is depicted in plate 5.2.

Plate 7.4 An intimate mixture of broadleaves and conifers on a fertile site. Regular thinning has produced valuable individual stems and a habitat of high wildlife value. Yorkshire, north east England. (Photo: J. A. Harris)

CHAPTER 8

Guidelines for Wildlife Conservation within Productive Woodlands

FOREST MANAGEMENT FOR WILDLIFE

Forests and woodlands provide suitable habitats for the majority of bird and animal species that are native in Britain and many insect and plant species as well. Comparatively minor modifications to forestry practices can therefore be of considerable value to wildlife conservation. Wildlife species using forests fall into two categories: (1) those species requiring relatively undisturbed conditions and which therefore benefit from selection forest systems, both single-tree and group-selection, because of the retention of woodland cover; and (2) those species exploiting disturbed conditions, taking advantage of clear felling systems and extraction routes. There are many more species in this second category. Conservation and management objectives must be defined for each woodland or forest because a diversity of forests, managed for different species, is more important in providing habitats than trying to create *maximum* diversity in all forests (Hunter 1990). Objectives will be limited by the size of woods, single purpose objectives being the most suitable for small woods.

The size of felling coupe is important, depending on the size of the wood, topography and avoidance of frost pockets, but an area of about two hectares has many wildlife and forest management advantages, of which protection and minimizing desiccation are major factors. Coppice systems, which are essentially perpetual thicket, are currently popular amongst conservationists because vegetation, especially bramble, and overhead cover is removed at regular intervals, allowing certain plants to flower each time the coppice is cut. Clear felling, carefully planned, can have the same effect.

Throughout the life of a forest crop, ground vegetation in plantations is dependent on the light intensity penetrating the tree canopy and reaching the forest floor. Heavy and frequent thinning will let light in to benefit plants, but on some soils too much light will favour aggressive bramble and bracken. As regular thinning is the basis of good forest management, conservation and production can go hand in hand.

A particular advantage of woodlands is that many animals and birds benefit from the protection and seclusion even if they do not themselves feed in woods. As management intervention for such activities as thinning and harvesting is infrequent, the value of woodlands as secluded places is very high.

If the highest conservation value is to be obtained from woodlands that are primarily producing timber, a conservation plan, preferably as part of the forest management plan, or at least a check-list of things to consider, is useful. The rest of this chapter is divided into the stages that ought to be considered in the preparation of a conservation strategy.

Stage 1: The evaluation of woodland in terms of economic production and conservation

The following points need to be considered:

1 The type of productive forestry to be practised and the ultimate harvesting objectives, such as long-term timber or short-term pulp rotations.
2 Assessment of the desirable diversification of the woodland for wildlife, against uniformity for ease of management. Should diversification be achieved by small, uniform blocks or by intimate mixtures? What forms of wildlife are to be encouraged? All woods cannot contain all species, so management needs to be planned accordingly, depending upon the geographical situation.
3 Visual effects on the landscape.
4 The amount of land to be left unplanted. Examples include wet areas, roads, rides, public footpaths, wayleaves (land under power transmission lines), uneconomic areas, frost hollows, stream and pond sides, clearings for deer control, deer or sheep downfalls (corridors between high and low ground), rocky areas.
5 The cost of conservation measures undertaken outside normal forest management; is the woodland to be self-financing or will some assistance be sought from the various grant-giving agencies?
6 The extent of public access, both statutory and by consent; in the latter case no-go conservation areas may need to be established.

Stage 2: Mapping and planning

Forest management always requires a 'stock map' which shows the forest divided into management units ('compartments'), whose boundaries can be identified on the ground by roads, rides and paths. The tree species present, usually their age (expressed as date of planting), and the area of each compartment is shown. Stock maps are useful for recording special conservation sites and for planning the management of areas with conservation potential, as they already show tree species and locations that can be identified. Forest working plans usually include separate records of each compartment, on which special conservation aspects and prescriptions can be recorded. Most important of all is to include specific conservation objectives in the 'objects of management' in the working plan. These objectives can often be part of the forestry objectives, such as early, heavy thinning to encourage both tree growth and diversity, but may sometimes require special measures, such as grading the edges of plantations. A section on conservation within the forest working plan is more likely than a separate plan to achieve integration of conservation throughout the forest, rather than merely in areas specially set aside.

The afforestation of bare land

Special consideration needs to be given to afforestation because major ecological changes will be caused. These need not be a disadvantage; indeed, there are usually both long- and short-term opportunities to add diversity through a whole new range of habitats that will be created. The following points should be borne in mind.

1 It is useful to establish and record the former land use before planting.
2 Whilst delineating compartments for management purposes, it is a good idea to record anything of special interest.
3 A map of soil types and their pH may be of even more value to conservation planning than to forest planning.
4 Make a note of the physical characteristics of the site: such as acid or basic soils, free draining or the presence of soil pans impeding drainage and angle of slope.
5 It is important to decide how much draining is necessary for tree establishment but to bear in mind that wet areas often have conservation value. Some of these can be left or developed further for conservation. Base-rich flushes (wet areas where flowing water appears above ground) in particular hold interesting plant populations.
6 Sites of Special Scientific Interest and other special features that have wildlife potential should be noted.
7 If there are potential conflicts with other interests, it is important to discuss their management or proposed treatment with appropriate people. In many cases conflicts can be resolved and arrangements made to take all interests into account.
8 Fundamental to wildlife conservation is a decision on the amount of open ground to be left in addition to forest rides and roads (see also Stage 1, item 4).
9 A consultant or local natural history society will be useful if further advice is required (see Stage 4 below). Every forest or unplanted area already contains significant wildlife and this should be evaluated carefully and thoroughly, before changes are planned.
10 From a forestry point of view it will be necessary to select species suitable for the geographical situations and soil types present and to decide on the proportion of broadleaves to conifers. Where several species can be grown, their variety will enhance the wildlife potential; but a wide range of species will inevitably reduce

the forest production, as there will only be a few that are optimal for many of the sites concerned (see chapter 7).

11 Plan to diversify large even-aged blocks by premature felling. Depending on markets, Christmas trees, poles or firewood can be sold instead of sawmill timber.

12 Finally, it will be necessary to decide upon the planting pattern; for example, groups, lines, intimate mixtures, mixtures in blocks of different species or monocultures. The last of these may be the best option from a forestry point of view, but diversity will then depend on the variety of age-classes alone.

Existing plantations

1 First, obtain a stock map showing the various species and ages, or create one if this does not exist.

2 It is important to calculate the present age-class distribution structure in order to decide on management policy so that eventually a normal forest is created. This will provide the maximum production on a sustainable basis and the maximum wildlife benefit (see chapter 4.) A bar graph with ten-year age-classes in columns, differentiated between broadleaves and conifers, will prove useful here (see figure 8.1).

3 A fundamental decision will be the proportion of broadleaves, conifers and mixtures, bearing in mind that broadleaves alone are seldom profitable and are a long-term investment; and also that mixtures often provide the richest habitats.

4 Finally, after the productive forest and wildlife potential have been assessed, it will be necessary to select appropriate silvicultural systems to meet the chosen objectives. Different methods may need to be applied to the same woodland; for example, manage part on a clear felling regime and part on a selection system.

Stage 3: Features with wildlife potential

The following features need special attention.

Streams

Retain some areas near streams as open grass or damp meadow and avoid shading them. In broadleaved areas, encourage scattered clumps near stream edges but avoid shading the whole length of the stream. In areas to be planted with conifers, leave any existing broadleaves near streams; do not plant conifers along stream sides in areas of acidic geology in order to reduce acidity from needle fall and to minimize acid runoff. Plan a mosaic of trees and unplanted areas to provide both shade and sun on streams. Where there are grazing animals, unplanted areas will be kept free of woody growth, otherwise some cutting may be required. Trees along stream sides provide cool water (in their shade) and food (insects) for fish, as well as nesting sites and otter holts. Avoid felling into streams. Where possible, protect headwaters by using long rotations and selective felling, thereby minimizing disturbance and silting.

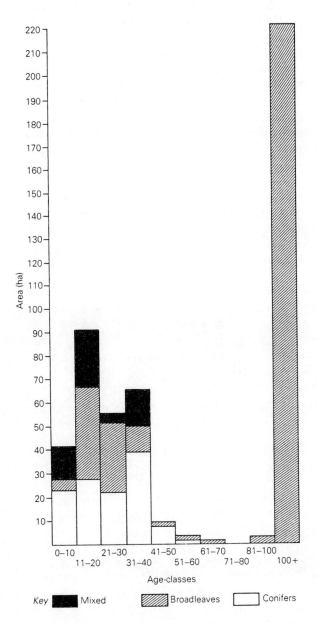

Figure 8.1 A typical ten-year age-class distribution diagram

Ponds

Keep ponds clear of trees on the south side to allow sunlight to reach the water; this helps plant growth and the recycling of nutrients, with all that follows from this. Avoid large, overhanging broadleaved or coniferous trees as excessive leaf fall chokes ponds and makes them *oligotrophic* (lacking in oxygen). Leave low bushes and small trees (such as willows and alders) near the water for animals and birds to use as cover when they come to drink and as resting places for insects. A large pond can be in both sun and shade and be deep and shallow, thus providing a variety of habitats. Alternatively, a number of small ponds can be scattered through a woodland, some deeper than others. Pond edges should be graded on the sunny sides, to provide warm, shallow water and banks with a gentle slope. Keep streams and the ditches flowing into them, and the ponds themselves, free of pollution.

Fire dams

If these are constructed for forest protection, they can be of great value for wildlife. Create them with graded edges on the sunny side. Site them away from large trees and do not plant too near to avoid shading them.

Plate 8.1 A pond in an upland coniferous forest in the English Lake District. The unshaded margin shelves gradually into the water and the artificial islands support nesting birds, including a colony of black headed gulls (*Larus ridibundus*). (Photo: E. H. M. Harris)

Wet areas

Swamps, flushes and ground with impeded drainage are worth retaining and should not be drained and planted up as they usually have good plant and other wildlife potential.

Drains

When forest drains are made they should not open directly into natural water courses, so that silt and chemicals are not carried into streams. In this way, scouring by flash floods is avoided and damage to aquatic life is minimized.

Slopes

Slopes on woodland edges, ride sides and roadside cuttings can usefully be developed for conservation. Drainage on slopes and banks is 'sharper', thereby reducing competition from the more aggressive plants that grow on flatter areas with deeper, richer soils. Sunny slopes facing south are particularly important for butterflies, insects and reptiles; shading of such areas should be minimized.

Rides

These should be wider than is necessary for forest management alone, although wide rides do have the advantage of being less shaded and thus keep drier for forest operations. At least 10 m width is desirable and sunlit rides running east and west can usefully be wider but aspect will also determine the amount of sunshine on hilly ground. Long straight rides should be avoided to minimize 'wind tunnelling'. Curved rides provide more shelter and a variety of sunny and shaded areas. Grass rides should be mown in the centre to provide dry areas; mowing is usually a part of normal management in woods where game rearing is carried out.

Ride sides

An abrupt woodland edge is of little conservation value. If ride sides are graded into the woodland with a border of low scrub or shrubs, their potential and interest will be enhanced. Such low cover on the woodland edge acts as a windbreak. The choice of shrubs should depend on the wildlife species it is hoped to encourage. Different sections of ride sides should be cut and mown on rotation every two to three years to avoid linear uniformity. Such borders do not need to be continuous and the forest edge can be graded by heavy thinning to encourage vegetation to grow under the tree canopy. Ride centres can usefully be mown annually to provide areas of short grass.

Firebreaks and power-line wayleaves

As these have to be left unplanted, there is an opportunity to plan mowing regimes that take account of a suitable time of year (after flowers have seeded) and frequency. Natural vegetation can be left, or wild flower mixtures can be sown.

Plate 8.2 A Scots pine (*Pinus sylvestris*) plantation where a variety of shrubs have been encouraged on the south facing edge to provide a graded margin beneficial to insects (especially butterflies) and birds. Hampshire, southern England. (Photo: E. H. M. Harris)

Glades, and open and bare areas

Open areas are of great value and can be established in plantations at ride junctions and such places as hilly and rocky mounds, and around ponds, stream sides and failed areas. Clear cut areas provide temporary open space and bare ground where timber has been extracted. These areas will provide grazing and resting areas for animals and will allow the growth of flowering plants. They will also provide areas for deer control.

Roads

The type of material used for forest roads needs to be considered; for example, limestone can increase plant habitat potential but may also change local conditions and introduce new plant species. Sloping verges enable annual plants to colonize. Roadside ditches provide wet habitats.

Groups of overmature, dead, windblown and windsnapped trees

It is important to retain some of these for hole-nesting birds, insects and fungi. The spreading branches of older trees provide nesting sites for birds of prey. Do not plant young trees under the canopies of overmature trees but allow the natural vegetation to grow there. Unwanted species, such as birch, in a crop can be ringbarked to provide dead wood and nesting sites. Trees do not need to be old to provide dead wood; stems as small as 20 cm in diameter are used by nesting woodpeckers.

Quarries and rocky outcrops
These are usually valuable wildlife areas and although they may not need much management, it is important to record them so that their value is not subsequently overlooked. Excessive shading of these sites should be avoided.

Bridges
Bridges over streams can be designed with ledges to provide nesting places for birds, such as wagtails and dippers.

Unusual shrubs or trees
Many woods, particularly in southern Britain, contain uncommon trees and shrubs, such as the wild service tree and spurge laurel, which should be left and encouraged to regenerate.

Localized habitats
Special habitats suitable for particular species should be noted and retained. Limestone flushes are often rich in plants. Blackgame have leks (gathering grounds for display and mating) in some upland forests which should not be disturbed. The nesting sites of birds of prey are often traditional and so are many badger sets. Ponds with dragonfly populations should be kept free of shade.

Stage 4: The evaluation of present and future wildlife potential
In most cases, the complete evaluation of the wildlife of an area is too complex to be carried out in depth. However, the occurrence of rare species should be noted and the larger animals, birds and butterflies recorded. Records should be made on a number of separate visits at different times of year. The records made by local naturalists may be particularly useful. Any assessment of wildlife conservation has also to take into account pest species; for example, grey squirrels, deer or rhododendron. Policies for control need to be developed that will not endanger other species, especially rarities.

When these assessments have been made it is possible to consider appropriate forms of management that will increase wildlife species numbers and diversity, especially the creation of a varied woodland structure. It is a good idea to give reasoned objectives for individual species. Also useful is a statement of objectives for specific areas, such as ponds, or south-facing slopes on basic rocks with a rich ground flora.

Stage 5: The wildlife plan
The next stage after evaluation is the preparation of a plan for wildlife management,

complementary to the forestry plan, or preferably integrated within it, with both revised regularly. There is a danger that plans that are expensive, complicated or time-consuming to prepare and carry out will not be followed after a change of ownership or even under the existing ownership. To be successful, the plan should avoid drastic or uneconomic measures, such as the removal of exotics without sound reasons, or coppicing where it is unsuitable; rather, it should seek to implement forest operations and silviculture of a type that will maximize both production and conservation. Different parts of the forest may be worked as discrete units with different objectives. At the time of writing the plan, suitable arrangements for monitoring the success of the plan need to be set up. It is worth considering using local natural history societies to carry out this work.

Stage 6: Conservation with little cost

Successful and useful conservation can very often be incorporated in the woodland management at no or little cost, and where this is the case proposals are much more likely to be implemented. Timing the conservation work to coincide with forestry work is the key to minimizing costs.

Rides and clearings

On areas to be afforested plan wide rides with irregular edges and curving shapes. Allow natural scrub to colonize ride edges or plant some of these with shrubs, especially fruit-bearing species, at the same time as planting the productive crop. Leave clearings at ride junctions. In established woods, widen ride junctions to form clearings at the same time as carrying out thinning. Widen rides when trees are saleable.

Fire dams

If dams are being created for firefighting purposes, there is little extra cost in grading out one end to encourage shallow water plants and access for animals. This should be on the side receiving the most sunlight, as the shallow water will then warm up quickly.

Features providing cover

There is no need for a forest to be very tidy and, indeed, unnecessary maintenance is costly. Leave old buildings and old trees if they are safe, walls, heaps of stones, piles of brushwood or thinnings. All of these provide cover for nesting, roosting, winter shelter and seclusion for birds, animals and insects.

Uneconomic areas

Cost savings and conservation improvements can both be made by leaving areas unplanted that would be uneconomic; for example, damp areas that would require a lot of draining but would still give poor growth, areas with soil pans near the surface, frost hollows and rocky areas. Leaving small fields unplanted adds a totally different habitat within the forest, but in the long term maintenance will be needed if these are not grazed. Keeping south-facing banks free of shade can be particularly beneficial. Consider not beating up (replacing trees that have died) failed areas.

Stage 7: Forest operations in relation to conservation

Weeding
Blanket weedkiller application the season before planting has been shown to kill extensive bramble and bracken and assist both young trees and the ground flora in following years. Spot weeding with herbicides round young planted trees, rather than complete weed control, is efficient, saves money and has no significant effect on the variety of vegetation. Keeping chemical applications to a minimum by careful timing is a saving too. Tree shelters provide the opportunity to leave natural growth between widely spaced trees. However, as widely spaced trees take a long time to close canopy, uniform vegetation of bramble or bracken may dominate the ground flora and will need control by cutting or herbicide application.

Providing diversity
Underplanting mature broadleaves with a conifer understorey, for example western red cedar or hemlock, can control excessive bramble growth as well as add variety. The cedar or hemlock can be removed and sold for poles later but in the meantime it provides a shrub layer for birds. The use of larch or broadleaves on the edge of evergreen conifer plantations has the effect of extending light into woodlands and encouraging the ground vegetation. The use of finely branched conifer provenances also lets more light into plantations. Perhaps the most significant forest operation at the time of planting is the pattern of mixtures. Observation has demonstrated that conifer/broadleaved line mixtures let more light reach the forest floor than groups of broadleaves in a conifer matrix. They are also easier to manage, though they are visually less attractive until most of the conifers are removed.

Brashing and thinning
Brashing is an expensive manual operation but it is useful to brash early to let in side light and thereby prevent the complete suppression of the ground flora. Leave brash and branches on the ground for shelter and nest sites but windrow (place in linear heaps) if possible to allow plant growth between. Heavy crown thinning rather than light, low thinning increases the amount of light reaching the forest floor and provides irregular patches of light, which is beneficial to the ground flora and insects such as the speckled wood butterfly. Thin the outer edges of planted crops heavily and early in order to allow rideside flowers to extend into both broadleaved and conifer plantations. Removing branches (high pruning) from final crop trees lets in light and facilitates movement under trees for birds such as the sparrowhawk and

owls. Sometimes it may be necessary to fence off sensitive wildlife areas or leave parts of plantations unthinned as barriers, if there is much public access. Non-timber trees that have grown up naturally can be left if they are not interfering with final crop trees.

Extraction
Felling and extracting thinnings disturbs and lays bare the soil, encouraging the germination of seeds and providing bare ground for species such as solitary bees, as well as for many flowering plants. This essential operation of productive forestry therefore has particular conservation value. The amount of ground laid bare and bramble controlled will depend on the extraction method. For instance, dragging logs out on the ground will bare it more than 'forwarding' produce out on trailers.

Regeneration
Forest edges can be worked on shorter rotations for species that benefit from clear felling; for example, merlin and curlew in the uplands, woodlark and nightjar in the lowlands. There is great scope within normal forest management practices to vary felling coupes in size, shape and distribution throughout the forest in order to provide a whole range of diverse habitats. Felling coupes in some cases need careful shaping to avoid frost pockets by allowing the cold air to move freely downhill. In felled areas, leave some single trees as perches for birds, such as tree pipits and owls. Coupes will attract different species depending on their management; slash (lop and top) can be windrowed (cover), burnt (bare ground), not replanted immediately and grazed, or given a pre-planting application of herbicide.

Timing of work
Forestry operations should be avoided at sensitive times of year for species affected by disturbance. This is especially important during the nesting season for birds.

Stage 8: Creative conservation work requiring extra expenditure
There will always be many opportunities for additional expenditure on measures designed specifically to meet conservation objectives. The following are some examples.

Ponds and flowing water
Available water is essential to all animal life. It is, therefore, particularly beneficial to make new ponds in areas lacking water and to clean out old ponds that have become filled with branches, leaves and silt. Flowing water is also valuable as it is well aerated and provides for a different range of aquatic wildlife.

Plants
There is considerable scope for sowing flower seeds to increase wild plants. Useful guidance is given by the NCC (NCC 1988). Pond plants can also be introduced.

Nest boxes
The erection of boxes for birds or bats, particularly in young woods, is always beneficial but some maintenance will be needed.

Rides and woodland edges
New or widened rides can usefully be established. The mowing of rides, clearings and stream sides on rotation is expensive but valuable. Woodland edge is an important area in the woodland. External edges often sustain a variety of wildlife as they are adjacent to other habitats. Planting hedges and shrubs, especially those bearing fruit, on the margins of woods encourages many species of wildlife. Such plantings will provide the greatest benefits in southern Britain.

In many areas, native trees and shrubs will appear spontaneously but if planting is needed, the choice of species should be made selectively, depending upon the wildlife species to be encouraged and the local geography. Unnecessary costs can be incurred by planting inappropriate species in unsuitable places. Many tree and shrub species are only of benefit to insects in the south of Britain. If birch is to be introduced, ground scarification and seed spreading may be much cheaper and more effective than planting, as birch does not transplant easily.

Stage 9: Sources of information and advice on conservation
Advice on particular aspects of conservation, or on an individual species, may need to be sought from a specialist national or local organization, or from knowledgeable individuals. Such assistance may be appropriate for habitat design at the beginning of a project, for advice on rarity, or for monitoring the success of the project. There are many sources of outside help and advice and these should be made full use of.

LITERATURE
Useful literature is available in the form of county records published by local natural history societies, and their specialist publications. There are books dealing with the local natural history of particular regions and these are usually available in local libraries.

Biological surveys, distribution atlases and other publications are produced by many organizations, especially The Botanical Society of the British Isles, The Mammal Society of the British Isles and The British Trust for Ornithology. The Forestry Commission's *Wildlife Conservation in Woodlands, Ranger's Handbook* and their many leaflets on wildlife are an essential basis of conservation management in woodlands, as are the former NCC publications. Publications are also available from the Farming and Wildlife Advisory Group and the Ministry of Agriculture. Information on bird boxes is provided by the RSPB and there is a useful book by L. Bolund (1987). Details of bat boxes can be obtained from the Mammal Society.

CHAPTER 9

Woodlands of the Future

We have discussed many ways in which wildlife conservation and forestry can be combined, and have shown that productive forestry is an excellent medium for a wide range of plant and animal species. The new forests created during the twentieth century in Britain are still young but as they come into full production and are managed on a sustained yield basis, habitat diversity will inevitably increase and the associated range of species will multiply, many becoming more common. It is unnecessarily limiting to see wildlife conservation only in the context of ancient woodlands and those woods where the production of timber has no place, as this reduces the opportunities for conservation. Indeed, in Britain there is very little primary woodland; that is, woodland growing on sites thought never to have been cleared by humans. The largest of these areas are the relics of the Caledonian pine forests in Scotland, remaining now in such places as Glen Affric in the Highland Region and the Black Wood of Rannoch in Tayside. Some broadleaved primary woodland exists in southern England; examples are parts of the Wyre Forest in Shropshire and the Forest of Dean in Gloucestershire.

All these relict areas, however, have been utilized for timber and forest products for many thousands of years and have been managed intensively for this purpose. The area they occupy is small and totally inadequate to supply domestic timber needs, so for generations Britain has imported large and expensive quantities of timber and wood products every year. Britain no longer has an empire to provide these imports. The countries that used to supply them are developing themselves so they need the timber at home, or for manufacture into saleable articles for export, thus providing themselves with work and increased revenue. At the same time, other traditional suppliers, Canada, the USA and Scandinavia, are realising that their natural forests need conserving. Only Russia still has extensive untapped supplies of softwood, but it is now accepted that even these are not inexhaustible. In the eastern states of North America, however, there are extensive stands of high-quality hardwood timber that have arisen on abandoned farmland when the descendants of the early settlers moved west. These are now yielding significant supplies for both home use and, increasingly, for export to Europe. This is not yet a fully utilized resource as no more than half the annual increment (see chapter 4) is being cut at present, with the result that a reserve of stock is being built up (Harris and Harris 1990). The principal species are various oaks, ash, black cherry and the sugar maples. Their availability from an economically stable region means that only the best broadleaved plantations will have an economic future in Britain.

Although there is not a long forest tradition in Britain, the climate is excellent for growing trees, which like the warm, wet conditions, particularly in western Britain. From Devon through Wales to western Scotland, there is land which is only marginally suitable for farming and it is sensible to grow trees on it. As the European Union develops, forestry should become more important and Britain should have an increasingly important part to play in this if imports are to be reduced so that budgets can be balanced. All western European countries, other than those of Scandinavia, import timber but none as much as Britain, yet coniferous timber trees grow faster and better in Britain than elsewhere in the EU and much faster even than in Scandinavia.

There is thus room for much more forestry in upland Britain and on a smaller scale in lowland areas too, where woodlands can be integrated into farmland. It is more than fortunate that an expansion of forestry, so very desirable from an economic standpoint, can also provide both ecological and recreational benefits of immense value without detracting from the primary objective of wood production. There is no other industry, except perhaps the water supply industry with its ability to support fishing, boating and birdwatching, which can enhance wildlife and recreation at the same time as providing a renewable raw material.

Woodlands provide refuges for animals, birds and plants that cannot survive in towns, industrial areas and against intensive modern farming practices. Many of these species hide and breed in woodlands, venturing forth when it is safe to forage in surrounding areas. Woodlands therefore have a much wider ecological role than just providing habitats for closed forest species. As a direct result of the increase of forestry plantations there are more deer in Britain now than ever before and also a wider range of species; the fallow, sika and muntjac have all been added to Britain's wild fauna. Indeed, deer are now so successful that they are major forest pest species in many areas. The native red squirrel has retained its foothold in Britain in the pine plantations of East Anglia and a few other places, without which it probably would not have survived in England. Pine marten, at one time almost extinct, is now common in the Great Glen and extending its range, entirely as a result of forestry. Polecats and wild cats are all increasing in coniferous plantations, adding exciting variety.

The crested tit, once confined to eastern Scotland, is increasing in Scottish plantations and the crossbill has colonized coniferous plantations in England and Wales. The nightjar and woodlark, which were both becoming rare, have found a new refuge in coniferous plantations, now their main habitat. The shy golden oriole is breeding in poplar plantations in East Anglia and the parrot crossbill nests in pine plantations in the same area. The firecrest from southern Europe is establishing a foothold as a breeding species in spruce in southern England and other species are certain to follow. For instance, it may not be long before the pine grosbeak, yellow browed warbler, pine bunting, mealy redpoll, arctic redpoll and brambling stay to nest, as all are already visiting Britain. Other likely colonizers include the nut-cracker, common in coniferous forests throughout Europe and Asia, and the black woodpecker, which has been spreading westwards in Europe and is now breeding just across the English Channel in France.

Woodlands are refuges for many meadow, hedgerow and roadside plants, as there is no heavy grazing or intensive cropping to prevent their establishment and chemical spraying is minimal. Rare woodland plants, such as the ladies' tresses and ghost orchids, are also found in managed forest. Many non-woodland insects can find suitable habitats and protection within such open areas of woodland as rides, roadsides, clearings, failed areas and replanted coupes. For these plants and insects, and for many birds and mammals, which species of trees form the woodland is not as important as the shelter and habitats that the trees provide. The age and structural variety of the woodland are important in determining the variety of habitats but too often, old, gladed, oak woodland has been favourably compared with woodland made up of other tree species, especially young coniferous plantations in the uplands, giving the impression that the latter have little conservation value. However, it is the age of the trees and the fertility of the site that are important, rather than the tree species present (Mitchell and Kirby 1989).

In previous chapters we have shown the potential that exists for the conservation of wildlife in managed woodlands, and in particular that active management can provide benefits for many species if normal forest management is carried out, because it maximizes habitat diversity by varying the age structure and species composition.

In chapter 1 we showed that Britain falls naturally into the taiga zone, the transitional mixed forest zone and the temperate broadleaved forest zone. Britain's man-made forests should follow this pattern if they are to maximize wildlife benefits. The upland coniferous forests will support species typical of the taiga, whilst mixed forest will support both northern and southern species. Conditions for growing good broadleaves improve southwards and eastwards, and it is in these regions that the conservation of southern species of plants and animals, especially insects, is important. No amount of broadleaved planting outside the climatic range and physiological limits of an insect species will improve its conservation. More woodlands of all types and as wide a diversity as possible are needed in order to improve and pass on Britain's wildlife heritage. It is fortunate that there is an economic need for these woodlands; without this stimulus there would be scant resources to provide wildlife habitats. Instead, Britain can capitalize on the need for more home-grown timber and the large sums that are being invested in new woodlands to bring back a more natural balance to the countryside, with many more mammals than were present in recent centuries and a much wider range of plants and birds.

The forests of the future will also increasingly provide opportunities to enjoy and study woodlands. This will be of immense value in Britain's industrialized society, now that over 90 per cent of the population lives in urban surroundings. There are material and social benefits to be derived from forests, including increased opportunities for leisure activities; urban living generates a need to come into contact with natural surroundings, and the 'wilderness' atmosphere of woodlands and forests provides for these needs. Woodland areas can accommodate informal car, picnic and walking visits to the countryside more easily than farmland, where damage to crops and farm stock is always a danger.

Woodlands and forests can absorb many more people than the open countryside and yet still provide solitude. Informal car parks can be provided in woodlands, and trails and paths arranged, without significantly reducing the timber-producing capacity, and well-placed caravan sites can be unobtrusive. All these developments can go hand in hand with wildlife conservation and need not conflict with it. Indeed, a greater appreciation of wild species by the general public is needed if conservation is to have its rightful place in an affluent society, and productive forestry will increasingly provide the means for studying wild species. The quiet pursuits of bird-watching and observing animals are well provided for in forests of all types, especially where there is a diversity of ages and tree species.

How will forestry practices develop as the new plantations mature? Sustained yield, the fundamental principle underlying forest management, was explained in chapter 4; the object is to go on producing a crop of trees and the timber they provide, permanently. This principle of sustained yield creates a forest, with all stages from establishment through thicket and pole stage to maturity, in which a range of habitats is provided. There are many sophisticated methods of forest management but the objectives are the same. For instance, in a woodland of mixed ages, a calculated volume of timber can be removed each year without reducing the overall yield. This 'allowable cut' can be calculated from growth records.

Forest management thus involves continuous growth of the trees and the regeneration of harvested areas. Regeneration can sometimes be by natural seedlings that have grown up under the mature trees before they were felled or that quickly fill the gaps left by felling. However, natural regeneration, although aesthetically appealing, is difficult to ensure and often is very slow to form new woodland. It can be supplemented by planting additional trees; more often, no attempt is made to wait for natural regeneration and the whole area is replanted. In Britain recently, taking a lead from the organization *Pro Silva* in Europe, interest has developed in so-called 'continuous cover forestry' or 'irregular forestry', which aims to mimic natural conditions by adopting regeneration systems under which the ground is never cleared, claiming this to be more natural. Referring to the provision for wildlife habitats,

Hart (1995) reports that 'Irregular forestry and the selection systems are clearly the better provider of these benefits' but gives no supporting evidence. As we have said in chapter 8, there are species which benefit from selection forestry but they are in a minority. Nor does continuous cover forestry approximate to natural forests in our northern temperate climates where windblow, even in lowland woods, is a common occurrence and the most usual precursor to truly natural regeneration.

Replanting, or artificial regeneration, has many advantages. The first is that there is much more certainty of success and the second is that the whole area is utilized, ensuring the maximum use of suitable land. Additionally, there is an opportunity to plant young trees from genetically improved stock or even to mix or change the species to include a more desirable species. Genetic improvement was first applied to the major commercial species, particularly to Sitka spruce. Recently, improved strains of our native Scots pine have become available and this will mean that this previously rather slow growing native species, although of high conservation value but often rejected on grounds of low yields, will be more readily used in primarily commercial plantations. Probably the greatest advantage is that much quicker early growth is achieved because the planted trees have been grown for two or three years in a nursery and are sturdy with well developed roots. All these advantages of artificial regeneration add up to increase the yield from plantations over natural forests by about two-and-a-half times; this will become more significant in the future as world timber supplies are expected to be outstripped by demand before the end of the twentieth century. Undoubtedly, technical improvements of this type will occur and will further speed up tree growth in the future. Higher growth rates and yield should not be seen as conflicting with conservation but as providing an opportunity to enhance it. Quicker growth means passing more quickly from one forestry stage to the next and more chance to diversify the whole woodland. Improved techniques will allow a wide range of tree species to be grown, again diversifying habitats; a recent example is the introduction of the tree shelter, which provides much improved opportunities for growing oak and other hardwood species.

The enormous potential of biotechnology, which can provide such benefits as faster growing trees with finer branches, narrow form and resistance to pest attack, is only just emerging. These improved trees will shorten the dark thicket stage, allow more light in and enable more species to coexist in the forest without excessive damage, and will also create opportunities to shorten broadleaved rotations and increase the diversity of species and variation in productive woodlands. Woodlands of the future will make use of the many successful introduced species, both broadleaved and conifer, as well as native species. They will be intensively managed to provide the timber and wood products that a highly industrialized society needs.

After a few years of protection from browsing animals and the weeding of the young trees, an established young forest soon reaches the 'closed canopy' stage; that is, the branches of the trees start to interlace as the trees grow outwards and upwards. From then on, the trees dominate the site and only shade-tolerant plants will grow under them. At this time the amount of wildlife is at a minimum. Very soon the trees begin to compete with each other and, if they are to go on growing well, they need to be thinned out by cutting the poorer trees and allowing more light to the better ones. The trees cut out are useful as stakes for fences, for firewood and for other small produce. This process of thinning is carried out periodically for the rest of the life of the crop, always with the object of benefiting the best of the trees. The aim is to produce a final crop of excellent trees, spaced evenly throughout the wood, so that they all have an equal chance to grow. The selection of trees to be favoured at each stage is judged by eye but the amount to be removed is calculated from tables based upon the maximum yield of the site concerned. As soon as thinning starts and more light enters the forest again, wildlife is tempted to return and, in a mature forest with large, widely spaced trees and an understorey, there is a variety of life exploiting the varied habitats provided.

Thinning is perhaps the process of greatest benefit to productive forestry and to wildlife conservation alike (Mitchell and Kirby 1989). It is essential to the success of both and entirely compatible with both, as the object is to let in the light. In the past, there were no good markets for small poles, but as Britain's forest industries grow, these are improving. Because the markets were absent, woods have been left unthinned, or were thinned very late; they have become dark and appear inhospitable. The greatest improvement for wildlife in woodlands in the future will be to thin early and regularly; but this will only happen if viable woodland industries are established, and these in turn depend upon a wider acceptance of forestry and enough productive woodlands to sustain them. Conservation and production can be expected to develop hand in hand in this way.

Plantations are often of one species because they are then easier to manage, since two species do not normally grow at the same rate. There is also usually a single most profitable species for a site. This 'monoculture' can, however, have dangers as disease can spread more quickly. In the very long term, the soil may be improved by a mixture of species or at least a species change when the crop is harvested. When bare land is being planted for the first time, particularly if it is poor and exposed, which has been the case on much of the new forest land in Britain, there are few choices as only hardy pioneer species can be used successfully. These are light-demanders that are the first species to occur in a natural succession but they do not produce the high yield of timber obtained from the more sensitive secondary species. The latter are more tolerant of shade and can grow under other trees where there is protection from wind and where the soil has been improved by the first crop. There is an opportunity, therefore, to add to the diversity of new forests in the second 'rotation' when the light-demanding pioneer species have enhanced the soil with their leaf fall.

It is wise never to clear large areas when the first rotation is harvested. Large-scale clearance would throw away the valuable woodland conditions that have been created with regard to the soil, control of weeds and particularly shelter from wind, this last being the single most important limiting factor to tree growth in Britain. The second-rotation forests are not being re-established on large cleared tracts of land, along the lines of the initial afforestation programme. The first rotation is felled piecemeal in comparatively small blocks by felling different stands as they mature. Those on the better sites mature earlier than those on poorer sites, even if they were planted at the same time. Different species mature at different ages. In large, uniform areas, some crops are felled prematurely, or will suffer from windblow, and others will be delayed. All this has the combined effect of evening out the yield and greatly diversifying the habitat.

The second rotation (well on its way in some older British forests now) is planted, or naturally regenerated, in small coupes (regeneration areas) scattered throughout the forest. There is a good deal of evidence from continental European as well as British forests that 2 ha coupes are an economic size to harvest and have advantages for habitat conservation, though in some cases smaller areas are appropriate. The Chilterns Plan for the regeneration of the ageing Chiltern beechwoods recommends not more than 3 ha coupes, but in some large upland forests bigger areas are appropriate. The early and mature stages of the forest sustain the richest wildlife so the incorporation of short- or long-term rotations within the forest plan can be considered, depending upon the species to be encouraged. Short rotations will favour open ground species. Long rotations will favour those depending on old growth and dead wood and should preferably be sited in protected valley areas where the risk of windblow is least (Peterken et al. 1992).

In upland Britain windblow has caused extensive premature losses in may large-scale afforestation schemes on poor and exposed sites. Started at Kielder Forest in Northumberland and now practised widely in all maturing forests, a system known as 'restructuring' is practised to overcome these difficulties. Shortly before the end of the first rotation, the plantations are divided into a series of separate felling coupes with relatively windfirm edges by using existing features, such as forest roads, rides and streams, in scale with the topography. The age class distribution is extended by manipulating the stands themselves so that in due course adjoining stands will be at different stages of development. Whilst the original objective was to create a more stable forest, in practice, restructuring has provided additional significant environmental benefits, i.e. an improved forest landscape and, of particular importance to nature conservation, an increasingly wide range of habitats adjacent to each other with differing ages and structures (Spencer 1995).

The phase of establishing large, new forests in Britain, particularly of conifers in the uplands and on poor soils, is largely complete. We are not likely to see many new commercial forests established, whilst further additions to Britain's woodland cover will be in small areas on surplus farm land, mainly in the lowlands, and most will have a significant broadleaved element. In addition, we are likely to see a slow improvement of the many abandoned small broadleaved woods by bringing them back under management. Both these developments and the regeneration of the upland forests by stages rather than in the massive, uniform blocks that they were first established in, will result in much more varied and intimate woodlands and forests. This will be increasingly significant and valuable for wildlife conservation and will enhance the appearance of the landscape of much of Britain.

In every case the forests of the future will be a mosaic of different age-classes spread throughout the forest area and constantly changing. At the same time they will be refuges for many forms of wildlife and will provide a variety of relaxation and recreational opportunities. Familiarity with all kinds of woodland, particularly with the trees they contain, will lead to their acceptance in the British countryside and perpetuate a varied tree heritage of much greater diversity than exists today. Conservationists need not fear these developments which will supplement the protected conservation sites and extend the range and numbers of both rare and common species. Forestry will provide an enormously valuable wildlife resource as part of good forest management, as an expanding forest industry develops.

Recent developments outside the commercial sector are the community forests around urban centres (where the aim is to create 'natural woodlands') and the New National Forest in the English Midlands. The latter is to cover 200 square miles eventually and will be about one third planted in roughly the proportion 3:2 of broadleaves and conifers. This is in an area with previously only 6 per cent woodland cover. In Scotland the Central Scotland Woodland Initiative aims to enhance the landscape between Edinburgh and Glasgow. All this new planting will also provide new wildlife habitats and constitutes significant developments from this point of view. None has wood production as a primary objective - public access and landscape enhancement are paramount - but in most cases it is of some importance. Where appropriate, wildlife conservation has an important place and, just as in primarily productive forests, its scope will increase as these community forests age and become more structurally varied. This will occur more quickly than in the large upland afforestation schemes of the past, both because these forests are being established on fertile lowland sites and because a wide variety of tree and shrub species are being planted. The New National Forest is to be dispersed through farm land rather than being a continuous area; thus it will have many external edges, the 'ecotones' so important to wildlife (see Introduction). All these initiatives will need continuous management if they are to achieve their complex objectives.

Another interesting and valuable initiative concerns the remnants of the ancient Caledonian pine forest in north Scotland. Attention was drawn to the importance of the Caledonian pine by Professor Harry Steven as long ago as 1957 in his book *The Native Pinewoods of Scotland* (Steven and Carlisle 1957) and was reinforced by a later visit he made to probably the most important relict, Glen Affric, in 1959. This led to the current awareness of the heritage value of the native pinewoods where some trees are over 300 years old, with uncommon species, such as wintergreens, ladies' tresses, twin-flower, lesser twayblade, black grouse, pine marten, wildcat, capercaillie, Scottish crossbill and crested tit. The forest is estimated to have covered 1.5 million hectares but the remnants are now reduced to a mere 16 000 hectares (Forestry Authority 1994a, 1994b) listed in *The Caledonian Pinewoods Inventory* (Forestry Authority 1994b). Encouragingly, 85 per cent of these remnants are currently under management (Taylor, C. 1994, Scottish Forestry 48,2). Recently the importance of these woods has been recognized by the imaginative New Native Pinewoods Grant available under the Forestry Authority Woodland Grant Scheme supporting the establishment of new woods of natural character outside existing pinewoods but within the former natural distribution of Scots pine in north Scotland. The objective is to establish new woods that emulate the native pinewood ecosystem by use of local Scots pine seed from seven strictly delineated regional provenances, and the inclusion of tree species naturally associated with the former Caledonian Forest, i.e. birch, aspen, willow, alder, rowan, juniper and holly.

Clearly the wildlife conservation value of this initiative will be of great significance; so will recently announced Woodland Improvement Grants (WIGs), also under the Woodland Grant Scheme, for unmanaged woods and for woodland biodiversity. These are in response to the report *Biodiversity: The UK Action Plan* (see below). One WIG project provides support to owners for bringing low value, unmanaged woods into management and includes selective felling, coppicing and conversion to high forest, respacing and rhododendron control. Another is designed to encourage owners to manage their woods for woodland biodiversity by promoting the conservation of habitats and species in semi-natural woodlands, coppice management for butterflies and dormice, and in Scotland, encouragement of the capercaillie. It includes the preparation and implementation of management plans to meet these objectives.

The future development of Britain's forests in the next few decades is likely to be much influenced by changes in the world scene. Growing concern over rapid destruction of tropical forests and natural coniferous forests in Russia, and a realization that the developed nations can no longer point the finger at forest exploitation in the developing world without setting the same standards at home, led to the Earth Summit held at Rio in 1992. Although the Rio Summit was not solely about forestry, a lot of attention since has been given to forestry in the responses to this world initiative. Thus major changes of emphasis have been proposed for British forestry which, if implemented, should have significant impacts on the increased provision of wildlife habitats. Two important principles have emerged: sustainability and biodiversity, both relevant to the theme of this book. Whilst these changes are in response to global concerns, they come at a time when Britain's comparatively small but maturing man-made forests can readily accommodate them and when our diminishing semi-natural broadleaved woods urgently need them.

Therefore, in response to alarm at the increasing rate of loss of forest cover world-wide, world leaders convened at Rio de Janeiro in 1992 and drew up The Rio Principles. These were followed in 1993 by the Helsinki Guidelines for the

Sustainable Management of Forests in Europe (*General Guidelines for the Sustainable Management of Forests in Europe*), which led to the UK Action Plan (*Biodiversity: The UK Action Plan*, HMSO 1994a) in four parts: *Climate Change, Sustainable Development, Biodiversity* and *Sustainable Forestry: The UK Programme.* The last, part of *The UK Sustainability Development Strategy* (HMSO 1994b), is intended to lead in Britain to a national forest strategy with environmental targets for forestry to meet, emphasizing multi-purpose forestry. Thus our forests will be seen as integrated ecosystems to be permanently maintained and this is to be the guiding principle behind future forest policy in Britain.

Key points in these documents relevant to forest wildlife conservation are:

1 Sustainable forest management
2 Provision of new habitats for plants and animals
3 Avoidance of irreversible degradation of forest soils, sites, flora and fauna
4 Conservation and enhancement of biodiversity
5 Management aimed at increasing the diversity of forest habitats.

The UK Steering Group on Biodiversity reported in 1995 (HMSO 1995) and, amongst other things, recommended changes to the Woodland Grant Scheme, laying emphasis on semi-natural broadleaved woodland and on yew woodland as important habitat types holding large numbers of threatened species. It recommended 'a new woodland initiative for management of semi-natural woodland to reflect the higher conservation and amenity value of such sites and the general lower timber production potential'. It also set national targets for 116 endangered species and 14 threatened habitats as well as recommending improved biological recording. A few of the endangered species are woodland species, in particular the red squirrel (dependent on conifers and for which the objective is to maintain current populations and re-establish where appropriate) and the narrow-headed ant (*Formica exsecta*) which occurs in Caledonian pine forest and the Plymouth pear. Included among the threatened habitats are the upland oak woods of western Britain, whilst the New Forest in Hampshire holds possibly the rarest fungus in Europe, the nail fungus (*Poronia punctata*), dependent upon pony droppings and once common throughout western Europe. Thus Britain has made a good start and indeed, in the words of The Wildlife Trusts, 'No other nation has come up with such a comprehensive programme for the conservation of its wildlife' (Natural World 1996).

At the same time, there are local initiatives emerging to supplement and assist with the implementation of biodiversity. Several wildlife trusts are working with local authorities supported by English Nature and encouraged by the Department of the Environment and the Royal Society for the Protection of Birds, to produce local guidelines for land managers, farmers, planning and highway authorities to address biodiversity issues at all levels, with both local and national targets.

'Sustainability' is beginning to be looked at in a much wider context than merely sustainable resources or sustainable forests but as sustainable communities. This is already accepted as an important principle of the sustainable management of tropical forests and their indigenous people by providing work for local people, and it can be applied in rural areas of the UK. For example, in response to unemployment between the wars, the Forestry Commission set up forest villages in the more remote afforestation schemes. Now that these forests are mature, their place in some rural communities is becoming much wider than the production of timber alone, to the extent that some may become community forests, owned and managed by the local community, as occurs elsewhere in Europe.

On the wider scene and following Rio, as mentioned previously, European nations adopted guidelines for sustainable development at Helsinki in 1993, whilst USA, Canada, Australia, Chile and Russia created the Montreal Process and the Santiago Principles. These initiatives should all lead to the adoption of international agreements and national standards. Of particular concern is the unscrupulous marketing of timber, particularly tropical timber, from forests where there is no attempt, and often no intention, to regenerate them. In response to this concern and prior to government initiatives, the Forest Stewardship Council (FSC) was set up in 1992 by the Soil Association as an international organization to certify products from sustainably managed forests by issuing Certificates of Responsible Forestry Standards under the 'Woodmark' label. It promotes the use of wood as a natural resource obtained from sustainably managed woodlands meeting the Soil Association's criteria of environmental protection, appropriate landscaping and biodiversity conservation alongside sustainable production. Similarly, the 'Qualifor' symbol accredited by the FSC is a certification of management quality rather than individual forest certification. The FSC, which now has its headquarters in Mexico, is working towards the preparation of 'principles of good forest management' which are to be applied through an internationally agreed accreditation programme. The membership of the FSC is broad but self-appointing and self-accountable. It is, however, the only accreditation organization with a global remit which is acceptable to environmental organizations, such as the World Wildlife Fund. It is also seen by large retailers in the UK as a credible organization to approve world-wide supplies of their timber products and thereby satisfy the environmental concerns of their customers.

While the overall FSC principles are universal, individual country assessments are being produced tailored to the conditions, needs and implementation methods most appropriate at national level. Thus there are a large number of accreditation organizations around the world which have been approved by the FSC to undertake timber accreditation to their standards. In the UK the Soil Association is an FSC approved accreditor and in North America there are several, all monitored and audited on a regular basis.

In several countries other groups have come together to produce approved systems in order to give confidence to consumers; for example, there are moves towards Scandinavian, Canadian, Australian and New Zealand 'Woodmarks' but these are not recognized by the FSC. In the UK, the Forest Industry Council of Great Britain (FICGB) launched their 'Woodmark' labelling scheme for British timber derived from well regulated forests. FICGB's 'Woodmark' authenticates wood products made from home-grown timber under government licence and thus subject to regeneration conditions, i.e. those imposed by the Forestry Authority through felling licences and replanting conditions. It depends upon Britain's commitment to sustainable forest management with an effective regulatory system and was established in response to forestry regulations at the Rio Summit.

Convergence of the two schemes in the UK, both using the 'Woodmark' label, is needed and inevitable, whilst taking account of the need for international compatibility, which is the aim of the Inter-governmental Panel on Forests. This is due to report to the UN Commission for Sustainable Development in 1997. Monitoring at reasonable cost will be the key to the success of these schemes.

Whilst the woodmark schemes attempt to set measurable standards of sustainable wood production and utilization, the concept of sustainability goes much further and extends to non-timber values as well. Foremost amongst these is wildlife conservation, not on its own but as an integral part of the whole. In Part I we have shown how long-term management with sustained yield as its basis, though not necessarily the primary purpose, provides a wealth of wildlife habitats. Thus with these recent developments the prospect for conservation of a wide range of woodland species, and indeed extension of the range of some rare species, is good in Britain. In Part II we look at their specific requirements.

Flora and Fauna in Productive Woodlands and Forests

CHAPTER 10

Plants

Productive forestry in Britain, particularly the use of exotic species, is often dismissed as of little value to the conservation of native flora. The opportunities provided by plantations and other managed woodlands are thereby undervalued. In this chapter we look at woodland plants in a wider context than Britain alone and go on to show that the native flora is both compatible with planted forests and by no means limited to broadleaved woodland.

THE ORIGIN OF WOODLAND PLANTS

The first vegetation to colonize Britain, as the glaciers melted, resembled that of tundra today. Mosses and lichens, together with bog and wetland species, were established early. The drier ridges were colonized by species with windborne seeds, including meadow flowers, grasses, and low shrubs such as willow *(Salix* spp.) and dwarf birch *(Betula nana)*. Some of the soils were already nutrient-rich and nettle *(Urtica dioica)*, tormentil *(Potentilla erecta)* and betony *(Stachys sylvatica)* are found in the early pollen record (Peterken 1981). It has been suggested that woodland soils had already developed into various forms of brown earth before the wet Atlantic period of about 5700 years BP (Dimbleby 1965), when leaching became more significant under the prevailing conditions. Pine *(Pinus sylvestris)*, birches *(Betula* spp.), juniper *(Juniperus communis)*, aspen *(Populus tremula)*, rowan *(Sorbus aucuparia)* and willows *(Salix* spp.) were the first trees to form woodland. Woodland ground vegetation must have developed then, as well as a fungal flora. As the tree canopy became more extensive, vegetation similar to that found in the taiga forests today developed and a varied woodland flora seems to have existed by this time. With rising temperatures, hazel *(Corylus avellana)*, followed later by oak *(Quercus* spp.), increased towards the end of the Boreal period (8000 years BP), when the climate was typically continental with warm summers, favouring such species as wood sage *(Teucrium scorodonia)*. The oak was probably sessile oak *(Quercus petraea)* as it thrives on drier and more acid soils than pedunculate oak *(Q. robur)* but the latter is more frost-resistant.

The Atlantic period saw the onset of warm, damp conditions which initiated the spread of the northern broadleaved forest (Zone 5:2; see chapter 1).Pedunculate oak may then have become an important component as it prefers deep, moist, fertile soils, although it was probably also introduced in preference to sessile oak much later when short-boled, heavy-branched timber was needed for house building and shipbuilding. It formed a forest composed of a number of species on the richer soils in the river valleys and low-lying land, mainly south of the Scottish border, together with alder *(Alnus glutinosa)*, lime *(Tilia* spp.) and elm *(Ulmus* spp.). This broad-leaved forest was by no means dominated solely by oak, as is often suggested but was similar to the remaining forests of northern broadleaves found today in Europe, Asia and North America.

During the Atlantic period, sessile oak was pushed further up the hillsides where the drainage was sharper and the soils were drier as the climate allowed a higher tree line than today (Anderson 1967). The elements of woodland flora that thrive in moist, deep soils found conditions favourable at this time. The changes in climate since then have caused an ebb and flow of vegetation types as fluctuations between cool and warm periods followed. The conditions prevailing at the climatic optimum in the Atlantic period have not been attained since; the cooling of the climate has rendered conditions less suitable for a number of plant species, with the result that many are now rarities in Britain. Unlike animals and birds, plants are unable to move physically as climatic conditions change, so patches of uncharacteristic plants remain, hanging on in unexpected situations where there are suitable microclimates. These are referred to as relict (or 'left behind') species. The 'Teesdale' flora, for example, contains a number of rare species, including a mixture of arctic and steppe plants which must have been early colonizers (Gilmour and Walters 1954).

The influence of microclimates can be observed in the distribution of many plant species and indeed in such small areas as gardens. The persistence of shade-bearing woodland plants on moorland is considered to be a reflection of the time that has elapsed since moorland was deforested, as they eventually disappear (Burnett 1964), but is also a reflection of suitable microclimates there. The lesser twayblade *(Listera cordata)*, a woodland species in other parts of its range, grows on damp sphagnum moorland in Scotland, where the conditions of shade provided by the heather or other plant cover are favourable to its survival. Western European or Atlantic species that require moist conditions, such as bluebell *(Hyacinthoides non-scripta)*, survive outside woodlands in the west because of the wetter climate and are not restricted to damp woodland as they are in the drier east of the country. Heather *(Calluna vulgaris)* finds conditions on the west coast too wet but it gradually increases eastwards; it then becomes confined to woodland situations on mainland Europe as conditions become warmer and drier, whilst it is not present at all in eastern Europe. Northern plants tend to occupy the northern slopes of hills with the more sensitive species on the warmer southern slopes (Salisbury 1961; Burnett 1964; Wang 1961). Similarly, the plants growing on the shaded sides of a woodland may be very different from those growing on the sunny side (Ellenberg 1988). Thus species which grow in a wide range of situations in the more continental climate of the European mainland are confined in Britain to the eastern side of the country and to warm soils, such as chalk, limestones and sands. All plants have an optimum range of conditions in which they thrive and many of them grow in a diversity of habitats if these conditions are fulfilled. However, they do not necessarily become dominant in areas where these optimal conditions occur if there is competition from other species. Without competition all woodland species could grow in all woodland types but competition and available plant nutrients determine the dominant vegetation in a particular situation (Ellenberg 1988).

The present plant composition of British woodlands consists of species that can be found in the taiga, the coniferous mountain forests of Europe, mixed forest and northern broadleaved forest (see chapter 1). Plants with a more southern distribution that occur in broadleaved forest are now only found in the warmer southern and eastern regions of Britain.

Many plant species colonized Britain before the Atlantic period; these comprise more than 300 out of the overall total of 500 species that have been identified from deposits (Godwin 1956). These are found today over a wide geographical range. Woodland plant species occur under a variety of forest types, their presence being dictated more by the conditions for growth beneath the canopy than by the tree species; thus, species typical of oak or beech forest in Britain are found in mainland Europe under other broadleaves and in mixed and coniferous forests.

The vegetation of Britain is therefore a complex mixture of species and has been subject to intensive investigation; a great deal of published literature exists. Numerous British classifications have been based on such things as geographical factors, the communities of the plants themselves, tree cover and the nature of the ground vegetation. Some rely on descriptions of the tree cover as the main classifying factor (Tansley 1949; Peterken 1981), whilst others use a wider range of factors but all are related to British conditions. Looking further afield, Poore (1955), Poore and McVean (1957) and McVean and Ratcliffe (1962) related British vegetation to that of the rest of Europe using methods and classifications developed by Braun-Blanquet, which have been developed into the National Vegetation Classification. Four volumes have now been published (1996). Volume 1 deals with woodlands (Rodwell 1991). The most recent classification following these lines was developed by The Institute of Terrestrial Ecology at Merlewood, Lancashire and has 32 classes based on Site Indicator Species Analysis developed by Hill et al. (1975). For this classification, plant species were selected by computer for dominance from 103 surveyed woodlands (Bunce 1982).

BRITISH WOODLAND PLANTS IN A WORLD CONTEXT

In very general terms, the woodland plants of Britain fall into the following world categories:

1 Northern species occurring in America and Eurasia.
2 Northern species occurring in Eurasia.
3 Northern European and west Asian species, including moisture-loving 'Atlantic' species.
4 Southern European and west Asian species, including moisture-loving 'Atlantic' species.

Table 10.1   World distribution of plants found in British woodlands

Boreal (or northern) species found in America and Eurasia	Boreal (or northern) species in Eurasia only (related species occur in North America)	Northern European and west Asian species	Southern European and west Asian species
Twinflower	Chickweed–	Blackberry	Bluebell
Lesser twayblade	wintergreen	Wood anemone	Yellow archangel
Ladies' tresses	Wood cranesbill	Dogs mercury	Wood spurge
Coral root	May lily	Herb Robert	Old man's beard
Wintergreen	Herb paris	Greater stitchwort	Bryony
Vaccinium spp.	Lily of the valley	Common cow wheat	
Wood sorrel	Solomon's seal		Teasel
Goldenrod	Ghost orchid	Figwort	
Nettle	Lady's slipper orchid	Wood avens	Carex strigosa
	Broad-leaved	Earthnut	Daphne spp.
Moschatel	helleborine	Sanicle	
Rose bay willow		Germander	
herb	Common twayblade	speedwell	
Deschampsia spp.	Butterfly orchid		
Luzula pilosa	Raspberry	Greater woodrush	
Bracken	Wood forget-me-not	Brome	
Lady fern	Hedge woundwort		
Male fern	Self-heal	White helleborine	
Hard fern	Bugle	Dark red helleborine	
Club moss spp.	Woodruff		
Carex pallescens		Violet helleborine	
Carex remota	Holcus lanatus	Deadly nightshade	
Wood horsetail	Carex pilulifera	Oxlip	
Bearberry	Buckler fern	Yellow pimpernel	
Crowberry	Bats-in-the-belfry		
Festuca ovina	Columbine		
Festuca rubra			
Cloudberry			

As it is not possible to deal in detail with all woodland plants, a representative selection has been chosen from these groups in the following section. British forests are still young but details of plant communities found in other countries come from long-established forests or semi-natural forests, both of which have been managed for a long time. They therefore give an indication of the potential that exists for plants in more recent British forests. Furthermore, in Britain the paucity of a woodland ground flora is particularly marked in woods planted on agricultural land, compared with those on old woodland sites where a woodland flora was already present. This is because colonization takes time and some species spread more rapidly than others. Plantations develop a flora dependent upon the site and geographical region in which they are planted and on the varying conditions for plant growth provided within them. Colonization and succession follow no set pattern and may be continuous or discontinuous. Depending on the conditions at the time succession commences, different plant communities may develop on similar sites (Peterken 1981).

Information in the following section is mainly derived from Summerhayes 1951; Clapham et al. 1952; Wang 1961; Perring and Walters 1962; Keble Martin 1965; Peterson et al. 1968; Ellenberg and Klotzi 1972; Wright et al. 1973; Schauer 1978; Changbai Mountain Research Station 1980; Polunin and Walters 1985; Mabberley 1987; Knystautas 1987; Ellenberg 1988; and from personal observations during Royal Forestry Society study tours, both in Britain and elsewhere, unless otherwise stated.

THE DISTRIBUTION OF WOODLAND PLANTS

Northern species occurring in America and Eurasia
The species in this group are found mainly in taiga, European and American coniferous mountain forest, mixed forest and northern broadleaved forest. They were early colonizers as the ice retreated. This group contains a number of rare species, some of which were confined to native Scots pine and birch forest in northern Britain until recently, although most have a much wider distribution in various types of forest in other parts of the world. These species are now extending into commercial plantations in Britain.

Twinflower *(Linnaea borealis)* is found abroad in spruce, pine and larch forest, creeping on the forest litter. It can now be found in spruce and pine plantations in Britain (Hill 1979). Lesser twayblade *(Listera cordata)* and creeping ladies' tresses *(Goodyera repens)* occur under conifers abroad, including pine and spruce, as well as in mixed forest and pure broadleaves, where the surface is acid and damp. Both now occur in plantations of both Scots pine and Corsican pine *(Pinus nigra* var. *laricio)* in Britain. Coral root *(Corallorihiza trifida)* and the wintergreens *(Moneses, Orthilia* and *Pyrola* spp.) are saprophytic on acid raw humus and are found under a wide range of trees, including spruce, fir, larch, beech, oak and birch, in other countries. In Britain they have colonized commercial pine plantations and can be found growing under Scots and Corsican pine. The wintergreens, like orchids, have small seeds so are easily and widely dispersed. This may explain the colonization of plantations that are at a distance from previously known stations and should enable populations to spread once established in a forest area. As with orchids, colonization may depend on disturbance exposing bare soil. Dustfalls in Britain originating in the Sahara indicate the distance over which some small seeds could be windborne.

More familiar plants with a wider distribution in Britain on acid soils include the various species of *Vaccinium,* which in other countries form an understorey in mature forests of spruce, fir, pine, larch, beech and oak, but occur more in the open in the wetter parts of Britain. Bilberry *(V. myrtillus)* is common in upland plantations and so is cowberry *(V. vitis-idaea).*

The species so far mentioned are confined to northern and upland regions in Britain but others have a wider distribution and are found in spruce, fir and larch forest throughout much of the northern hemisphere as well as in mixed and broadleaved forest; wood sorrel *(Oxalis acetosella,* synonymous with *O. montana* in the USA), golden rod *(Solidago virgaurea),* nettle *(Urtica dioica),* moschatel *(Adoxa moschatellina)* and rose bay willow herb *(Chamaenerion angustifolium)* are familiar examples. In Britain willow herb is common on clear felled areas and wood sorrel appears under trees after thinning and in mature stands. Grasses in this group include tufted hair grass *(Deschampsia caespitosa)* and wavy hair grass *(D. flexuosa),* both now common in upland plantations. The hairy wood rush *(Luzula pilosa)* too is found frequently in plantations and so is wood sedge *(Carex sylvatica).* Of the ferns, bracken *(Pteridium aquilinum)* and male fern *(Dryopteris filix-mas)* are found under a wide range of tree species in many countries and depend on the amount of shade cast by the canopy. In Britain, both are shaded out in young, dense Sitka spruce plantations but survive under larch (Hill 1979; Bamford 1987). Hard fern *(Blechnum spicant)* tolerates heavy shade and survives well under spruce (Bamford 1987). The club mosses *Lycopodium annotinum* and *L. clavatum* occur in open spruce and fir forest in Asia and America; both colonize ground ploughed for planting in Britain and are limited to upland areas.

Northern species occurring in Eurasia

These plants are found in many forest types, including taiga, mountain coniferous forest, and mixed and broadleaved forest. Many are represented by related species on the American continent. On the whole, they are found on less acid soils than the previous group. Many are on the edge of their range in Britain as summer temperatures are not high enough for optimum growth. As a result, species found on a wide range of soils in other countries are confined to warm soils in Britain, such as those over chalk and limestone, though they may have had a wider range at the climatic optimum.

Some species are only found in northern Britain. Elsewhere, chickweed wintergreen *(Trientalis europaea)* is found growing on acid humus under larch, spruce, fir and pine and in acid oak and birch woods. In Britain, at the beginning of the twentieth century chickweed wintergreen was found mainly in native pine woods in Scotland or on damp moors that had once been under woodland, but it has now extended its range into many upland plantations (Hill 1979). Wood cranesbill *(Geranium sylvaticum)* and golden saxifrage *(Chrysosplenium alternifolium)* also have a mainly northern distribution in Britain and grow in spruce and fir forest in the taiga elsewhere. Both can be found now in Britain's upland plantations, the saxifrage where damp conditions prevail.

It is noteworthy that a number of plants which are found under a wide range of trees in other countries are thought of in Britain as being restricted to broadleaves. However, it is likely that they were early colonizers with pine, birch and oak and are now restricted to broadleaved woods because of the lack of native conifers. The May lily *(Maianthemum bifolium)*, which is now local in Britain, occurs more widely elsewhere on loams with an acid surface under mature spruce, larch and mixed woods, as well as under birch, aspen and other broadleaves. Herb Paris *(Paris quadrifolia)* and lily of the valley *(Convallaria majalis)* occur in spruce, fir, larch, pine, birch, aspen and other broadleaved forests in many countries but have a predominantly eastern distribution in Britain on warm soils. Solomon's seals *(Polygonatum odoratum* and *P. verticillatum)* are found in mixed woods in their wider range, but a predominantly coniferous plantation on limestone in the Lake District, which is well thinned and south facing, has a ground flora containing lily of the valley and Solomon's seal, as well as orchids. This single example may indicate the increasing floral variety that can be expected as introduced conifers, managed with conservation in mind, take their place in Britain.

Plants that always excite attention include the orchids, which in Britain are regarded as typical of beech woods, but it is probably the light regime rather than the tree species which favours them. The very rare ghost orchid or spurred coralroot *(Epipogium aphyllum)*, found occasionally in beech and oakwoods in Britain, is also found elsewhere under pine, spruce and fir, as well as in mixed forest; it is saprophytic on the leaf litter and requires damp conditions as well as warmth and shade. The ghost orchid was probably more widespread in the past, as was lady's slipper orchid *(Cypripedium calceolus)*, which is found under a range of trees in Europe and Asia and occurs under pine in Scandinavia on base-rich soils (Aune 1977). Other orchids with a northern distribution and growing under a wide range of tree species include the broad leaved helleborine *(Epipactis helleborine)*, recorded in mature pine plantations in Britain, the common twayblade *(Listera ovata)*, recorded in both young and mature pine and larch plantations, and the butterfly orchids. The lesser butterfly orchid *(Platanthera bifolia)* and the greater butterfly orchid *(P. chlorantha)* both grow in lightly shaded areas under both conifers and broadleaves.

There are other northern Eurasian species that have a more widespread distribution in Britain. These include the raspberry *(Rubus idaeus)*, common in plantations on clear felled areas and as an understorey after thinning, wood forget-me-not *(Myosotis sylvatica)*, hedge woundwort *(Stachys sylvatica)*, self-heal *(Prunella vulgaris)*, bugle *(Ajuga reptans)* and woodruff *(Galium odoratum)*. All these are found in plantations, especially on road and ride edges, and in well-lit places under the trees. Grasses in this group include Yorkshire fog *(Holcus lanatus)*, which is common under trees where there is enough light; pill sedge *(Carex pilulifera)* too is common in plantations. Common buckler fern *(Dryopteris dilatata [austriaca])* is found under spruce and fir in Europe and Asia as it stands shade. Both Hill (1986) and Bamford (1987) comment on its present abundance in spruce plantations in Britain.

Northern European and west Asian species, including moisture-loving 'Atlantic' species

This group contains plants that have a northern and western distribution in Europe but that are not found in the taiga (northern coniferous forest). In most parts of their range they can be found growing under spruce, pine, fir, larch and mixed forest, as well as under broadleaves. They include many common species that are regarded as plants of broadleaved woodland in Britain. On acid soils, bramble *(Rubus fruticosus)* is common in all types of forest and requires control rather than conservation. On less acid soils, wood anemone *(Anemone nemorosa)*, dog's mercury *(Mercurialis perennis)*, herb robert *(Geranium robertianum)*, greater stitchwort *(Stellaria holostea)*, cow wheat *(Melampyrum pratense)*, figwort *(Scrophularia nodosa)*, wood avens *(Geum urbanum)*, earthnut *(Conopodium majus)*, sanicle *(Sanicula europaea)*, germander speedwell *(Veronica chamaedrys)*, greater woodrush *(Luzula sylvatica)* and slender false brome *(Brachypodium sylvaticum)*, can all be found in mixed and open coniferous plantations in Britain. As with the northern flora, a number of species need warm conditions in which to flower and many orchids are confined to warm soils; again, these are found under a wide variety of tree species throughout much of their range especially where coniferous forests grow on basic soils, but are at present mainly confined to beech woods in Britain. For example, white helleborine *(Cephalanthera damasonium)*, dark red helleborine *(Epipactis atrorubens)* and violet helleborine *(Epipactis purpurata)*, all occur under pine, spruce and fir on basic soils in many countries. A beech plantation in Hertfordshire, established with conifer nurses, now holds a good population of the violet helleborine and such examples can be expected to become more widespread. Deadly nightshade *(Atropa belladonna)* is a species with a similar distribution to these orchids in Britain, extending as far north as the border counties. Oxlip *(Primula elatior)* in Britain is regarded as dependent on broadleaved woods for survival but it occurs in coniferous and mixed woods elsewhere in its range. Its restriction to the eastern side of Britain is related to temperature rather than tree species.

Southern European and west Asian species, including moisture-loving 'Atlantic' species

The plants in this group do not extend into coniferous forest; they require warm temperatures and rich soils and therefore grow in climates where mixed and broadleaved forest occur. They were probably the last group of plants to colonize after the ice had retreated, when conditions had become warmer and damper in the Atlantic period. The Atlantic species, such as bluebell, which thrive in cool, damp conditions are widespread, especially in the west, growing in mixed and broadleaved woods as well as under well-thinned conifers. The range of southern European species in Britain is restricted, particularly in Scotland where they only have a local lowland distribution. Their limit in Britain is perhaps an indication of what would be the northern limit of the mixed forest zone if it were present naturally. The distribution of native tree and shrub species supports this, as more than half of them are confined to England and Wales (see chapters 2 and 5). Yellow archangel *(Lamiastrum galeobdolon)* and wood spurge *(Euphorbia amygdaloides)* are typical of this group and become dominant when coppice is cut but they can also be found growing in plantations. Other species include old man's beard *(Clematis vitalba)* and bryony *(Tamus communis)* as well as the rarer mezereon *(Daphne mezereum)* and spurge laurel *(Daphne laureola)*.

In addition to species that arrived in Britain naturally, there is now a large number of introduced species. Many of these are arable weeds introduced inadvertently in the past. Most are of southern origin (Salisbury 1961). Plants introduced into gardens have escaped and naturalized, whilst others have been deliberate introductions (Hyams 1979). In addition, a few very localized species are only found in Britain but are local forms of more widespread European species that retreated because of the cooler climate since the Atlantic period. Whitebeam *(Sorbus aria)* is such a southern species; found mainly in central and southern Europe, it is now confined to southern Britain though a number of localized and isolated forms survive in scattered localities in the north and west (Clapham et al. 1952). The various 'dune helleborines' *(Epipactis* spp.) now represent a number of isolated local forms occurring on sand dunes but which may once have been more widespread as the type. It is of interest that the damp conditions provided by pine plantations on sand dunes have provided a suitable habitat for these orchids.

British plants therefore comprise a mixed collection of species with varying origins and occupying a wide range of habitats. Many find suitable microclimates within woodland in which to survive although they are not strictly woodland species; in fact only 76 British species can be regarded as mainly woodland plants (Ratcliffe 1977). The greatest differences in ground vegetation occur between very acid soils, which carry a very restricted range of species, and more basic soils. Of the 236 species of plants found in woodland that are listed in *A Nature Conservation Review* (Ratcliffe 1977), only 66 are associated with acidic soils; 15 of these are ferns and 9 are grasses, leaving a total of 42 flowering plants. Plantations growing on acid sites will therefore never hold such a wide range of species as those on the more basic soils. Nor do semi-natural woodlands; for example, the ground flora under beech on the flat, acid plateaux of the Chilterns is limited compared with that found towards the foot of the hills where the soil is basic.

THE SIGNIFICANCE OF ANCIENT WOODLAND SITES

In his book *Woodland Conservation and Management,* Peterken (1981) makes a distinction between ancient or primary woodlands and woodlands arising from plantations: 'the flora of primary woodland contains an intimate mixture of both early and later plant colonisers accumulated over time'. In broad terms this is true, but many have taken it to mean that these species require the present broadleaved forest cover for survival, whereas the examples given above illustrate that climate, the amount of light, soil type (acid or basic, wet or dry) and available nutrients are far more important. An examination of Peterken's ancient woodland indicators shows that a number of species found widely in the taiga abroad are included. In Britain these probably colonized with pine, birch and aspen, as all occur under this type of tree cover in the taiga. Examples are wood sorrel, pale sedge *(Carex pallescens),* remote sedge *(Carex remota),* wood horsetail *(Equisetum sylvaticum),* the butterfly orchids, lily of the valley, May lily, herb Paris, woodruff, bats-in-the-belfry *(Campanula trachelium)* and columbine *(Aquilegia vulgaris).* Other ancient woodland indicators given by Peterken are southern in distribution and are found only in the mixed and broadleaved forest zones (Zone 6 and Zone 5; see chapter 1). These are small teasel *(Dipsacus pilosus),* yellow archangel, *Carex strigosa,* small leaved lime *(Tilia cordata)* and wild service tree *(Sorbus torminalis),* all of which do not grow naturally in Scotland and were probably late colonizers. Peterken's remaining ancient woodland species occur in coniferous, mixed and broadleaved forest in Europe, but some of these species have a local distribution in Scotland.

In the Adirondacks, New York State, studies have shown that the 'non-alpine' spruce/fir forest 'has mostly a ground vegetation identical with that of the mixed hardwood and softwood forest at lower elevations . . . Linkola (1924) found a similar situation in Switzerland. In the softwood belt extending above 1200-1300 m in elevation, he found the ground vegetation of the lower part to belong mostly to the same types as did those of the hardwood belt below . . . Thus many characteristic species of the ground vegetation do not have the same climatic ranges as the tree species usually associated with them' (Heimburger 1934).

Peterken lays much emphasis on the slow colonizing ability of ancient woodland plants, 'the species with a strong affinity showing little or no ability to colonise secondary woodland'. He points out that many of these require damp woodland in the eastern counties for survival and fragmentation of woodland restricts their colonization, recent woodlands adjacent to old woodland being colonized more quickly than isolated woods (Peterken 1984). Elsewhere he also says that coniferous planting on ancient woodland sites does not reduce the range of species present but the number of plants of a species may be reduced (Peterken and Game 1962). However, it needs to be pointed out that colonization is by no means slow further west and that insufficient attention has been given to his statement that 'few (ancient woodland) species appear to be strongly associated with ancient woods throughout their range'. For example, mixed woodland in east Wales on fields planted up in

1960 contained wood anemone, yellow pimpernel *(Lysimachia nemorum)*, woodruff and yellow archangel, plants typical of ancient woodland, only 26 years later (personal observation). Another locality further west consisted of fields that had been planted with mixed woods in 1957. By 1986 yellow pimpernel, figwort, early purple orchid *(Orchis mascula),* earthnut and common dog violet *(Viola riviniana)* were present. In south Scotland, Sitka spruce plantations 25 years old, planted on land far from existing woods and which had been grouse moor for a long time, were already being colonized by wood anemone (personal observations). Peterken (1986) has stated that in Scotland 'habitat continuity is less important as an ecological factor than it is in lowland England'. We think that even in England the colonizing ability of many species has been underestimated, especially where their physiological requirements are met. A paper by Whitbread (1990) strengthens these views. Certainly, many northern species found in the taiga are species adapted to the changing conditions that occur in such forest (see chapter 6) and this may also hold true for many northern and general species found in mixed forest.

Work in plantation forests in Europe, quoted by Ellenberg (1988), has shown that planted forests can be rich in plant and animal life and that this is not universally less diverse than in natural forest. It has been demonstrated that pine planted on former degraded agricultural land takes at least three generations to develop woodland communities. Mosses are dominant in the first rotation, followed by berried shrubs in the second and *Vaccinium* forming the ground layer in the third. This process takes longer in some cases than in others, however. Plantations of Douglas fir *(Pseudotsuga menziesii)* on better soils were shown to have developed a ground flora similar to that of broadleaved forests. Vegetation developed more rapidly under spruce than pine, and typical woodland communities developed in some cases in the first rotation, especially species thriving on the acid litter layer, such as ladies' tresses and twinflower. The type of community that developed depended on soil conditions. On acid soils the difference between plant communities under conifers and broadleaves was not considered significant, whereas conifers grown on dry limestone develop a mixed community. Acid-loving species root in the thin humus layer present in such situations and deeper rooting plants exploit the soil beneath to create a community not found under broadleaves. On most soils, the differences in ground vegetation depend on whether the canopy is composed of light-demanding or shade-tolerant trees. Thus a similar 'open ground' flora tends to develop under light-demanding oak and pine (Ellenberg 1988; Kirby 1988).

WOODLAND PLANTS

Woodland flora consists of a mixture of species depending primarily on climate and soil type, the latter being more important to plants than the tree species forming the canopy above them (Ratcliffe 1977; Hill 1979). Ellenberg (1988) states that in Central Europe 'there is not a single case where a higher plant is exclusively associated with a particular tree species'. Mitchell and Kirby (1989) also state: 'once site

(a)

(b)

(c)

(d)

heterogeneity and difference in growth stages and stocking are allowed for, no ground flora species or vegetation type can definitely be associated with any one tree species, or even with just conifers and broadleaves'. Although pure plots of coniferous and broadleaved tree species laid out on mineral soils to study the development of woodland conditions by Ovington (1953, 1954) showed large differences in the ground flora at 22 years, he concluded that 'there is little doubt that tree canopies by controlling light intensity and rainfall at ground level are of primary importance in determining the seral stages of ground flora'. When these sample plots were studied at 44 years by Anderson (1979), there was less difference between them, and the vegetation resembled that of native oak woodland. The variation in floral composition of the individual mature plots was the result of earlier canopy differences. Similar conclusions were reached about soil changes, the effects of tree species over the whole rotation assuming less significance than the effects found in the earlier stages (Anderson 1987). These studies illustrate the importance of the time factor, as it is the age of a stand that influences the ground flora. Hill (1986) points out that mature and well-thinned stands of Sitka spruce can support quite a rich ground vegetation. A heavily thinned 80-year-old stand of Sitka spruce observed in Northumberland had a similar flora to oak woods nearby and many other examples have been recorded (personal observations). The greatest changes in the ground flora are seen at canopy closure and it is at this stage that the effect of different tree species is greatest. Spruce eliminates almost all the ground flora but larch and pine allow enough light for some plants to survive so that recolonization is more rapid as soon as thinning starts (Hill 1986). Ratcliffe (1977) states that 'the woodland field layer may show little relationship to the particular tree species overhead but its composition is greatly influenced by the intensity of shade cast by the canopy'. The light requirements of plants increase as the acidity of the soil increases (Ellenberg 1988) and this is a further factor determining the type of vegetation .

Extensive closed woodlands are not as rich in flowering plant species as other habitats because the majority of plants require sunlight and open conditions. Neglected dense coppice when compared with a conifer plantation near by had a similar flora, as many of the flowers present in the open early coppice stages had been shaded out (Hill 1987b). This illustrates that light rather than tree species is the controlling factor. It is often stated that open-ground flora within forests is of little value because it is divorced from natural communities. This contention ignores the fact that in natural forest many of these species occur in wet and boggy areas and natural clearings. The communities present on open moorland in Britain today have developed from such open areas that have been extended by human activity. Nevertheless it must be recognized that 'grouse moors, deer forests and sheep walks

Plate 10.1 Plants typical of native Scots pine (*Pinus sylvestris*) woods which have colonized pine plantations of both Scots and Corsican (*P. nigra* var. *laricio*) established on non-woodland sites. The flowers illustrated were all taken in plantations in north Scotland. (a) Common wintergreen (*Pyrola minor*) in flower; (b) serrated wintergreen (*Orthilia secunda*) in fruit; (c) creeping ladies' tresses (*Goodyera repens*) in flower; (d) lesser twayblade (*Listera cordata*) in flower. (Photo: J. A. Lambert)

have acquired a considerable wildlife value in their own right' (NCC 1986), although they are man made.

Woodland plants comprise species that are able to thrive in deep shade and others that prefer partial shade on woodland margins (Brown 1981). The advantage of this strategy is the minimizing of competition from more aggressive species but woodland plants can also be found growing in open conditions wherever competition is minimal, for example on hedge banks. The amount of shade experienced by ground flora depends on the tree species and their density but both these factors can be manipulated to a large extent in managed forest by the routine operations of brashing, pruning and thinning, all of which allow overhead and side light to reach to the ground. Flowering plants are shaded out when light penetration falls below 12 per cent, and mosses below 8 per cent (Hill 1979). The *type* of light reaching the ground varies under different species. The amount of red light penetrating under spruce is half that under oak, but the amount of blue and violet light under spruce is double that of oak (Coombe 1957). The *intensity* of light is also important. Bursts of sunlight reaching the ground in the form of 'sun flecks' penetrate dense canopies and enable ground vegetation to survive (Robertson 1988).

Species that can survive under dense tree canopies reproduce by underground storage organs as well as by seed. They are thus able to withstand the alternating light and dark phases that occur in a maturing woodland, provided that neither phase lasts for a long time. Plants increase in numbers and diversity during the lighter phases, and then decline as shading increases. Many shade species produce seed in the light phases, surviving by vegetative or self-fertilizing means during the dark phase. The presence of deciduous trees lets full light reach the ground in spring and favours plants flowering then but the shade cast later in the season reduces light intensity to low levels. Even broadleaves can prevent 80 per cent of light penetrating the canopy and beech can intercept up to 95 per cent. As a result, ferns 'appear to be fairly indifferent as to whether the trees are deciduous or evergreen' (Ellenberg 1988) and this seems true of other summer flowering plants. Many plants flourish in the early stages of coppice woodland when shading overhead cover is removed. The same plants are found in mature and thinned woodland, but because the density is not so great and the flowering is not synchronized they do not excite as much attention. However, under clear cutting systems abundant flowering occurs at the time of regeneration and this is a very positive conservation advantage of even-aged forestry over selection forestry.

Plants of the forest edge (marginal species) and quick-growing species requiring nitrogen that colonize clear felled coupes rely mainly on seed for dispersal. This either blows in from elsewhere, or is buried seed that germinates when conditions become suitable. Buried seed survival has been studied under tree crops of various ages. The studies have shown that viability varies according to both the length of time the seed has been buried and the species concerned (Hill 1986). Some pessimistic views have been expressed about the seed persistence of some species in the uplands, suggesting poor survival in acid peats (Hill 1979). Waterlogged soils inhibit germination but preserve some seed. Seed viability lasts longest at low, even temperatures with moderate humidity. Attempts have been made to germinate buried seed but these experiments have been short-term and have met with varying success. Under natural conditions germination occurs sporadically over long periods, whenever the conditions are suitable; the length of time the seed is viable is different for different species, some persisting longer than others (Hill 1986). After germination, a small number of individual plants of species that produce copious seed can increase their numbers rapidly whilst conditions are favourable. Further seed shed as conditions become unsuitable enter a dormant phase and form a 'seed bank' ready to be activated when suitable conditions return. Such conditions are often created cyclically by normal forest operations. Soil disturbance encourages germination of seed banks, as germination is inhibited by high concentrations of CO_2 in the soil (Salisbury 1961). The value of disturbance is particularly well illustrated by the reappearance of the fen violet *(Viola persicifolia)* in East Anglia, thought to be extinct, after 70 years, when peat cutting was resumed (Moore 1982). Another example is the persistence of henbane *(Hyoscyamus niger)* on medieval archaeological sites. Henbane used to be grown medicinally and seeds of it and other species, ranging from 100 to 600 years old, taken from soils of known age, have been successfully germinated. Disturbance on such sites often leads to the reappearance of this plant (Odum 1965; Moffat 1987).

The amount of light and humidity under the tree canopy determines the ground vegetation present and this changes with the age of the crop. The influence of the ground vegetation itself on the soil and humus layer is still little understood and may be more important than the covering tree species (Miles 1986). The effect of tree species on soil structure certainly varies and is most significant on poorly buffered soils, particularly with species such as pine, which produces a mor humus. On such sites birch produces a mull humus whilst oak and beech are somewhere in between (Miles 1982). On better soils the differences are not as marked. Work in the USA has shown that sugar maple *(Acer saccharum)* and paper birch *(Betula papyrifera)* increase the pH of the surface organic layer; red oak *(Quercus rubra)*, beech *(Fagus grandifolia)* and white pine *(Pinus strobus)* decrease the pH (Art 1989).

OPEN-GROUND PLANTS THAT OCCUR IN WOODLAND
The plants that find suitable habitats in the open parts of a forest can be divided into three groups: (1) swamp and wet flora; (2) grassland species, including species from moors, lowland heath and meadows and (3) species colonizing bare ground.

Swamp and wet flora
Species favouring wet or swampy conditions can be conserved within forests by leaving wet areas undrained and unplanted. Indeed, such sites are an important refuge for these plants as so much farm land has now been drained. Wet 'flushes'

(a)   (b)

Plate 10.2 Plants associated with pure broadleaved woodland photographed in beech plantations established in mixture with conifer nurses. (a) Spurge laurel (*Daphne laureola*) in flower; (b) twayblade (*Listera ovata*) in flower. (Photo: E. H. M. Harris)

which are richer in base minerals need particular attention as they hold interesting wetland flora. Damp areas can also be found near ponds and streams, all of which support their own flora; these sites need to be in sunshine for much of the day, so shading by crop trees should be avoided and some trees may need to be removed. Many wet-ground species are found growing in boggy areas within natural forest and should not be eliminated at the time of planting. Sundew *(Drosera rotundifolia)* and butterwort *(Pinguicula vulgaris)* are good examples. Sundew occurs in sphagnum bogs in spruce and fir forest in the taiga. Both sundew and butterwort freely colonize wet roadside banks in upland forests in the west of Britain.

Grassland and other similar species
These species grow best in open conditions. Areas can be left unplanted to provide suitable habitats but unless these are grazed or cut regularly there is danger of the desired species becoming swamped by more aggressive species. A number of experiments have been carried out to evaluate the effects of fencing out stock; this has

always been shown to result in rank herbaceous growth and tree regeneration (Peterken 1986). Herbaceous competition is less on infertile soils and on steep slopes. In the uplands many plants are restricted to sites inaccessible to grazing animals, and the reduction of grazing pressure, especially by sheep, could assist the spread of such species if rock outcrops and small 'alps' are left unplanted.

Roadsides, wide rides and widened ride junctions provide useful habitats. Roadside banks, where the drainage is sharp and the soil poor, allow the less competitive species to grow but on flat ground and in the absence of grazing, mowing may be necessary on rides to preserve flowering plants. Regimes for this have been worked out for lowland forests and support a wider range of plants than those on acid soils in the uplands (Anderson and Carter 1987). Even in flat areas, the creation of mounds facing south will encourage flowering species, an effect that can be seen on a small scale on mole and ant hills as well as on some archaeological remains. Many meadow and heath species invade clear felled areas: in upland forests, heath bedstraw *(Galium saxatile)* and tormentil are common, whilst trefoils and clovers *(Trifolium* spp.) occur in lowland forests. The persistence of meadow and moor species on clear felled coupes depends on the intensity of management, whether the brash is burnt or windrowed, and on the time taken for the next crop to close canopy. Management techniques can make use of this. For example, chemical treatment with a degradable herbicide on a felled area in the Lake District to kill all herbaceous growth the summer before planting was followed six years later by a second treatment confined to small spots around the planted trees. A varied ground flora was thereby encouraged, which included both meadow and woodland species. In particular, bramble had been controlled, thereby allowing other plants to thrive (Royal Forestry Society 1987). Treatment of this kind is both economic forestry and excellent conservation, as the sensitive ground flora is assisted to flourish when it would otherwise be suppressed in competition with quick-growing aggressive species.

Species colonizing bare ground

A wide range of species belong to this group, including many weeds. Most of them have windborne seeds and they colonize bare areas quickly, soon invading disturbed ground beside roads and in clear felled coupes. Rose bay willow herb is a particularly rapid colonizer in forests throughout the British Isles, and indeed throughout the northern coniferous and broadleaved zones. Foxglove *(Digitalis purpurea)* is a common colonizer in lowland areas. Violets *(Viola* spp.) in woodland benefit from any disturbance that produces bare ground, such as timber extraction and small-scale disturbance, including 'rooting' by animals such as badgers and, on the European mainland, by wild boar.

THE CONSERVATION OF PLANTS ASSOCIATED WITH WOODLANDS

There is considerable scope for the encouragement of a wide range of plant species in managed forests. Many of the newer upland forests of pine and spruce are on poor and degraded acid soils or peat. Initially they have a sparse flora which is similar to that of the moors and grasslands from which they were derived where centuries of agricultural use has destroyed most of the woodland flora. The potential of pine forests on the dry soils in eastern Scotland, and on sands, for encouraging rare and uncommon plants has been mentioned earlier in this chapter. The widespread colonization of new forests by birds and butterflies is an indicator of the parallel but inevitably slower colonization by plant species; work in various parts of Europe has indicated future trends. Many upland forests will have to be grown on short rotations because of windblow and this will favour open-ground and early successional species, as already described in chapter 9. Where conditions allow longer rotations and where thinning is carried out, a diverse ground flora will develop. In many places this process has already started; it resembles natural succession, and once pioneer crops are replaced in the second and subsequent rotations, stands of varying ages will carry a diverse flora and provide further seed sources. The value of brash as well as 'lop and top' from thinning and felling for releasing nutrients, especially by burning, is already recognized in the coniferous forests of the USA (Washburn 1973).

Lowland forests are grown on a wider variety of soils and the potential for plant conservation is much greater because a wider range of plant species are favoured. In such forests new plant communities containing a wide variety of individual species have the chance to develop. Positive management can favour different species at particular stages in the rotation. Evidence of the occurrence of plant species under various forest types around the world, given earlier in this chapter, shows that a wide range of flora can be sustained in both coniferous and deciduous woodland. Ovington's plots demonstrate that similar floras can be sustained under a range of tree species in the later stages of the rotation, although there are differences at first. Early brashing, thinning and pruning increase side and overhead light and provide sensitive means of control over the ground flora. The periodic disturbance and increased light penetration provided by thinning is particularly important; after thinning, the ground flora temporarily increases then gradually recedes until the next thinning. Crown thinnings favouring the final crop trees give a more varied pattern of light reaching the ground than a uniform low thinning, which is therefore of less benefit to the ground vegetation. The selection of tree provenances for new planting can also exert an effect; for example, light-branched trees are beneficial. Extraction is of considerable value in reducing aggressive ground flora such as bramble and the sympathetic use of weedkillers on rides is useful to control aggressive and dominant species. Clear felling releases nutrients and favours plants requiring light and nitrogen.

In lowland conditions, there are opportunities for much longer rotations and the production of large timber trees, both coniferous and broadleaved, resulting in more stability of the ground flora. Increased stability allows selection forestry to be practised, which is beneficial to shade-bearing species, including a small number of rare, shade-tolerant plants. On these better sites, short-rotation broadleaved tree crops also have a place; sycamore and cherry, for instance, provide an early return, and sycamore on basic soils sustains a diverse ground flora. Plant species found under sycamore in Europe include wood anemone, ransoms *(Allium orsinum),* yellow archangel, dog's mercury, woodruff, Solomon's seal, twayblade, sanicle, herb Paris, strawberry *(Fragaria* spp), bugle and woundwort (Ellenberg and Klotzi 1972). All these species occur in two well-thinned stands of sycamore in Hampshire. Mixtures of conifers and broadleaves also sustain a diverse ground flora; this is discussed more fully in chapter 7. Although less aesthetically pleasing, line mixtures allow more ground vegetation to survive than groups as more light reaches the ground in the deciduous lines than in small deciduous groups. Many mature hardwood crops which were established as conifer/broadleaved mixtures contain plants of high conservation value. This was in fact the traditional way of establishing beech in the Chilterns.

All managed forest provides the potential for the conservation of meadow, heath and wetland species within the forest if suitable areas are managed for this purpose. Indeed, areas of reasonable size are more likely to be preserved within the forest than outside it; except in nature reserves, the potential for damage on outside sites is high. Altogether the degree of management and the site potential are more limiting factors than the tree species planted, as far as the conservation of plants is concerned.

In considering plant conservation in managed woods and forests, the opportunities for sympathetic introductions should not be ignored. Inherent variability will ensure that many plant species will adapt to a wide range of conditions (Sutherland and Watkinson 1986).

FUNGI

Fungi have no chlorophyll with which to manufacture carbohydrates in sunlight as green plants do and so depend on other sources for their nutrients. Some are parasites and are dependent on their hosts for survival. Others are saprophytes, feeding on dead and decaying material. Some saprophytic fungi are able to enter into intimate relationships with other plants; lichens, composed of fungi and algae, are one example and mycorrhizas (both internal and external), which aid the development of orchid seeds and trees, are another. Some fungi are associated with insects for their dispersal; for example, the fungus that causes Dutch elm disease. Fungi range from small species (microfungi), such as rusts, to those with easily recognized fruit bodies (macrofungi), such as the common mushroom *(Agaricus campestris).* Over 3000 species of macrofungi occur in Britain and some have a world-wide distribution. Honey fungus *(Armillaria mellea)* and the fly agaric *(Amanita muscaria)* are familiar world-wide examples. Many fungi cause disease and damage (pathogens), others are important decomposers in the soil, especially in acid soils.

Most of the life of a fungus is spent in the non-fruiting stage as bundles of threads (hyphae) which are known as the mycelium. Sexual reproduction by means of fruit bodies is erratic and depends on temperature and humidity. As the right combination of these factors occurs in the autumn, many fungi fruit at this time. Woodlands are particularly rich in fungi because of the humid conditions under the tree canopy. Soil type may also have some influence on fruiting. Identification of fungi is only possible when fruits are produced. The mycelium may be present but undetected until conditions occur that stimulate the production of fruit bodies (Watling 1974; Alexander and Watling 1987; Bon 1987). Experiments in a conifer plantation on a base-poor soil showed that the addition of fertilizers and soil conditioners influenced the production of fruit bodies (Hora 1959, 1972). Waterlogged soils and soils with a poor oxygen supply inhibit Sitka spruce mycorrhizas, and fruiting does not seem to take place under the heavy shade of this species in the later stages of the rotation, although the mycorrhiza is present (Walker 1987). For some species of fungus the soil pH appears to be more important than the trees under which they are found. *Boletus luridus* occurs under oak and hazel but is also found associated with willow in Sutherland. The common factor in its occurrence is that of base-rich soils (Watling 1974).

The distribution of fungi in relation to climate, soil and vegetation type is little understood. The same fungus species may occur under different tree species in different parts of the world. Competition between fungi may be a factor, different associations possibly being formed in the absence of more competitive species. For example, *Rozites caperata* is found under oak in America and Europe, under birch in Greenland and under Scots pine in Scotland. *Sarcodon imbricatum* too is found under oak in America but occurs under pine in Scotland (Watling 1974).

Fungi probably first colonized Britain when vegetation spread after the ice age, open grassland species being the first to arrive, followed by those now associated with pine-birch woodland. Most of these are northern species and the more southern and western species arrived later. *Amanita virosa* is a northern species associated with birch in northern Britain but it occurs with conifers in Europe and was recently recorded with Sitka spruce in Scotland (Alexander and Watling 1987). *Amanita phalloides* replaces it in southern and western Britain, where the latter occurs under broadleaves and conifers. Many other fungi appear to have the ability to thrive under both broadleaved and coniferous trees, whilst others are only found under either broadleaves or conifers. Some species, especially of parasitic fungi, appear to be restricted to particular tree host species. Where saprophytic species occur in association with particular trees, the dividing line between different strains within a fungal species associating with particular tree species, and distinct species of fungus, is difficult to draw. The influence of leaf litter and ground vegetation on the suitability of the humus sustaining fungi is still largely unknown. Young trees of the same species may show genetic variation in their ability to form mycorrhizal associations, some seedlings using one species of fungi and some another (Mason and Last 1986), whilst some clones are more successful than others (Walker 1987). Few if any fungal genera are restricted to one tree species and very closely related species are found in association with different tree species. Again the distinction between strains and species is not precise. Saprophytes growing on wood, however, appear to be more specific than saprophytes living on humus (Watling 1974), though amongst these species there are some found on the rotting wood of both broadleaves and conifers, especially in the later stages of decomposition.

Woodland fungi can be divided into the following categories:

1 *Grassland fungi.* These species are found in woodland on ridesides, clearings, areas with an open canopy and other open spaces. They are not dependent on the tree species forming the woodland.
2 *Parasites on trees and plants.*
3 *Saprophytic fungi decomposing wood.*
4 *Saprophytic fungi decomposing vegetation.* Many of these species enter into mycorrhizal associations, particularly with trees; some do so more readily than others.

Different fungi occur in different stages of forest succession. Mason and Last (1986) state that 'fungal successions . . . seem to be commonplace within forest ecosystems'. Decomposing fungi in soil show a succession from primary species to secondary species. Saprophytic fungi on dead and dying wood also show an orderly succession (Mason and Last 1986). The species colonizing twigs and branches are different to those attacking the trunk (Watling 1974). Pathogenic fungi vary with the stage of the rotation as tree growth modifies the environment within the canopy (Murray 1979). Saprophytic fungi on humus, including those forming mycorrhiza, also follow successional patterns as woodland matures (Mason and Last 1986). Over 2000 species are now known to form mycorrhizal associations and the succession of genera in conifers and broadleaves is similar in different parts of the world. Where trees colonize or are planted on non-woodland sites, especially on mineral soils, the fungal species associated with them are pioneer fungi such as *Hebeloma* and *Laccaria*. At canopy closure the diversity of species falls and *Amanita, Cortinarius, Lactarius, Russula* and *Tricholoma* spp. become dominant (Mason and Last 1986).

Where woodland follows woodland and the soil is thus richer in humus, fungi typical of the later stages of succession are able to compete and colonize young trees. There is also evidence of saplings obtaining nutrients through mycorrhiza from older tree roots (Mason and Last 1986). It is thought that this may aid the establishment of shade-bearing species. Pine and larch form mycorrhizas which are not inhibited in growth by heather and these are able to transfer nutrients to Sitka spruce when grown in mixture through the roots colonized by the fungi (Walker 1987). This fact has been exploited in commercial forestry by planting pine with spruce on sites where it is difficult to establish the latter species.

Many fungi can now be found in exotic conifer plantations. One hundred and twenty-four mycorrhizal and saprophytic species have already been recorded in association with Sitka spruce, many of which are normally associated with birch and pine (Alexander and Watling 1987). Their number will probably increase. A number of species have been recorded in plantations of lodgepole pine *(Pinus contorta)*, another species from North America. An uncommon fungal species, *Suillus flavidus,* usually found under Scots pine on *Sphagnum* moss, has also been found in plantations of mountain pine *(Pinus mugo)*, a Mediterranean species (Watling 1986).

The conservation of fungi

Many fungi cause damage and disease, so for these species control is more important than conservation. However, many harmless and useful species can find habitats within woodland. Fungi typical of heath, grassland or bare areas (including fire sites) find habitats within woods of all kinds. Suitable areas include rides, ridesides, clearings, meadows and felled areas. A wide range of fungi occur in woodland itself, as it provides humid conditions in which fungi thrive. The species occurring depend upon the stage of succession in the woodland, the geographical situation (northern species being found more commonly in the north and vice versa), the type of woodland (conifer or broadleaved) and in some cases upon the tree species present. Relatively few fungi are restricted to a single tree species but many are associated with either broadleaves or conifers only, whilst others occur under both. Further species are associated with a wide range of vegetation types. Most fungi are found on the more acid soils but a small number of species occur on basic soils, mainly in southern Britain.

Normal forest management providing a succession of age classes will therefore produce habitats for a wide range of species of fungi and mixed woods of conifers and broadleaves will hold the widest range. Dead trees, felled logs, unwanted thinnings and brashings can be left to provide dead wood for a range of harmless saprophytic fungi. Fungi reproduce by spores which are readily dispersed and resist adverse conditions, often remaining dormant for many years. Recolonization of woodland after felling is therefore rapid within the normal range of a species. Outside that range, fungi persist in the vegetative state, sexual reproduction only occurring under very favourable conditions. As many fungi have a world-wide range, more associations with exotic trees are to be expected in the future. Because fungi like a humid atmosphere, woodlands are the main habitat of the majority of species and normal management alone will provide a succession of habitats for them.

MOSSES AND LIVERWORTS (BRYOPHYTES)

Although some mosses are adapted to living in dry, open conditions, the majority grow in moist and shady habitats. Mosses have no true roots with which to take up moisture, so are dependent on humid conditions around them. Some absorb moisture all over the leaf surface, whilst others take it in through hair-like structures that function like roots. They are able to tolerate lower light intensities (2 to 9 per cent of full illumination) than vascular (flowering) plants (10 per cent upwards) (Hill 1979), and therefore exploit shady situations free from other plant competition. The majority of species, including the rare 'Atlantic' species, are found in the north and west of Britain where humidity is high and they can be found there growing on rocks in open conditions; mosses and liverworts are confined to damp woods in the drier east. Mosses and liverworts do not depend on the species of tree under which they grow but on the degree of shade and humidity provided. Mosses growing on bark are dependent on whether the bark is acidic or not, not on the tree species. Many species grow equally well in the shade of rocks. Conditions are more favourable for mosses under conifers as the needle litter does not smother them as much as the heavy leaf fall of broadleaved trees (Ellenberg 1988).

Many mosses and liverworts are northern species and mostly occur in acid situations but some species are confined to basic substrates. Mosses are a characteristic component of taiga forest, where the dense shade favours their development. In the taiga, mosses form carpets or 'feather moss' communities as a result of the high tree density and shade. Where pH, soil type, tree species and other factors are similar, stands with fewer trees support herbaceous or shrubby vegetation, whilst 'feather moss' communities develop under more densely stocked stands. Under very heavy shade even moss communities are shaded out (Larsen 1980). Typical mosses occurring in feather moss communities are *Hylocomium splendens* and *Pleurozium schreberi*. Both are circumboreal and are found in coniferous plantations in the uplands of Britain. Although a wide range of moss and liverwort genera are found in the natural taiga, little work has been done on the species that colonize coniferous plantations in Britain. Hill (1979) has recorded species found commonly in upland plantations, but the full range of species present has not yet been explored. However, considerable potential exists for the encouragement of a wide range of such species under the shade of coniferous plantations.

The conservation of mosses and liverworts

Most mosses have northern and western distributions and it is in these areas that managed forest can provide habitats for many species. Circumboreal mosses and liverworts colonizing bare ground, such as *Polytrichum commune, P. piliferum* and *Marchantia polymorpha,* take advantage of bare ground on burnt and clear felled areas, colonizing by means of spores. Some woodland species are able to survive long periods of desiccation, activity being resumed when conditions once more become suitable. These species are able to survive the removal of woodland cover by clear felling. Some mosses are sensitive to desiccation and cannot withstand drying out; this tendency becomes more important in the dry, eastern parts of Britain. Thus sensitive species that might survive a clear fell in the humid west may need the permanent protection of woodland cover in the east. In such areas, group regeneration or selection forestry will maintain the woodland cover and conditions that are needed; coppicing would cause desiccation.

ALGAE AND LICHENS

Algae are small plants growing in filaments or as single cells containing chlorophylls which determine their colour. They can be found growing on most trees, especially in polluted areas where they replace lichens. Lichens are not single plants but combinations of various fungus and algal species that grow together for mutual benefit, exchanging nutrients and providing protection for each other. There are about 1400 species of lichen in the British Isles. The majority of lichens are northern species, growing on bare soil or rock, but some grow on trees and many range throughout the Boreal region. Lichens are characteristically pioneers of new habitats that are poor in nutrients and were early colonizers after the ice age. Light and humidity are the most important factors affecting lichen growth and different species show marked preferences for the particular moisture conditions in which they are found. The richest lichen communities grow in oceanic regions and on mountains where mist and cloud provide moisture. For example, *Lobaria* and *Sticta* in Britain 'grow on a wider variety of species and age classes (of trees in the northwest) than in less oceanic districts' (Ratcliffe 1977). In other parts of Europe *Lobaria* is found growing on both broadleaved trees and conifers and this is true of other lichens, the trees acting only as a suitable substrate (Ellenberg 1988).

Shade causes lichens to die out and to be replaced by the more shade-tolerant bryophytes (mosses and liverworts). The effect of shading can be seen on individual trees; the lichens that tolerate drier conditions grow on the top side of the upper branches, whilst those requiring more moisture are found on the lower branches and the trunk. Similarly, the more shade-tolerant species occur on the north and east sides of the trunk (Harding and Rose 1986); these in turn give way to mosses near the base of the tree, where shading drastically reduces light and increases humidity. Soil lichens are adversely affected by the accumulation of litter on the ground. Many lichens are sensitive to pollution and die out in heavily polluted areas. However, since the Clean Air Act of 1954, lichens have been recolonizing in the London area (Rose and Hawksworth 1981).

Natural succession of lichen species takes place on trees, rocks or soil, until an equilibrium is reached. Lichen communities of 'climax' species may then persist for long periods. Individual lichen species in successions are not long lived and most develop rapidly and reach maturity in 15 to 20 years. In the Boreal forest, if lichen woodland is burnt, the bare ground is colonized by a succession of different species, eventually leading to 'climax' reindeer lichens in 30 to 80 years as tree cover increases again (Rowe and Scotter 1973). Similar successions take place as a tree grows from sapling to maturity, open-grown trees providing a wide range of habitats in their crowns.

Lichens grow into different shapes and are broadly divided into crustose (crusty), foliose (leafy), or fructose (branching) types. Some lichens will colonize a wide range of substrates, whilst others are more specific. The majority grow on soil or rocks but some grow on trees, particularly on the bark. The acidity of the bark of trees varies, conifers having a lower pH (3-5) than broadleaves (4-6), so different species develop on them. For example, *Cetraria* species occur mainly on conifers and many *Parmelia* species on broadleaves. Lichens found on broadleaves also vary depending on the smoothness of the bark. Although oak supports a wide range of lichen species, 'few, if any lichens are strictly confined to oak' (Harding and Rose 1986). Other broadleaves, such as sycamore, ash and beech, also support a large number of species. The range of species supported by various trees is shown in table 10.2. No indication is given of the division between northern and southern species in the table or how many are found on conifers other than pine.

Table 10.2 Lichen species associated with tree species

Tree species	Number of lichen species recorded
Oak	326
Ash	265
Beech	213
Elm spp.	200
Sycamore	194
Willow spp.	160
Birch	134
Pine	133
Rowan	125
Alder	116
Lime	83

Source: Figures from Harding and Rose 1986.

The conservation of lichens

Lichens are pioneer species, exploiting harsh and extreme conditions where competition from other plants is minimal. Many grow on a wide range of substrates such as rocks, soil and trees, especially in the north of Britain and opportunities exist in the uplands for providing suitable habitats in forests. Existing communities of unusual species of soil and rock lichens need to be kept free of excessive shade which would otherwise allow mosses to dominate. Ploughing clear felled areas, especially extensive areas which will dry out, encourages colonization by soil lichens in humid, upland areas. Little information is available concerning the occurrence of lichen communities in plantations, but species such as *Usnea hirta* are frequent on plantation conifers in the uplands. Trees, especially on the edge of plantations and woods, sustain a range of species as succession proceeds; the retention of some trees beyond rotation age will provide habitats for the later successional species. In the less humid areas of the country, some lichen communities are confined to sites in pasture woodland that contain old, isolated broadleaved trees. Agroforestry systems in the future may provide more open-grown trees and these will provide suitable habitats for a range of lichens. In areas with low humidity, many northern and western lichen species only survive vegetatively in undisturbed situations and are unable to spread and colonize new substrates by spore dispersal, as they do under the more favourable humid conditions in the west.

CONCLUSION

Managed forest, including introduced conifers, is by no means a restraint on plant conservation, as is often supposed. Indeed, forests provide a wide range of very diverse habitats and diversity increases with age. The full potential of managed woodland as a plant conservation resource in Britain has by no means been fully demonstrated yet.

CHAPTER 11

Mammals

Successive ice ages removed plant cover and animal life from Britain as the separation of the islands from the mainland of Europe, by the formation of the English Channel, brought to an end recolonization by land mammals (Yalden 1982). Bats were an exception as they can fly and so migrate over considerable distances. As a result, only about thirty species of land mammal arrived by natural migration to mainland Britain and fewer still to Ireland. Most of these are northern species that colonized early. Five of them have since become extinct (Corbet and Southern 1977). These were the wild ox *(Bos primigenius)* and the forest-dwelling mammals, bear *(Ursus arctos)* and beaver *(Castor fiber)*, which had gone by the twelfth century; and with the increasing destruction of woodland, wild boar *(Sus scrofa)* and wolf *(Canis lupus)*, which had disappeared by the middle of the eighteenth century (Fitter 1959).

During the Roman period and perhaps earlier, other mammals were introduced deliberately and are now integrated into Britain's wildlife. Their successful establishment may be the result of the number of unoccupied habitats that were available (Fitter 1959). Rabbit *(Oryctolagus cuniculus)* and fallow deer *(Dama dama)* were amongst the first and were imported for their food, fur and hide value. Although originally kept in warrens and parks, both species subsequently escaped and established themselves in the wild.

Rodents found their way to Britain with human assistance. They included both species of rat *(Rattus* spp.) and the house mouse *(Mus musculus)*. The house mouse appeared early (its origin probably being Persia) and it is thought to have found its way to Britain with Neolithic farmers (Berry 1981). The black rat *(Rattus rattus)*, also from Asia, is usually thought to have followed in the twelfth century but rodent remains found in York suggest it may have arrived much earlier, with the Romans (Rackham 1979). As human populations increased, overseas trade flourished and rats were carried as unwelcome passengers on ships, taking with them bubonic plague. Towards the beginning of the eighteenth century another Asian rodent, the brown rat *(Rattus norvegicus)*, also found its way to Britain on ships (Fitter 1959).

As living standards rose and leisure increased from the seventeenth century onwards, many animal introductions were made, both for their sporting and amenity value. Roe deer *(Capreolus capreolus)* from Germany were released in southern England about 1800 to repopulate areas that had lost their indigenous populations in medieval times (Fitter 1959). Other species of deer were introduced as exotic animals to many parks and some eventually escaped into the wild. Only two species have established themselves in any numbers: sika *(Cervus nippon)* and muntjac *(Muntiacus reevesi)* are both still expanding their range (Dansie 1970; Horwood and Masters 1970). The grey squirrel *(Sciurus carolinensis)* was introduced as a curiosity to a number of parks in the late 1800s and is now a familiar sight throughout England and is still expanding its range (Lloyd 1983). It is a severe pest of broadleaved trees.

Recent attempts at reintroductions of extinct native animals have included the release of wolf and beaver in confined areas (Pinder 1979). The release of wolves on the Isle of Rhum has been suggested as a means of controlling the natural red deer *(Cervus elaphus)* population there (Yalden 1986).

Mammals come into direct conflict with man more than does any other group of animals except insects. Rodents and grazing animals compete for food, whilst predators take a toll of domestic animals and game birds. As a result, human influence on the mammalian fauna of Britain has been twofold, by habitat modification or destruction, and by persecution (Langley and Yalden 1977; Yalden 1986). The initial modification of the forest environment by Neolithic farmers created larger open areas than those within natural forest and provided more diversity of habitat. A similar situation occurred when the virgin forests of the USA were cleared for agriculture in the nineteenth century and some species of mammals increased in numbers (Trippensee 1948). Work in Africa has also shown an increase in rodent numbers after forest clearance (Gorman and Robertson 1981). Burning in southwest Australia increased open-ground species of animals, such as kangaroo *(Macropus fulinginosus)* and wallaby *(M. rufogriseus)*, as well as certain bird species (Christensen and Kimber 1975).

As the human population grew, the area of woodland in Britain shrank and became fragmented, resulting in the local extinction of the larger mammals. Roe deer had gone from medieval England by the early 1700s and after the destruction of the Scottish forests in the eighteenth century, both roe and red deer took to living on the open hills, as many still do today. Wild cat *(Felis silvestris)* and pine marten *(Martes martes)* retreated to remote areas and suffered extensive persecution, almost to the point of extinction at the beginning of the twentieth century (Langley and Yalden 1977). As shooting game for sport became more widespread as a leisure activity in Victorian times, the smaller predators, such as stoat *(Mustela erminea)* and weasel *(M. nivalis),* were ruthlessly destroyed. The enormous numbers recorded in old estate records give some idea of the destruction.

The reduction in such persecution during the two world wars, and the expansion of forest cover since the 1920s, have resulted in a steady rise in the numbers and distribution of many species. Polecat *(Mustela putorius)*, pine marten and wild cat are now steadily expanding in the cover provided by the new coniferous forests, which has also halted the decline of the red squirrel *(Sciurus vulgaris)* (Corbet and Southern 1977; Velander 1983; Yalden 1986). The pine marten is present in some areas in such large numbers that its impact on other wild life has given conservationists cause for concern (pers. comm.). All species of deer have benefited from the increase in both coniferous and broadleaved woodland area and have found conditions so ideal that control is now necessary to prevent excessive damage. Thus the conservation of mammals in woodland is a matter not only of the encouragement of species in order to maintain viable populations but also of the ability to control numbers by management, based on ecological principles. When conflicts with human interests arise, a control strategy needs to be chosen that will keep damage to an acceptable limit and allow humans and animals to coexist (Corbet and Southern 1977). In this chapter, therefore, we consider control as well as conservation.

In order to consider conservation of mammals in productive woodland, in the widest sense of both preserving and controlling by management, an understanding of the basic requirements of each species is necessary. Factors of importance include suitable climatic conditions; the availability of food; the presence of cover for protection during breeding, resting, or evading enemies; and sufficient suitable habitat to provide a viable home range in which the animal and its offspring can survive. Population densities and the amount of competition, both between species and between individuals of the same species, are also important.

Climate and latitude both have an effect on animal populations and distribution. Some species require warm conditions in which to breed, whilst others can survive in a cooler climate with a long summer day length and short winter days. For this reason, fifteen species of bat breed in the south of England but only the pipistrelle *(Pipistrellus pipistrellus)* occurs throughout the British Isles (Corbet and Southern 1977). Although other species of bat can be found further north in other countries, the cool summers in Britain, the result of the Atlantic climate, limit insects available for food, compared with the warm summers on the European mainland. The dormouse is a species with a southern distribution in continental Europe whilst in Britain it is restricted to the south of England (Hurrell and McIntosh 1984). The mountain hare *(Lepus timidus)* on the other hand, survives best on the higher ground in northern Britain; although it has been introduced to parts of Wales and the Pennines, it has never become numerous in those regions (Fitter 1959). Most British mammals are, however, very adaptable, with wide distributions and the ability to thrive in a variety of habitats, including all types of forest. All British land mammals use woodland to some extent, even if only for brief periods; for example, hares normally live in open areas but seek food and shelter amongst trees during bad weather. Indeed, in Britain, a heavily populated country, woodlands provide the main refuge for most mammals (Steele 1975).

The most important factor controlling an individual animal's survival is an all-year-round supply of food. Winter is the testing time for those mammals that do not hibernate; the weather during this critical period controls population levels, prolonged adverse conditions causing dramatic falls in numbers in some species. It is during the winter that the forest manager must ensure that there is sufficient food in the woods for the mammals that are being encouraged, or must supplement it artificially. At the same time, the total numbers of any species causing damage can be reduced most effectively towards the end of the winter when natural mortality has already taken its toll. Intensive control measures at this time most effectively reduce populations as they are already at their lowest ebb. However, feeding game during the winter may benefit unwanted species, such as the grey squirrel, enabling it to maintain good condition and to produce large, early litters and allow time to breed again the same year (Kenward 1983).

In order to ensure a supply of food, many mammals take up territories or home ranges, often delineating the boundaries by scent markings. The extent of the home range varies according to the available food supply, which may vary in abundance in different parts of the range (Belovsky 1983; Karnil et al. 1987). In areas where food is plentiful ranges are small, increasing in size as food becomes harder to find (Corbet and Southern 1977). Larger mammals, especially the carnivores, require larger home ranges to sustain themselves and in many instances the different sexes occupy discrete but overlapping ranges (ibid.). Individual animals patrol their ranges and know them intimately.

Closely linked with the home range of female mammals is the choice of a breeding site. Here, cover, food supply for the mother and suitable conditions for rearing the young successfully are vital factors. Cover is important in providing protection from the weather, freedom from disturbance and protection from predators. In contrast, open areas are important in providing food for some species, and also dry lying-out places and play areas for young animals.

Population studies and a better understanding of the social structures peculiar to a particular species are subjects still requiring extensive research. The development of radio tracking has revolutionized these areas of study and the results of such studies will assist the forest manager to devise means of manipulating animal populations within woods to produce viable communities in balance with other interests. This will ensure that damage is not excessive and that instead a wide range of interests is maintained.

The following brief accounts highlight those features of importance that enable particular animals to make full use of woodland habitats. Identification is not included. Except where stated, information has been derived from Harrison Matthews 1952, Southern 1964 and Corbett and Southern 1977.

As mammals are also dependent on climate, they fall into four main groups, just as plants do:

1 Northern species found on all Continents (NAE)
2 Northern species found in Eurasia (NE)
3 Northern Europe and West Asian species (NEWA)
4 Southern Europe and West Asian species (SEWA)

Table 11.1 World distribution of mammals occurring in British woodlands

Boreal (or northern) species found in America and Eurasia (NAE)	Boreal (or northern) species found in Eurasia (NE)	Northern Europe and west Asian species (NEWA)	Southern Europe and west Asian species (SEWA)	Other introduced species
Stoat	Common	Hedgehog	Wild cat	Rabbit (SEWA)
Weasel	shrew	Mole	Dormouse	Edible
Mountain hare	Pigmy shrew	Polecat	(mixed and	dormouse
Fox	Water shrew	Water vole	broadleaved	(SEWA)
	Pine marten	Yellow-necked	forest)	Brown rat (Asia)
Extinct:	Otter	mouse	Lesser	Fallow deer
Elk	Badger	Pipstrelle bat	horseshoe	(Asia)
Bear	Brown hare	Barbastelle	bat	Muntjac (China)
Beaver	Red squirrel		Leisler's bat	Chinese water
Wolf	Field vole		Bechstein's bat	deer (China)
	Bank vole			Sika deer
	Wood mouse			(China,
	Harvest mouse			Japan)
	Red deer			Mink (North
	Roe deer			America)
	Long-eared bat			Grey squirrel
	Noctule bat			(North
	Daubenton's bat			America)
	(Bats occurring in the southern region of the zone)			
	Greater horseshoe, Whiskered, Natterer's Mouse-eared, Serotine			
	Extinct: Wild boar			

INSECTIVORA: ANIMALS SUBSISTING MAINLY ON INSECT PREY

Hedgehog (*Erinaceus europaeus*)

Distribution Throughout the British Isles. Beyond Britain, the hedgehog is found mainly in mixed and deciduous woods in western Europe. NEWA.

Habitat requirements The hedgehog dislikes dense woodland and wet ground. Optimum sites provide cover near to feeding patches, especially grassland. These include woodland edges, open mature woodland and small blocks of woodland of all kinds (though hedgehogs are not as common in pure coniferous woods), scrub, hedgerows and neglected patches of ground. Thick cover (bramble, nettles, bushes) is used for daytime resting in the summer but no permanent nest is occupied other than during the breeding season. Compact winter nests of leaves, dried grass and moss are built in protected places where temperature extremes are buffered, such as woodpiles, holes in banks, holes between tree roots, animal burrows, thick bramble and hedgerow bottoms.

Density This varies with habitat quality and with season, being high in the autumn when the young of the year augment the populaton. Densities of one animal per hectare have been recorded. Home range, movements and social structure have been little studied.

Food The hedgehog selects food to maximize energy intake and minimize energy expenditure and thus does not necessarily eat the most abundant prey. In one study, earthworms and carabid (ground) beetles provided 50 to 80 per cent of energy but only comprised 10 to 14 per cent of diet. Soft-bodied prey was preferred (slugs, caterpillars and insect larvae) (Wroot 1985). Spiders, flies and other insects are eaten, as also are small mammals and birds, carrion and eggs (Yalden 1969).

Conservation Although generally considered harmless, the hedgehog is still controlled in areas where game rearing is important. A study in Norfolk showed that eggs were taken from 23 per cent of nests in hedgerows and field boundaries. Where game bird nest density was high, more nests were found and destroyed, so that some temporary control may be needed in such situations in the nesting season (Tapper et al. 1982). Otherwise in the managed forest no special provisions need to be made.

Mole (*Talpa europaea*)

Distribution Occurs throughout the British Isles, except for Ireland. Elsewhere, is found in coniferous, mixed and deciduous forest. NEWA

Habitat requirements The mole dislikes shallow or stony soils because they are difficult to tunnel in; also waterlogged or very acid soils as few earthworms are present in them. The optimum habitats are permanent pasture and deciduous woods, where burrowing is easy and earthworms are plentiful. The mole is present but scarce on heather moors and in coniferous woods, though in the latter it is sometimes present in moderate numbers (Mellanby 1981). Dense trees of any kind are avoided, so moles mainly occur on restocked areas, open mature woodland, clearings and rides.

Density Eight animals per hectare have been recorded in winter and sixteen per hectare in summer on grassland.

Food Earthworms are the main food source; they form 50 per cent of the diet in summer, whilst 90 to 100 per cent of winter feeding is on worms that have been collected in the summer and stored. The other 50 per cent of the summer diet consists of insects, slugs, beetle larvae and fly larvae, the proportion taken rising during dry weather when worms are difficult to obtain.

Conservation The mole is mainly harmless in woodlands but it can sometimes cause damage on restocked areas by burrowing under young trees. Mole hills do, however, provide bare earth for plant colonization. No special conservation measures need to be taken as suitable stages of the rotation will be used, in addition to clearings and rides.

Common shrew *(Sorex araneus)* and Pigmy shrew *(Sorex minutus)*

Distribution Both species occur throughout mainland Britain but the common shrew is not found in Ireland. Elsewhere, both have a wide distribution in all forest types. NE.

Habitat requirements Both shrews dislike areas devoid of ground cover. They are found in all types and ages of woodland, both coniferous and deciduous, where low cover or forest litter is present. Runways are made in the forest litter or in shallow burrows below ground. The common shrew is the more abundant species within woods and spends 80 per cent of its time underground in winter, compared with 50 per cent for the pigmy shrew. The pigmy shrew is found more often on woodland edges and in rough grassland. Both species make nests of grass in trees, holes in banks, under logs and underground. Shrews also climb and search for food above ground level.

Density Population density is related to food supply. In one case the number of pigmy shrews on blanket bog increased when this habitat was improved (Lance 1973). Home range is usually from 900 to 1000 m². Six animals per hectare have been recorded in larch in August (Don 1979). Small numbers of the common shrew have been recorded in a 40-year-old Sitka spruce (*Picea sitchensis*) plantation and in the clear felled areas there (Thompson 1986).

Food Earthworms, insects and larvae found in soil litter. Spiders and some species of beetles are preferred. Pigmy shrews tend to prefer different species of woodlice to the common shrew. Some vegetable food is eaten and seed can be important in pure conifer woodland when coning begins. Shrews have been recorded eating Douglas fir (*Pseudotsuga menziesii*) seed in large quantities in the United States.

Conservation Shrews are beneficial as they consume many insects. No special measures need to be taken to encourage them in woodland.

Relation with other animals Shrews form an important prey of several predators, especially owls.

Water shrew *(Neomys fodiens)*

Distribution Throughout mainland Britain but not in Ireland. NE.

Habitat Although found mainly near water, the water shrew also occurs some distance away. In Poland it is found in coniferous, mixed and deciduous woodland.

Density No information.

Food In woodland it is recorded as taking adult flies, earthworms, rove beetles, millipedes, centipedes and woodlice (Churchfield 1979).

Conservation The water shrew is harmless. The presence of a clean water supply, in the form of a pond or stream, will help to create suitable conditions, though this does not appear to be essential.

Bats (Chiroptera)

Distribution The fifteen species of bat that occur in Britain are found mainly in the south, some of them with a very local distribution. Only four species are found commonly in Scotland. Eight of the fifteen British species have a more northern distribution in Asia because summers there are continental in type and conditions warmer, thus insect prey is more readily available.

Habitat requirements Suitable roosting sites are of prime importance. Sheltered roosts are required in winter where temperature and humidity fluctuations are small and where there is little air movement to dissipate body heat during hibernation. Such roosts are found in tree holes in wooded valleys, caves, mines, tunnels, pot-holes, disused icehouses and in some buildings, such as church porches and old barns (Racey and Stebbings 1972). Summer roosts include holes in trees, particularly old woodpecker holes, buildings, lofts and behind cladding. Breeding females congregate together in nursery colonies. A particular site may not be occupied continuously, the bats leaving if conditions become unsuitable, for example a roof space becoming too warm. Very few species of bat are able to live completely under the tree canopy. In Bialowieza in Poland, the majority of species were found to be dependent on open space for hunting food (Harris and Harris 1988a).

Food Most accounts list small insects, beetles and moths. Cockchafers and dor beetles are eaten by most bats. A study of the brown long-eared bat showed that moths formed 40 per cent of diet in the summer. Beetles, caddis flies, spiders, bugs and earwigs were also eaten. Moths were not taken indiscriminately, however, 94 per cent of those caught belonging to the Noctuidae or owlet moths which fly at night. They included the large yellow underwing *(Noctua pronuba),* the silver Y *(Autographa gamma),* the lesser yellow underwing *(Noctua comes),* the cabbage moth *(Mamestra brassicae),* the dark arches *(Apamea monoglypha)* and the common rustic *(Caradrina blanda).* All these moths are common and have large, soft bodies that provide a good food supply (Thompson 1983).

Conservation Most bats are at the extremity of their range in Britain, the number of species declining from south to north. Bats take a wide range of insects. The use of some pesticides, such as pentachlorophenol, is lethal to them, the chemical being ingested from the fur during grooming. Synthetic pyrethroids are safe to use where bats occur, such as in house lofts and buildings (Racey and Swift 1986). Bats use bat boxes readily if provided in pole stage and mature woodland, mainly for summer roosting, but some boxes are used for overwintering. A number of bat box designs are available from the Forestry Commission, the Mammal Society and the Fauna and Flora Preservation Society. Early thinning will assist some species to hunt amongst the trees but most species prefer to hunt along woodland edges and along roads, rides and in clearings. The presence of water is important as it provides a source of insects and open space in which to fly. All bats are protected in the UK under the Wildlife and Countryside Act 1981.

CARNIVORA: ANIMALS THAT ARE MAINLY MEAT EATERS

Fox *(Vulpes vulpes)*

Distribution Occurs throughout the British Isles. Elsewhere, the fox occurs in all forest zones in both North America and Eurasia. NAE.

Habitat requirements The fox is highly adaptable and exploits a wide range of habitats, including towns. It can be found in any deciduous or coniferous woodland with ground cover and is most abundant where the habitat is varied, for example where small woods and fields are adjacent. The choice of site for breeding earths is dependent on soil type, as freely draining and easily excavated soils are needed. Vegetation cover and topography are also important (Insley 1977). Other breeding sites used include holes in rock scree, holes under tree roots, badger setts, enlarged rabbit holes and artificial holes and tunnels.

Density This varies according to the food supply. A New Forest (Hampshire) estimate gave an overall density of 2.18 foxes per km^2 with a minimum density of 0.75 adult foxes per km^2, compared with an overall density of between 2.06 and 2.61 adult foxes per km^2 in London where food is freely available. Breeding dens were found to be closer together in unbrashed plantations (Insley 1977). A full account of social behaviour is given by Macdonald (1987).

Food The range of animals, birds, insects, vegetable matter and carrion taken is extremely wide and varied. This is why the fox is so successful in a wide range of habitats. For an account of the extensive range of prey taken by urban foxes, see S. Harris 1977. Like other mammals, individuals have food preferences; in two studies, voles were being taken rather than mice (Macdonald 1976) and rabbits rather than birds (Frank 1979). Large numbers of voles are eaten in young plantations and at this stage, foxes live and feed wholly within the forest (Hewson and Kolb 1975). Earthworms are eaten by vixens with cubs (Jefferies 1974).

Conservation and control Foxes have always been in conflict with humans, preying on both domestic and game animals. Damage, when it occurs, is excessive as foxes destroy more prey than they eat, especially when the prey is confined. However, in woodlands where game management is unimportant, the presence of foxes may be beneficial. It has been claimed that natural regeneration is helped by foxes who keep small seed-eating mammal populations in check (pers. comm.). Foxes are important as potential carriers of rabies. Research has shown that young foxes move long distances in the autumn, especially where densities are high; the spread of rabies has been halted by concentrating on the destruction of young animals, leaving a core of sedentary adults in residence. Work in the USA shows a death rate of 50 per cent where populations are high. At densities of one fox per km^2 the disease is maintained but at densities below 0.5 foxes per km^2 it dies out. Conventional control methods in Britain seem to exert little influence on fox populations. This is partly because of the social structure, which ensures that there is a pool of non-breeding vixens ready to fill a gap left by the death of breeding females (S. Harris 1977; Macdonald 1987).

Plate 11.1 Pine marten (*Martes martes*). A rare mammal formerly confined to north Scotland, the Lake District and north Wales, which has benefited from the habitat provided by upland coniferous plantations and is rapidly expanding its range. (Photo: K. & M. Bullen)

Pine marten *(Martes martes)*

Distribution Occurs in north-west Scotland, Galloway (from releases in 1981 and 1982), the Lake District, North Wales, Yorkshire and Ireland. Early in the twentieth century the pine marten was at a very low level but has been spreading steadily south-east in Scotland since 1959 (Velander 1983) as well as in Wales, entirely as a result of coniferous planting. In other countries the pine marten occurs in all forest zones. NE.

Habitat requirements At the beginning of the twentieth century and as a result of deforestation, the pine marten was found on open ground. Areas with some tree cover and some ground cover, such as rough grass, rushes and tree litter, are preferred. Recent records show a move away from open habitats into the new coniferous forests (Velander 1983, 1989). Similar trends exist in Ireland. Here, one study showed 16 per cent of animals recorded in deciduous woodland, 46.4 per cent in deciduous woods underplanted with conifer, 6.2 per cent in mature conifer, 16.4 per cent in thicket stage conifer and 15.5 per cent in scrub (O'Sullivan 1983). Breeding dens are made in hollow trees, under logs, in rock crevices, in old squirrel dreys and in large birds' nests; such dens are usually located near fresh water. Tracks are used regularly, typically along forest rides (Velander 1983). Large nest boxes erected for birds are sometimes taken over by martens (pers. comm.).

Density This depends on the availability of prey, which in turn is controlled by the stage of forest succession. Soviet work has shown 0.6 marten per 1000 ha in young plantations, 2.8 per 1000 ha in middle-aged plantations and 3.7 per 1000 ha in mature plantations.

Food Small animals, especially field voles but also rats, rabbits and squirrels, are taken. Red squirrels are the main food in Sweden and Finland. Small birds, frogs, beetles, wasp grubs and other insects and larvae, together with fruits (blackberries, raspberries) when in season, make up the rest of the diet. The pine marten will eat carrion and can therefore be easily poisoned (Hurrell 1968).

Conservation The marten was persecuted in the past by gamekeepers and farmers and is still destroyed. Out of 230 animals where the cause of death was known, 70 per cent of deaths were the result of trapping, snaring, shooting and poisoning (Velander 1983). Reduced persecution and the seclusion of new forestry plantations has allowed the pine marten to increase in numbers and to extend its range. The reintroduction of the marten to further aid its spread has been advocated by O'Sullivan (1983) and Yalden (1986). In some parts of Scotland there are now good populations and, although martens are protected, fears have been expressed by conservationists concerning the effect their predation will have on other species. However, it would seem ironic to control an animal that a few decades ago was on the verge of extinction. No special measures need be taken to encourage these mammals where they already exist in forest areas.

Stoat (*Mustela erminea*)

Distribution Throughout the British Isles. In other countries the stoat is found in all forest zones. NAE.

Habitat requirements The stoat may be found in a very wide variety of habitats, its presence being dictated by food supply and suitable cover. When hunting, the stoat usually follows regular routes along hedges, fences, ditches and walls. Stoats are good climbers. Breeding nests are made in holes in walls, trees, rocks and burrows.

Density This varies with the availability of prey, and the larger male occupies a larger territory than the female. In a young conifer plantation where voles were numerous (up to 540 per ha), one male stoat had a territory of 20 ha (Lockie 1966).

Food Rabbits were the main prey before myxomatosis, after which stoat numbers fell considerably (Hewson 1972). Small rodents and birds are eaten as well as insects, earthworms, birds' eggs and fruit.

Conservation and control The stoat takes game birds and their eggs during a limited season in spring when some trapping may be necessary; otherwise it does no harm (Tapper et al. 1982). No measures need to be taken to encourage it in woodland.

Weasel (*Mustela nivalis*)

Distribution Occurs throughout mainland Britain but absent from Ireland. Elsewhere occurs in all forest types. NAE.

Habitat requirements The weasel occurs in a wide variety of habitats if there is suitable prey and cover. Runways are used regularly (Lockie 1966). The weasel hunts and travels along ditches, drainpipes, hedges and walls bordering woodland, and follows features within the wood; it is an expert climber and takes nestlings from nest boxes. Dens are made in holes in walls, banks and trees, under piles of stones and in log-stacks.

Density Like those of other carnivores, numbers are related to habitat quality and the available food. The larger male holds a bigger territory than the female. In young coniferous plantations where voles are numerous (up to 540 animals per ha), males held territories ranging from 1 to 5 ha (Lockie 1966). In deciduous woodland, where prey numbers ranged up to 39 per ha, male territories were between 7 to 15 ha and female territories from 1 to 4 ha. There is some evidence that at low rodent densities, below 10 animals per ha, breeding does not take place.

Food The main prey taken is voles and mice (50 to 75 per cent of diet). Birds, especially young song birds, comprised over 25 per cent of the diet in May, June and July (Tapper 1976b). Rabbits, rats and shrews formed part of the diet in some studies. As a rule, individual weasels favour one type of prey (Day 1968; Walker 1972).

Conservation and control The weasel is for the most part beneficial to humans. Some game bird chicks are taken and make up between 2 and 5 per cent of diet, mainly in the critical fledgling period during May to July (Tapper 1976b), so control can be limited to this period. However, King (1973) found that baiting traps with pheasant eggs was unsuccessful, as the weasels found them too tough to open. Reports of egg predation by weasels therefore need verification. No measures need be taken in woodland for their conservation.

Polecat *(Mustela putorius)*

The polecat was confined to Wales before 1950 but is now steadily expanding its range into the border counties of Shropshire, Gloucestershire and Herefordshire (Walton 1968). It has spread recently into Derbyshire, Warwickshire and Oxfordshire *(Mammal News,* 1995). In other countries it occurs in all types of forest. NEWA.

Habitat requirements The polecat is found in a wide variety of habitats. Young conifer plantations provide suitable areas because of the number of voles available (Poole 1970; Blandford 1987). The polecat breeds and lives in underground holes which it may excavate under tree roots and in hedge banks. It will also use fox earths, rabbit burrows, holes under rocks and in walls, and old buildings.

Food A wide range of prey is taken. Mammals comprise 35 to 70 per cent of the diet and include hares, rabbits, voles, mice, shrews, moles and hedgehogs. Small birds and their eggs are eaten (6 to 14 per cent of prey). Other prey includes insects, worms, slugs, frogs and carrion (Poole 1970).

Conservation The polecat is beneficial to humans on the whole but some birds are taken and domestic poultry is sometimes eaten. Polecats are still trapped and poisoned and many become road casualties but no special measures need be taken in woodland for their conservation.

Mink *(Mustela vison)*

Distribution The mink was introduced from North America in 1929 for its fur. It is now naturalized and occurs throughout most of the British Isles.

Habitat requirements In North America, the mink is found in both the deciduous and coniferous zones and extends into the tundra. It usually lives near rivers, streams and lakes where there is good vegetation cover on the banks. Existing holes, including rabbit burrows, are used for dens. Trees are occasionally climbed.

Food A variety of prey is taken, including birds, mammals, fish and invertebrates.

Conservation The effects of mink on other forms of British wildlife are still being studied. On the whole, it appears that the amount of damage that was expected has not been realised and was overestimated. It seems that the mink and otter can co-exist as the mink is less aquatic and the reduction in prey population numbers is less than anticipated. However, control is necessary in woodlands where any game bird rearing is undertaken, also where domestic poultry and ducks are kept and on fish farms. As with other predators, damage can be high where the prey is concentrated in numbers and easily available.

Otter *(Lutra lutra)*

Distribution Occurs throughout Britain. Numbers have declined because of river pollution. NE.

Habitat requirements Otters occur in all types of woodland, provided there is fresh water in the form of streams, rivers or small lakes. Holts for breeding are made amongst tree roots, and horizontally spreading roots provide particularly suitable places. Sycamore and oak are especially good in this respect (Mason et al. 1984), so is ash. Where there is a shortage of natural sites, artificial holts can be supplied and these have been used with success.

Food Items eaten include fish, frogs, birds, small mammals, insects, worms and Crustacea.

Conservation If otters are to be encouraged in a woodland, good water quality is essential to provide a suitable food supply, especially of fish. Artificial holts can be provided if natural sites are scarce. Where there are fishing interests otters may cause some damage as more fish are taken than eaten, but this loss must be accepted as a conservation cost. Successful reintroductions of this once-declining species have been carried out on some rivers. However, it is now increasing naturally in many southern rivers.

Badger *(Meles meles)*

Distribution The badger is found throughout the British Isles. In other countries it is found in all types of woodland. NE.

Habitat requirements The results of the National Badger Survey organized by the Mammal Society (Neal 1972; Clements et al. 1988) showed that the badger has a preference for deciduous woodland and mixed woodland (49 per cent of recorded setts), but a wide variety of other habitats are used. Setts are also found in coniferous woodland, and in Roxburgh, for example, where such woodland predominates, 34 per cent of the setts were recorded in conifers. Coniferous woods are important habitats where other woodland is scarce (Thornton 1988). The surveys found that whatever the woodland type the following factors were important.

1 Enough cover to enable the badger to leave the sett unobserved.

2 The type of soil. This needs to be freely draining to keep the sett dry; it must be easy to excavate but firm enough to prevent the tunnels collapsing.

3 Food needs to be within reasonable distance, thus few setts are found in the centre of large woodland areas unless there is enough open ground in the form of clearings, rides and meadows (Neal 1972). Badgers do, however, travel long distances when foraging. Because these conditions are important, setts tend to be concentrated in suitable areas, other tracts of ground being unoccupied. The majority are found on the edges of woods near their food supply, and suitable soil types on the sides of hills are favoured (Dunwell and Killingsley 1969; Neal 1972).

Density An average density is about 12 adults per 1000 ha but in very favourable conditions much higher figures have been reported. In one study where the average population was 9 animals per 1000 ha, the mean territory size was 104 ha, the larger groups holding territories of 131 ha and the smallest group occupying 54 ha (Cheeseman et al. 1985).

Food A wide range of animal and plant food is taken and varies according to availability and the time of year. Carrion is also eaten. Earthworms constitute the most important item of diet and provide an easily exploited food source; this is particularly important to lactating females who have to find a considerable amount of nutritious food quickly whilst rearing the young (Jefferies 1974). Studies in Italy showed earthworms were only a small part of the diet there, insects and fruit being the most important (Pigozzi 1988). Regurgitation of grass and other food for feeding the young has been recorded (Howard and Bradbury 1979).

Conservation Badgers are generally beneficial to humans as they eat rodents and destroy wasps' nests. Some poultry and game are eaten but the amount is not significant, the badger being a relatively slow-moving animal compared to some other predators. Damage to fences can be avoided by the installation of badger gates on well-used routes (Neal 1955; 1972). In some areas the badger population is infected by *Mycobacterium bovis* which causes tuberculosis in cattle and, if the raw milk is drunk, also causes a form of the disease in man. There is some evidence that at low badger densities the disease is unable to persist in the badger population (Cheeseman et al. 1985). No special measures need be taken to encourage badgers in woodlands. Felled and restocked areas provide foraging grounds until the trees close canopy but setts can be found in quite dense woods.

Wild cat (*Felis silvestris*)

Distribution The wild cat was restricted to northern Scotland at the beginning of the twentieth century, having been driven to live on open ground as a result of deforestation and persecution (Langley and Yalden 1977). With the increase in woodland area and the relaxation of persecution, the wild cat is now expanding its range steadily southwards into southern Scotland. In other countries it occurs in all forest types. SEWA.

Habitat requirements The wild cat is found in woodland and scrub on the European mainland; the new forests in Scotland have provided it with a similar habitat to colonize, as they provide cover, food and freedom from persecution. Lairs are made in holes in screes, rock caves, trees, in foxes' earths, under bushes, in bracken and in old birds' nests. Like the domestic cat, it likes to bask in the sun on dry, open areas. It is also a good climber.

Density Territories are recorded of 60 to 70 ha, but little is known about densities. As with other predators, the size of territory required may vary with the quality of the habitat.

Food Small rodents, especially voles, rabbits, hares, birds, carrion and insects (cockchafers and grasshoppers) have been recorded. Outside Great Britain, mice, shrews and squirrels are taken and no doubt constitute a food source in Britain too.

Conservation Within woodland, wild cats eat many small rodents and do not need controlling. Some damage may be caused to poultry. No special measures need to be taken to encourage this species in woodland where it is free of persecution. Wild cats are protected in the UK under the Wildlife and Countryside Act 1981.

LAGOMORPHA: SMALL HERBIVOROUS ANIMALS

Rabbit (*Oryctolagus cuniculus*)

Distribution The rabbit was an early introduction from the Mediterranean region. It is now generally distributed throughout the British Isles, though it is not as numerous as it was before myxomatosis, a virulent virus disease introduced in 1953. Before that, it was a major agricultural and forestry pest, its numbers having risen steadily since 1800 for various reasons (Sumption and Flowerdew 1985). SEWA.

Habitat requirements The rabbit is able to live in all types of woodland: coniferous, mixed and deciduous. It makes burrows in banks and hedgerows near the woodland edge, alongside rides and in clearings near a food supply. Burrows are found in all soils, except the very wettest. If the ground cover is thick enough, rabbits will 'lie-up' above ground and will breed there.

Density The numbers of rabbits were very high before myxomatosis but few areas support large populations now as the disease exerts periodic control. Myxomatosis is now present in a number of forms, some more virulent than others. Genetic resistance to the disease has been demonstrated by Ross and Tittensor (1986).

Food A wide range of plants are eaten, including agricultural crops and young trees. The bark of older trees is chewed at the base, mainly in winter, and if ringbarking encircles the tree it dies.

Conservation and control In the last 150 years the impact of the rabbit on the British countryside has been widespread. Before 1800, trees could be established without fences but, as rabbits increased, this became impossible until the introduction of myxomatosis. In some areas the population is rising once more (Gill 1992), and fencing is becoming necessary again (Ross and Tittensor 1986). This is due to inherited resistance, now occurring throughout Britain (Kolb 1994). In areas where there are game interests and thus predator control, populations are higher than those without (Trout et al. 1986). The impact of the rabbit and the effect it has had on both vegetation and animals were reviewed by Sumption and Flowerdew (1985). For the first few years after reduction in numbers many plants flowered profusely but as the ungrazed vegetation began to follow through the natural succession to woodland (which had been arrested by the rabbit), many animals and plants were adversely affected. The increase in the area of broadleaved woodland shown in the Forestry Commission census of 1984 was almost certainly the result of this natural regeneration in the absence of rabbits.

Brown hare (*Lepus capensis*)

Distribution Occurs in mainland Britain in lowland areas, introduced to the north of Ireland. Although widely distributed in Europe and Asia, the brown hare is found at lower latitudes than the mountain hare and in more open country. NE.

Habitat requirements The usual habitat of the brown hare in Britain is agricultural land, but the increase in vegetation cover and the reduced competition from rabbits led to an increase in numbers after myxomatosis (see the rabbit above) and a movement into woodland. A similar movement occurred in France (Sumption and Flowerdew 1985). Hares also move into woodland of any kind in the winter, seeking food and shelter. Recent surveys showed hares present in many woodlands, especially those with a ground cover of grass (Wray and Harris 1994).

Density The hare is only found at low densities within open woods. Hares have been slowly declining in Britain since 1960 for reasons not understood; a similar decline has occurred in the rest of Europe (Krebs 1986).

Food A wide range of plants are eaten, especially cereals and grasses.

Conservation and control Hares can cause damage in young plantations by stripping bark and nipping off the tops of young trees. Characteristically they do this down a whole line of newly planted trees. They will jump rabbit fences but tree shelters do provide protection against them.

Mountain hare *(Lepus timidus)*

Distribution The mountain hare is found on high land in Scotland and Ireland; there are small populations in North Wales and the Pennines. In other countries it has a more northern distribution than the brown hare and is found in taiga forest as well as on the open tundra. NAE.

Habitat requirements The mountain hare is usually found on heather moorland in Britain but forests are used during the winter for food and shelter.

Density The mountain hare is rare in granite areas but commoner on land over base-rich schists. Numbers were found to increase on improved blanket bog and when moorland was converted from grassland to trees (Watson and Hewson 1973).

Food Mainly heather and cotton grass. Young trees, bark and cones are eaten in hard weather. In one study, willow and rowan were selected in preference to birch, whilst all species of conifers were eaten indiscriminately (Hewson 1977).

Conservation and control The mountain hare seeks the shelter of woodland in the winter and may then cause damage to young trees. As well as browsing, hares will work down a line of young trees, nipping out the tops and leaving them on the ground.

RODENTIA: RODENTS

Red squirrel *(Sciurus vulgaris)*

Distribution The red squirrel is found mainly in coniferous and mixed woods although its range does extend into deciduous woods. By the seventeenth century in Britain it had become scarce and reintroductions were made in the seventeenth and eighteenth centuries. It is now found mainly in Scotland, Ireland, the Lake district, Northumberland, Wales and East Anglia, with small populations elsewhere, such as south-west England, the Isle of Wight and Brownsea Island, Poole (Lloyd 1983). Surveys that have been carried out in Scotland 'provide a good basis for what might prove to be a period of expansion of the red squirrel north of the Caledonian canal where conifer plantations of spruce and pine, now maturing, are widely dispersed' (Lloyd 1983). NE.

Habitat requirements The red squirrel is predominantly a squirrel of coniferous forest. It adapted to living in deciduous woodland in Britain as a result of the retreat of coniferous forest in the warm Atlantic period. It has recently retreated from areas now occupied by the grey squirrel and various hypotheses have been put forward to explain the inability of the two species to coexist. The most popular have been disease and the driving out of the red squirrel by the grey. It now seems most likely that coexistence is not possible in deciduous woodland because both squirrels have a similar daily activity and social structure, with no strict territories and with home ranges of individuals overlapping. Both species also use the canopy and ground layer in a similar manner. The larger and more prolific grey squirrel, adapted to living in *deciduous* woodland, is therefore at an advantage. Red squirrels digest acorns less efficiently than greys. They prefer ripe hazel nuts but greys will eat unripe nuts, thus depriving the red of this food source *(Natural World* 1994). Normally when closely related species occur in the same habitat they use it in different ways (MacKinnon 1978). The survival of the red squirrel therefore almost certainly depends on large areas of coniferous forest (Gurnell 1987). Dreys are usually built next to the trunk of conifers but squirrels also breed in tree holes, old birds' nests and nest boxes. The drey is constructed of pine needles, grass *(Deschampsia)* and mosses *(Hypnum, Hyloconium, Pleurozium)* (Tittensor 1970).

Density Red squirrels are most abundant in conifer areas of over 50 ha. In such places, densities of 0.8 squirrels per ha have been recorded. Life expectancy is up to 10 years. The size of the territory is related to the density of squirrels, 1 per ha being an average figure (Gurnell 1987).

Food Conifer seed (of pine, spruce, Douglas fir and larch) and mast (beech, oak, chestnut, hazel) form the main food as a result of their high calorific value. The endosperm of conifer seeds is extracted, that of spruce being preferred in Europe as the seed is easier to open than pine and has a higher calorific value (Moller 1983). Squirrels feeding on Norway spruce show a lower mortality rate from disease than those feeding on pine (Gurnell 1983). The food value of Sitka spruce, which has a smaller seed, has not yet been investigated (Gurnell 1987). In late spring and early summer when these foods are scarce (though green pine cones are eaten) and there is a seasonal food shortage, flowers of pine, larch and spruce are eaten, together with birds' eggs, insects, fungi, bulbs and roots, whilst tree bark may be stripped for the nutritious phloem beneath.

Food is sought during two main feeding periods, morning and afternoon, except in the winter when feeding activity takes place in the morning and is of short duration (Degn 1974; Moller 1983). Degn (1974) noted a period of activity in the evening during September and October. This may be related to the laying down of autumn fat (work on humans has shown that more fat is laid down during the night after an evening meal than during the day after breakfast). Feeding occurs on the ground in winter but more time is spent in the canopy than by grey squirrels. Foraging occurs in the tree canopy in May, June and July (Kenwood 1983; Tonkin 1983).

Conservation and control The red squirrel has been a forest pest in the past, causing damage to larch, Scots pine, spruce, silver fir, birch, beech and sycamore, by bark stripping. The eating of buds and shoots distorts the shape of trees, especially if the leading shoot is damaged. This damage has to be weighed against the desirability of conserving what has been a diminishing species in many parts of Britain, bearing in mind that the species is protected under the Wildlife and Countryside Act 1981. Squirrels use the pole and mature stages of forest as well as open ground. Sitka spruce plantations are colonized at about 25 years when coning begins. In areas where both grey and red squirrels occur, if the latter are to be favoured, the planting of large-seeded broadleaves such as oak and beech should be avoided and planting should be confined to mixed species of conifer so as to spread food availability (Gurnell 1987; Gurnell and Pepper 1993).

Grey squirrel *(Sciurus carolinensis)*

Distribution The grey squirrel, introduced from North America during the nineteenth century, is now found throughout most of England south of Cumbria and Northumberland, having been introduced and translocated within this area since 1876. In Scotland it is found mainly in the central lowlands and Deeside (Lloyd 1983). It has also been introduced to Ireland but is not found elsewhere in Europe.

Habitat In North America, the native grey squirrel occurs in areas of mixed broadleaves that contain a wide range of species with large seeds. In Britain it is found in many habitats, including mixed broadleaves and conifers, and in pure conifer areas where these are not too extensive. The grey squirrel's general behaviour is similar to that of the red squirrel, with no strict territories and individuals having overlapping home ranges (Don 1983). Dreys are built in either broadleaves or conifers and are composed of twigs and branches (often with their leaves on) and lined with grass and moss. Holes in trees are sometimes used as dens, whilst nest boxes of a suitable size have been used for breeding (Rowe 1983).

Density The grey squirrel is found at much higher densities in Britain than the red, ranging from 1 to 13 per ha.

Food Oak (acorns with less tannin being selected (Moller 1983)), sweet chestnut, hazel, beech mast, sycamore samaras and a variety of fruits form the main items of diet, together with roots, bulbs and fungi, so that foraging is mainly on the ground during the autumn and winter. Squirrels are difficult to trap in the autumn when food is plentiful (Rowe 1980). In years when there have been heavy seed crops, squirrels may stay feeding on the ground longer than usual (Moller 1983). Then more survive over winter and females come into breeding condition earlier, especially in areas where food is provided for game (Kenward 1983).

In spring the squirrel starts feeding in the tree canopy, selecting high energy foods from those available; thus flowers and catkins are eaten in preference to leaves and shoots. Poplar catkins are taken in April and oak flowers in May. June and July are months when there is an energy gap and a range of food is eaten in the tree canopy. Phloem is obtained after stripping the bark from the more vigorously growing trees (10 to 40 years old), 2 to 3 years after canopy release by thinning. Such attacks are related to phloem width and to body weight of squirrels (Kenwood, Parish, Holm and Harris 1988a, 1988b). Squirrels move into areas with a good supply of food and they will congregate on particular trees that are fruiting heavily. If food is in short supply, breeding does not take place (Gurnell 1983). As with the red squirrel, feeding activity is confined to two main periods of the day, morning and afternoon, separated by rest. Feeding in the winter is restricted to the morning (Tonkin 1983).

Conservation and control The grey squirrel gives many people pleasure in urban situations but it is a forest pest which severely restricts the growing of broadleaved trees. Buds and leading shoots are eaten, causing deformation, and trees are killed by girdling when the bark is stripped. A number of theories have been put forward to explain bark stripping but it seems generally agreed that it takes place during a time of food shortage and therefore high trapability. The supply of high-energy mast and spring flowers comes to an end in late spring, when the new supply of nuts and mast is not yet available (MacKinnon 1978). Late spring damage by bark stripping is more severe where populations are high (often because of winter feeding of game and livestock) and where the habitat has a limited supply of foods to fill the 'hungry gap'. The amount of damage also seems to be correlated with local and annual variation in sap flow (Kenwood 1983, 1988). Grey squirrels have been recorded stripping bark from fourteen species of deciduous trees, including sycamore and other maples, beech, oak, birch and ash, whilst cherry is left alone. Conifers regularly attacked include pine, spruce and larch. A detailed table of squirrel damage is to be found in Gill 1992. Conservation cannot be justified as the grey squirrel is much the most serious pest of broadleaved trees. Poisoning is the most effective means of control available at present (see Forestry Commission publications). A new trap has been developed that is activated by grey squirrels only.

Field vole *(Microtus agrestis)*

Distribution Occurs throughout mainland Britain but is absent from Ireland. Elsewhere, the field vole has a wide range and can be found in all forest zones. NE.

Habitat requirements Optimum habitats are those with rough grass and therefore plenty of cover. Human activity has helped to increase the habitat for the vole through the clearing of woodland and scrub. There is evidence that populations persist in good habitats, less favourable areas being colonized when the numbers rise during the breeding season (Tapper 1976b). Young forestry plantations, both conifer and broad-leaved, and especially in the first rotation, provide a good habitat because of the cessation of grazing. The field vole is also found in mature woodland that is not dense and that therefore has sufficient ground cover, in hedgerows, on blanket bog, scree and moor. Shallow burrows are made just beneath the surface, or tunnels are made above the ground in long grass. Nests are built at the base of grass tussocks or in burrows.

Food Leaves and stems of grasses are the primary food, particularly in the spring and early summer when their food value is high. Herbs and seeds are eaten later in the year (Tapper 1976b). Roots, young trees and bark are also eaten.

Density High densities are reached in favourable situations giving rise to vole 'plagues'. This aspect has been reviewed by Tapper (1976b). High populations occur in the early stages of afforestation, up to 500 per ha, and attract a wide range of predators (Newton 1983). In the second rotation, on a felled area of Sitka spruce where the lop and top was left on the ground, densities of 31 per ha were recorded (Thompson 1986). Usually only young animals overwinter and few adults survive into the second autumn.

Conservation and control Voles can cause considerable damage to forestry interests, both by eating tree seeds and by gnawing and ringbarking young trees at ground level and below the surface. Trees that are not completely girdled may survive but are distorted in growth. Conifers may recover if side buds replace leading shoots that have been damaged (Brambell 1974). Control of voles is seldom effective and is expensive but tree guards or tree shelters are the best defence. Damage is less with overall weeding. Weeding round trees also reduces damage as voles do not like crossing bare ground (Davies and Pepper 1987).

Voles provide food for a wide range of predators and in that role are an important factor in the maintenance of rare predator populations. Such predators include owls, hawks, harriers, eagle, polecat, pine marten and wild cat. During phases of rapid vole increase predators may exert little control on population levels but an abundance of prey leads to a build up of predator numbers. As the peak in vole numbers begins to decline the increased numbers of predators exert an effect (Tapper 1 976b).

Bank vole (*Clethrionomys glaroelus*)

Distribution The bank vole is found throughout mainland Britain. It was discovered in Ireland in 1964, where it was probably introduced, and is now spreading slowly.

Elsewhere it extends through all forest zones. NE.

Habitat requirements The bank vole is found in both coniferous and deciduous woodland, in scrub and in hedgerows. Ground cover, such as bracken and nettles, is important, few voles being found where the forest floor is clear (Eldridge 1968). Like the field vole, runs are made above ground and burrows below ground. Nests are built close to the ground or in burrows where thermal conditions are stable (Cotton and Griffiths 1967). Bank voles climb up small trees and shrubs more readily than field voles; they seldom venture far from hedgerows or the woodland edge (Eldridge 1968, 1971).

Density Few adults survive over winter, the animals breeding the following spring being the young of the previous year. The extent of the home range varies with habitat. Ranges of 50 m have been recorded in woodland and 120 m long in a linear habitat (Healing 1980). Densities range from 4 per ha to 74 per ha, depending on habitat and the time of year. Densities of 6 per ha were recorded in larch with little ground cover except on ridesides and in small areas where thinning had let more light through the canopy (Don 1979). Densities of 15.7 per ha have been recorded in 40-year-old Sitka spruce, falling to 11.7 per ha on recent clear falls (Thomson 1986).

Food Tree and shrub leaves and grasses are eaten in spring and early summer; fruits and seeds are taken in the autumn. The flesh but not the seed of rose hips and holly is eaten, as also are acorns, hazel nuts, ash and yew seed (Eldridge 1969; Smal and Fairley 1980). Sitka spruce seed was the most important food item in a 40-year-old plantation. Fungi are eaten, especially in autumn (Thomson 1986). In one study, the bark of yew and ash was taken but not that of hazel or oak. Voles, unlike wood mice (see below), prefer not to eat hard seeds if other foods are available. Ferns and woodrush are not taken (Smal and Fairley 1980). Insects and worms supplement the diet. The cocoons of forest moths are eaten, including those of *Bupalus piniarus*, which is a serious pest of pine.

Conservation and control Damage is similar to that of the field vole on young trees but usually occurs higher on the stem. Tree seed is taken and damage may occur in nurseries. Like the field vole, the bank vole is an important prey species for a wide range of predators. Control measures are the same as those for the field vole.

Water vole (*Arvicola terrestris*)

Distribution Occurs throughout Britain but not in Ireland. NEWA.

Habitat requirements Usually near water, but outside Britain lives underground away from water.

Conservation and control There is little information on this species. The presence of fresh water appears important. Water voles can cause damage to willows.

Wood mouse *(Apodemus sylvaticus)*

Distribution Occurs throughout mainland Britain and Ireland. Elsewhere it may be found throughout all forest zones. NE.

Habitat requirements Found in most woodland types, except in very dense, young plantations (Yalden 1971), though it may occur on the edge of the latter. Studies in broadleaved woodland found that a ground cover of bramble and bracken was more important than a shrub layer (Corke 1977). It is more of a woodland edge species than the yellow-necked mouse when they occur in the same area (Corke 1977). The wood mouse has been recorded in Britain high on mountains well away from woodland (Wilkinson 1987).

Runways are made in woodland litter and burrows are made below ground. These are short in clay soils but extensive in sands or loams, with nesting and food storing chambers (Jennings 1975; Gurnell 1979). The wood mouse is a good climber and arboreal runways are used to avoid predation (Montgomery 1980). An extensive bibliography of wood mouse studies can be found in Judes 1982.

Density Few adults survive over winter, the young of the previous year forming the main breeding stock. Numbers are at their lowest in March and April, highest in September and October. Seven to 100 individuals per ha can be found in good habitats and after good seed years in oak and beech. Nineteen to 28 per ha have been recorded in conifers (Forbes and Lance 1976); 34 per ha in larch (Don 1979); and 6.2 per ha in dense 40-year-old Sitka spruce (home range 63 m), with 4.5 per ha in the clear falls (home range 42 m) (Thomson 1986). Home ranges vary up to 1000 m2, the male occupying a large area (Montgomery 1979). Wood mice range more widely than bank voles in their search for food and, because the species is nocturnal, continuous cover is not so important (Flowerdew 1976).

Food Insects, buds and shoots, larval lepidoptera including sawfly larvae *(Cephalica lariciphila)* (Don 1979) are taken; grain, seeds, fruits and nuts are taken in autumn and winter. Sitka spruce seed was the most important item in a 40-year-old plantation (Thomson 1986).

Conservation and control The wood mouse eats tree seed in forests and nurseries and may require control in the latter by poisoning and trapping. Mice provide food for a wide range of predators, including owls and hawks.

Yellow-necked mouse *(Apodemus flavicollis)*

Distribution In Britain, the yellow-necked mouse is only found in southern and eastern England. A full discussion of its restricted distribution is given by Corke (1977). NEWA.

Habitat requirements The yellow-necked mouse prefers to live in woodland rather than on the woodland edge. In other countries it is found in dense coniferous forest (Van den Brink 1967). It is recorded at higher altitudes than the wood mouse in France and Norway. There appear to be no reports of the occurrence of this species in lowland coniferous woods in Britain. A dense, low canopy is favoured and high populations were found in a wood of yew, beech and hazel where windblown trees formed dense entanglements near ground level (Montgomery 1979). Nest boxes provided for birds have been used.

Density Eight to 40 individuals per hectare have been recorded (Montgomery 1979).

Food No data available.

Conservation and control As for the wood mouse.

Dormouse *(Muscardinus avellanarius)*

Distribution Mainly found in the south and east of Britain. SWEA.

Habitat requirements The dormouse is an animal inhabiting either the edge of woodland or woodland in the early stage of succession with thick vegetation. Dense woodland which has no shrub layer or boundary hedge is not suitable. Bramble and honeysuckle are important elements in the habitat. Both are used for nesting in and also as a food supply. Nests can be found 10 m above ground as the dormouse is a good climber. Birds' nests and bird nest boxes are sometimes used (Hurrell and McIntosh 1984). Nests for hibernating are usually at or below ground level.

Density No data available. Old records noted high numbers in hazel coppice, which will have provided an ideal habitat as coppiced woodland never gets beyond the thicket stage.

Food Hazel nuts are the most important element, with sweet chestnut, beech, acorns and fruits (blackberries, crab apple) taken in the autumn when the animals are storing up fat for the winter. In spring, leaves (of honeysuckle), flowers and pollen are eaten, together with Lepidoptera and aphids (Richards et al. 1984).

Conservation The dormouse is harmless and is a protected species in the UK under the Wildlife and Countryside Act 1981. Because of the reduction in habitat (particularly hazel coppice), together with a change in climate, it is not as common as it was at the beginning of the twentieth century. Conservation is thus required. Woodland edges in the southern and eastern counties should be provided with hazel hedges and thick cover to encourage this species (Hurrell and McIntosh 1984). The thicket stage of both coniferous and broadleaved plantations provides a habitat but sufficient seed or fruit bearing trees also need to be present to provide autumn food. Nest boxes can be provided for summer use.

Harvest mouse *(Micromys minutus)*

Distribution Mainly in the south and east of England. Elsewhere, the harvest mouse has a wide range in mixed and deciduous forest, extending into the taiga. It seems to require continental-type summers. NE.

Habitat requirements Although associated with corn-growing areas in the past, the harvest mouse can be found in any situation where tall vegetation provides cover. It has been suggested that the harvest mouse 'today is thriving best in marginal habitats, often between wet and dry ground, along river courses and in marshes, probably as it did in primaeval times' (Glue 1975). A survey carried out by the Mammal Society found nests in hedgerows (13 per cent), bramble (11 per cent), field edges and ditches (10 per cent and 9 per cent respectively), scrub (7 per cent), woodland border (4.5 per cent) and young coniferous plantations (4 per cent), as well as a variety of other situations (S. Harris 1979). Winter nests are constructed near the ground at the base of grass tussocks and in piles of logs or stones. The harvest mouse is also found nesting on ridesides in mature coniferous plantations. Motorway verges are colonized where the vegetation consists of tall grasses, willowherb *(Epilobium* spp.), teasel *(Dipsacus fullonum)* and thistles *(Cirsium* spp.) (Trout et al. 1978); that is, vegetation very similar to that which can be expected on clear cuts in woodland in southern Britain.

Density This varies with habitat. The harvest mouse has a short life span so the population turnover is rapid, the greatest numbers being found in November when densities can reach 200 per ha. The mean home range of male mice is 400 m^2 and of females 350 m^2, with ranges probably overlapping (Trout et al. 1978).

Food Insects, fruits, seeds, berries and grain are recorded but there is no detailed work on the diet (Trout et al. 1978).

Conservation No significant damage is caused by this species. The harvest mouse forms prey for hawks, owls, stoats and weasels and is vulnerable to a wide range of predators, being active throughout most of the day and night (Trout et al. 1978). No special measures need be taken in woodland for the conservation of this species. Restocked areas and other sites with rough grass will provide suitable habitat. Spot weeding with herbicides around planted trees rather than complete weeding will leave food and vegetation for harvest mice.

Plate 11.2 The edible dormouse (*Glis glis*), a protected species, introduced from Europe, causes considerable damage to both broadleaves and conifers in the Chilterns, southern England. (a) Edible dormouse; (b) a characteristic 'window' gnawed in pine bark. (Photos: (a) Forestry Commission, (b) E. H. M. Harris)

Edible dormouse *(Glis glis)*

Distribution Found only in the Chilterns, having escaped from captivity in Tring Park, Hertfordshire. Its range is expanding, mainly because of unlawful releases. The edible dormouse was thought to have been introduced originally by the Romans for food but recent archaeological work in York has shown dormouse remains there to be those of the garden dormouse (*Eliomys quercinus*) (O'Conner 1986), a European species absent from Britain.

Habitat requirements The edible dormouse is found mainly in young woodland and pole stage crops.

Density Densities of 1.7-1.8 per hectare have been recorded in mature and pole stage beech (Hoodless and Morris 1993). Up to 30 individuals have been found in domestic attics in winter.

Food Fruits, nuts and bark.

Conservation and control This introduced species is protected under the Berne Convention (see chapter 3 above). Where it occurs in Britain, however, it is responsible for extensive damage to both conifers and broadleaves between 10 and 30 years old by gnawing the bark of the stem in the crown and forming characteristic rectangular 'windows' in it. The trees later break at this point during high winds. In parts of the Chilterns it seriously damages trees in both broadleaved and coniferous woodland. At present this animal is protected and it is illegal to trap or kill it without a licence. Trapping is the preferred control method and nesting boxes have been used to catch edible dormice in Europe (Jackson 1994).

ARTIODACTYLA: MAINLY LARGE HERBIVORES

Red deer *(Cervus elaphus)*

Distribution Found mainly in Scotland, with herds occurring in the Lake District, East Anglia, the New Forest and the south west; as well as small numbers in other places near old deer parks. In Ireland found in Donegal and Wicklow. In other countries the red deer has a wide distribution and occurs in all forest types. NE.

Habitat requirements Although the red deer had to resort to the open hill in Scotland, with the destruction of forest, it is primarily a woodland animal and has therefore found an ideal habitat in the expanding area of coniferous forest in Scotland and northern England. A combination of cover near to suitable grazing areas is sought in woodland; a condition provided by the 'normal forest'. The thicket stage is used for resting and cover. In winter, hinds graze different areas to stags on the hill, selecting the more calcareous grasslands (Staines and Crisp 1978; Watson and Staines 1978).

Density Home ranges on the open hill have been recorded as 400 ha for a group of hinds and 800 ha for a group of stags. Large herds are found on some deer forests. The size of the range may be a reflection of the amount of shelter and food available within an area (Staines 1974). Deer in woodland occur in smaller groups, some herding taking place in winter. Woodland deer have heavier body weights and a higher breeding success than those on the hill (Mitchell and Grant 1981; Staines 1986). Red deer may live for 20 years.

Food The red deer grazes rather than browses and feeds on grasses as well as

Plate 11.3 Upland forestry plantations, established in the twentieth century, have created an ideal habitat for the red deer (*Cervus elaphus*) necessitating expensive fencing. (Photo: E. H. M. Harris)

heather, shoots of trees and leaves. Bark (including that of yew, and also yew leaves) is stripped and eaten. Food quality is important and deer are selective in their choice of food (Staines 1974; Jackson 1974) but as the deer is a ruminant, roughage is also necessary for proper digestion.

Conservation and control The expanding woodland cover in upland Britain has favoured the roe deer which in some places now does considerable damage by browsing trees and stripping bark. There is some evidence that bark is used as roughage at certain times of the year and it may also provide minerals. Many deer move into woodland for shelter during the winter and trees may be browsed for their moisture content (Jackson 1974). Willow and aspen are stripped in preference to other broadleaves, and Norway spruce and lodgepole pine are the preferred species amongst conifers (Staines 1986). Control is usually carried out by shooting and thus is difficult in woodland. Numbers need to be kept well down if damage is to be avoided (Ratcliffe 1987, 1988). For details of damage see Gill (1992).

Sika deer (*Cervus nippon*)

Distribution Introduced from China and Japan. Scattered feral herds occur throughout Britain.

Habitat requirements Sika use all types of forest. The thicket stage is used for cover and resting.

Food The sika deer is a grazer like the red deer, feeding mainly on grasses and shrubs.

Conservation and control Damage is similar to that of red deer. Where the ranges of these two species overlap, some inbreeding occurs and hybrids are common. Hybridization with the red deer also occurs in the sika's natural range (Harris 1987).

Roe deer (*Capreolus capreolus*)

Distribution Mainland Britain from Yorkshire and Lancashire northwards but absent from Wales and Ireland. Roe deer became extinct in England in medieval times and were reintroduced to Dorset and East Anglia in the nineteenth century. They are now spreading as a result of reafforestation. In other countries the roe deer occurs in all forest types. NE.

Habitat requirements Any type of woodland with sufficient cover can hold roe. In Scotland the roe deer has become adapted to feeding and living on moors with deep heather that has followed tree removal, but the reafforestation programme has helped it to recolonize many parts of Scotland. Roe deer usually live in small groups that hold territories from April to August. The boundaries of territories are scent marked, rides, streams and banks being used. Dry, sunny areas in which to lie up are an important constituent of suitable habitat.

Density The average size of breeding animal territories is 7 ha; that of a male may overlap those of two or three females.

Food Roe deer are browsers rather than grazers and the main food taken depends on availability. In young conifer plantations grasses, herbs, heather and conifer foliage are eaten, and in older plantations, conifer foliage, grasses and bramble. In deciduous woods ash, oak, hazel and other shrubs are browsed. Bracken and other ferns, rushes, sedges, fungi, corn, turnips, clover, fruit and roses are also eaten. Bramble is particularly important in winter (Henry 1981; Hosey 1981). Yew is browsed as high as it can be reached.

Conservation and control Roe deer damage trees by browsing, severely restricting growth (Welch et al. 1988). Young trees are frayed to mark territories in the summer months (April to August) and to remove velvet from antlers (April to June). At high densities roe deer cause considerable damage. Culling needs to be heavy, as twins are produced regularly in woodland areas.

Fallow deer *(Dama dama)*

Distribution Originally introduced, probably by the Romans, and now occurs in scattered herds throughout Great Britain, having escaped from parks.

Habitat requirements Any type of woodland, provided that there is some cover available in which to hide during the day, though this is not so important where there is little disturbance. Fallow deer may then choose to rest in a dry sunny place out of the wind, often in full view on the edge of rides and woods.

Density The number supported by an area depends on food availability within the wood and the surrounding fields.

Food Fallow deer are grazing animals rather than browsers (Jackson 1977). Grasses, rushes and foliage are eaten. In the autumn, acorns, beech mast, apples, chestnuts and blackberries form a major part of the diet. Bramble and mosses are taken during the winter and almost anything is nibbled in late winter, particularly dog's mercury, bluebell leaves, ivy, yew, holly and ferns.

Conservation and control Fallow deer damage young trees by browsing, nibbling almost anything and removing protective tree guards unless these are very securely fixed. Fallow deer also damage trees by fraying during the autumn rut and when removing velvet from the antlers (from late summer to early autumn). Bark is also stripped from trees. Walnut trees are particularly attractive in the rut and small trees are thrashed into the ground. Culling needs to be heavy if numbers are to be kept at a reasonable level.

Muntjac *(Muntiacus reevesi)*

Distribution Was introduced from China but has escaped from parks, Woburn in Bedfordshire being one of the main centres. It is now found throughout south-east England and is spreading rapidly north and west (Chapman et al. 1994).

Habitat requirements Thick cover is required, especially bramble. Occurs in deciduous woodland, young coniferous plantations and mixed plantations.

Density No information is available but muntjac appear to hold territories like the roe, with male and female ranges overlapping.

Food The muntjac is mainly a browsing animal (Jackson 1977). Bramble forms over 50 per cent of the food eaten. Herbs, leaves and grasses are also important. Apples and other fruits are eaten in the autumn. Ivy, dog's mercury and bluebell

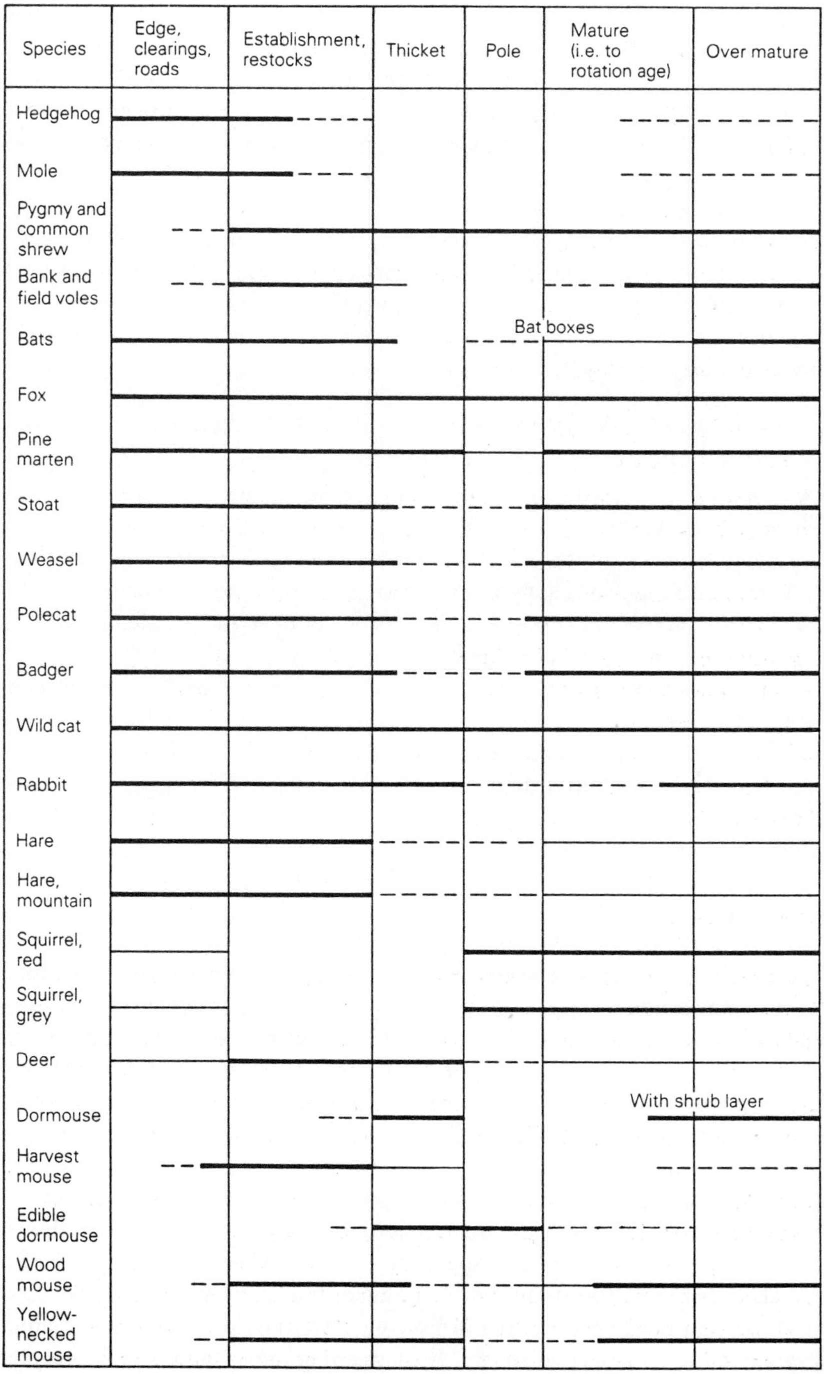

Species	Edge, clearings, roads	Establishment, restocks	Thicket	Pole	Mature (i.e. to rotation age)	Over mature
Hedgehog						
Mole						
Pygmy and common shrew						
Bank and field voles						
Bats				Bat boxes		
Fox						
Pine marten						
Stoat						
Weasel						
Polecat						
Badger						
Wild cat						
Rabbit						
Hare						
Hare, mountain						
Squirrel, red						
Squirrel, grey						
Deer						
Dormouse					With shrub layer	
Harvest mouse						
Edible dormouse						
Wood mouse						
Yellow-necked mouse						

Figure 11.1 The forest stages used by mammals

leaves are also eaten (Jackson 1977). Conifers eaten include Douglas fir, Scots pine, Norway spruce, Sitka spruce, western hemlock and yew; amongst the broadleaves, oak, birch and cherry are eaten. The diet of the muntjac in China includes rhododendron leaves (also recorded in Britain), *Vaccinium,* ivy, roses, oak, pine and yews (*Cunninghamia* and *Cephalotaxus*) (Jackson 1977).

Conservation and control As numbers increase some damage to young trees is being reported. The male muntjac strips the bark up to a height of 1 m with his teeth, sometimes ringbarking the tree and killing it, but the damage is usually on one side, causing distorted growth. Cherry, a species left alone by the grey squirrel, is frequently attacked by fraying.

THE CONSERVATION OF MAMMAL SPECIES IN WOODLAND: SOME GENERAL GUIDELINES

All the mammals discussed use deciduous, mixed or coniferous woodland in some or at all stages of the rotation. The tree species themselves do not influence the presence of many species, but the conditions provided by the woodland do (see figure 11.1). Woods on fertile soils support higher numbers than those on poor soils. The wide range of foods eaten by most mammals is one of the principal factors enabling them to exploit a wide range of habitats.

Mammals using woodland can be assigned to four categories from a conservation point of view:

1 Rare species requiring protection;
2 Predators;
3 Mammals neither rare nor doing damage;
4 Successful mammals doing damage, and therefore requiring control.

Rare species requiring protection

The following species require special protection so that, wherever possible, management should aim to encourage them in the parts of the country where they occur naturally. These species include the dormouse (a southern species near the extremity of its range), the harvest mouse (a northern species requiring a more continental climate), the red squirrel and all bats. The last fall into two groups, northern species requiring a more continental climate and southern species at the northern edge of their range in southern Britain.

Dormouse .A woodland edge animal requiring a deciduous, scrubby edge or hedge. Thick patches of bramble, especially in south-facing sheltered places, are important, together with fruiting hazel (the main food) or fruiting oak, beech or sweet chestnut. Shrubby conditions following clear falls provide suitable situations, provided there is a food source. Spot weeding rather than overall weeding of new plantations will allow a wider range of food. Nest boxes can be provided for breeding.

Harvest mouse An animal requiring thick vegetation. Found along ridesides in all types of woodland, young plantations and clear cut areas. To encourage harvest mice, vegetation should be left uncut and in recently planted areas spot weeding round the trees rather than complete weeding should be practised.

Red squirrel The principal factor required is a large area of coniferous woodland. Here the grey squirrel is unable to compete successfully to the extent that it does in broadleaved woods. In suitable areas, broadleaves (especially those with large seeds) should be kept at a minimum and a continuous supply of conifer seed needs to be available. Such conditions obtain once the trees are half-way through the first rotation, as there will then always be trees coning.

Bats The provision of bat boxes in all types of woodland beyond the thicket stage helps all species but their siting is important. Sheltered situations not too far from woodland edges should be chosen. Artificial caves have also been built and have proved successful for hibernation. The provision of a water source encourages insects and provides a feeding area for bats.

Predators
These include the still rare pine marten, polecat and wild cat. Although these predators need protection as a priority, all having been at the point of extinction by the beginning of the twentieth century, the following remarks apply to predators in general.

Human beings are the main enemy of any predator; after that the availability of food is the principal limiting factor. Because of the great variety of prey taken, predators can be found in a wide range of habitats, including all types of woodland. A 'normal' forest (see chapter 4) provides the combination of cover, shelter and food supply needed. However, all predators, even the rare ones, are still extensively persecuted in the interests of game and, in some cases, agriculture. Undoubtedly some domestic animals and game species are taken by predators, especially game when their numbers are maintained at an artificially high level. However, there is always a 'floating', non-breeding population of predators ready to fill gaps created when a resident population is wiped out. It is for this reason that the total number of predators killed on some estates remains remarkably constant from year to year. Control during the game-bird nesting season may be necessary for a short period but the number of rodents and rabbits taken by predators during the rest of the year is beneficial. Total predator destruction leads to increases in the numbers of rodents and rabbits, which in turn then need controlling. The essential conservation message with regard to predators is therefore to cease persecution and, where necessary, to practise sympathetic control. Culling can be limited to certain times of the year, and will be more effective than continuous and indiscriminate killing.

Mammals neither rare nor doing damage

All these mammals can be found in woodland of all types at the appropriate stage of the rotation, provided the conditions given under each species are met. They do not therefore require any special measures.

Successful animals doing damage and therefore requiring control

These include, deer, grey squirrel, vole, mice and rabbit, all of which are increasing as a result of afforestation. Rats may need control at times, as they often move into woodland during the summer.

The control of pest species is complex, and detailed recommendations on control must be sought elsewhere from the extensive literature available. Pest mammals can be either small or large. The smaller mammals breed rapidly and, if conditions are favourable for survival, they reach pest proportions very quickly, even if the overwintering populations are small. Work has shown that the number of mice and bank voles in a wood is related to the abundance of the tree seed crop. Numbers in a survey were highest in mixed hardwoods and hardwood/conifer stands (Mallorie and Flowerdew 1994). Trapping and poisoning provide a check on numbers but control by predators will have a useful influence when predator numbers have had time to build up. The factors that aid the rapid breeding of pest species, such as food supply and shelter, can be removed or fenced off.

Rabbits and squirrels breed early in the year if they have wintered successfully, so their body weights are high in the spring. More young are then born but their individual body weights are less than in smaller litters. Early breeding also allows more litters to be born during the year but the rate of survival will depend on the prevailing conditions. Grey squirrels are usually poisoned, trapped or shot; rabbits can be gassed, though this may kill other species that shelter in their burrows. The vulnerability of plantations to attack by squirrels needs particular attention (Kenwood et al. 1996).

Deer have increased rapidly in all types of woodland in the last few decades and now cause considerable damage. Fencing planted areas keeps deer out for some time. Tree shelters are only effective against muntjac. Shooting is the usual method employed to control numbers; it is particularly important to keep the female population at a low level and at the same time to cull a proportion of males. When food is short only the older females breed but these are more successful in rearing their young than inexperienced juveniles. *Forestry Commission Bulletin number 71* examines control methods in relation to red deer and emphasizes the importance of forest design for deer control by providing open space where animals can be shot cleanly.

The position of the red squirrel, grey squirrel and pine marten in present-day woodlands and forests epitomizes the conservation opportunities they are now providing. The native red squirrel had adapted to Britain's largely broadleaved woods, like so many other species normally living in the coniferous and mixed forest zones, but suffered almost to the point of extinction with reduction in woodland cover. By contrast, and typical of an introduced species, the grey squirrel has been very successful. The red can only compete where large conifer forests give it an advantage. The pine marten suffered even more severely than the red squirrel but is now recovering as a result of afforestation. As the natural predator of the red squirrel, the two species will come into balance, but if the pine marten started to control the grey squirrel too, broadleaved forestry would benefit greatly. Perhaps the only way to control the grey squirrel would be to introduce the pine marten (or the related beech marten from the European mainland) into lowland areas.

CHAPTER 12

Birds

The pattern of recolonization of Britain by birds after the first ice age is similar to that by other animal groups, but because of the greater mobility of birds it was more rapid. The initial treeless arctic tundra conditions would have encouraged species that were able to make use of such habitats. A few of these tundra species still breed in Britain where conditions are suitable for them, mainly in the north of Scotland. The majority of tundra species are summer visitors, exploiting the short summers and moving away to avoid the harsh winters.

With the warming of the climate, grasses and tree species began to colonize the denuded soils, which provided habitats similar to those in the forest tundra zone today (a zone with scattered trees). Lakes and swamps occupied the lower ground with willows on their margins, whilst birch, aspen and pine grew on the drier sites. Gradually tree cover increased and the mixture of tree-covered and open areas diversified habitats further. At this stage the landscape resembled the taiga zone today; that is, a mixed landscape of trees, mainly conifers, and wet areas, similar to that which now covers vast regions of Russia and the North American continent (see chapter 1). Many British bird species inhabit the taiga zone elsewhere in the Northern hemisphere, both in Asia and America, where conifers are the principal species. It is reasonable to assume that birds living in the taiga zone today originally colonized Britain when conditions were similar, although some of the continental tree components were absent, especially spruce and larch. The addition of spruces, firs and larches, particularly to the new upland forests, has increased the number of tree species present and provides the wider range of tree species that occurs in the taiga throughout the rest of the northern hemisphere (see chapter 6).

Following the warming of the climate after the ice age, the forest cover that developed in Britain included a mixture of broadleaved species which are typical of the few remaining mixed forests in Europe and those of Asia, where human influence has been less than elsewhere. The mixed forest zone provides a habitat in which elements of the taiga zone (with lower temperatures) and the broadleaved zone (with higher temperatures) intermix (see chapter 1), providing habitats for a wide range of species. There are no woodland species of British birds so totally dependent on pure broadleaved forest that they cannot exist in mixed forest, and all British species are typically inhabitants of the natural mixed forest zone. There are only about a dozen British woodland birds that are naturally restricted to the mixed and broadleaved zones only. These are species with a distribution related to higher summer temperatures than prevail in Britain and whose ranges are of a more southerly or more continental type, such as the nightingale. Species that extend into the natural coniferous zone, and these are the majority, do not depend on pure broadleaves for survival, other factors such as forest structure and the availability of food and nesting sites being more important. A number of these species have adapted to man-made conditions; for example, many woodland birds have found the equivalent of woodland in gardens, parks and hedgerows. The lack of conifers in Britain in the past has resulted in many species typical of taiga or mixed forest conditions modifying their behaviour and adapting to wholly broadleaved habitats. The British coal tit, for example, has a thicker bill, capable of dealing with a wider range of seeds, than its counterpart on the European mainland, which does not normally extend into the broadleaved zone. The nuthatch, thought of as a typical representative of broadleaved woodland in Britain, is found in the taiga zone elsewhere, feeding on insects and conifer seed.

Although a few British birds have a distinct preference for either coniferous or broadleaved trees, especially when the trees themselves provide the main direct food source in the form of buds and seeds, most species are to be found in a wide range of forest types. When considering the conservation of birds in woodland, it has to be remembered that all species do not occur in all parts of Britain. Birds with a more northerly breeding distribution (for example, redwing) do not nest in the lowlands and the south and are on the southern edge of their range. Conversely, the nightingale is confined to the south of Britain and is on the northern edge of its range. The different climate prevailing in the west of the country supports different communities to those in the east. The presence of birds in Britain is due primarily to factors other than tree species, including altitude, topography, size of wood and woodland structure (Newton 1986a; Bibby 1987; Fuller 1995). Size has been studied by Ford (1987), who showed that a shrub layer is more important than woodland area in creating species diversity. The structure of woodland has been analysed in relation the presence of song birds in Aberdeenshire by French et al. (1986). They concluded that structure is more important than the tree and shrub species composing the woodland and that 'variation among tree types does not indicate tree species effects but rather effects of differences in the cover profile characteristic of these types'. They found that the 'best' woods were those containing the highest diversity of habitat, which in turn was related to age. Open space of 10 per cent to 35 per cent and a good shrub layer were important elements and the latter was most beneficial if evergreen. The 'best' woods had similar structures to each other, whilst the tree species forming the canopy was not significant. French and his colleagues noted that their findings 'partly contradict the generally held view that broadleaved trees always support more species than conifers'. Their recommendations for improving woodland for birds include: diverse structure combinations (at least 4 structure types in 10 ha); at least one third mature and old trees, together with other age classes; open land over 20 per cent of the area; tall shrubs and dominant trees occupying one third each, if possible overlapping, or if not, the tall shrub areas next to pole stage or mature forest; retention of some dead wood; avoidance of large even-aged blocks. French et al. state that 'plantations need not be poorer than "natural" woods in song birds and can indeed be considerably richer, if the above guidelines are followed'. Their recommendations can be fulfilled by straighforward forest management without expensive commitments.

Food supply is another important factor. Most species eat a wide range of foods and different food items are exploited in different parts of their range. Foraging by animals and birds is directed towards saving energy, so larger prey is selected in preference to smaller items if it is available (Belovsky 1983; Karnil et al. 1987).

The mixed forest zone sustains the widest range of species as it includes the elements of both the taiga and broadleaved zones. As we explained in chapter 1, Britain lies mainly in the mixed forest zone, with elements of the taiga zone on high ground and in the north of the country and a gradual transition to the broadleaved zone in the south. As the majority of British forest birds are very adaptable and are to be found throughout the natural taiga, mixed and broadleaved zones elsewhere, it is reasonable to suppose that colonization by birds took place rapidly once tree cover began to appear after the ice age. The few broadleaved zone bird species (about a dozen; see below) that do not extend into the taiga zone, were later arrivals after the climate had ameliorated in the warm Atlantic period. As we mentioned above, it is a widely held misconception that most of Britain's woodland birds are dependent on broadleaved woods for survival. This is by no means true and it has to be remembered that Britain's natural tree flora is poor in species, especially conifers, so that the habitat types available were limited in the past. Pure broadleaved woodland has been shown to support higher numbers of birds per unit area than conifer woodland, but this is mainly the result of the warmer conditions (lower altitudes) and better soil fertility in which such trees are grown. Leaf fall and faster nutrient cycling due to increased light penetration are also factors. Not surprisingly, territories held by birds are smaller on more fertile soils (as food availability is higher), so the number of birds per unit area is higher (Newton 1986a). What this reflects is the carrying capacity of the environment, which in turn influences the total numbers of a particular species able to utilize it. However, mixed stands support higher total numbers and greater species variety. The poorer soils and climatic conditions with lower nutrient cycling in the taiga zone are reflected by the lower overall numbers of birds that are able to use the available food supply; but soil fertility does not affect the range of species present (French et al. 1986; Newton 1986a).

It is useful to consider the natural forest zones inhabited by birds and the situations in woodlands used by them, including birds' use of all types of managed forest, which in many cases reflects their use of natural forest. This will show that the application of good forest management, together with some positive modifications, provides conditions in all kinds of managed forest that will suit most woodland birds. (The more complex problem of open-space forest birds we discuss later.) Conservation recommendations must take account of basic requirements, such as sufficient food supply for both survival and breeding, nesting sites and the suitability of territory. As a general principle, small birds require small areas to fulfil these needs, whilst many predators need large, well spaced territories. Most birds take a wide range of food and are not specific feeders. A considerable amount of research on birds in forests (large areas) and woodland (small areas) is in progress, and information of practical value is now coming forward. One interesting result comes from a study by Lack (1988), which indicates that small areas of woodland are of more value than linear hedges.

The forest zones described in chapter 1 as applied to Britain are Zone 5 (temperate broadleaved deciduous forest), Zone 6 (mixed coniferous and broadleaved forest) and Zone 7 (boreal, northern coniferous forest or taiga). For simplicity these are referred to as broadleaved, mixed and taiga forest. The stages a maturing forest goes through have been defined in chapter 4 and this information has been put together in the following section to describe habitats used by birds and how they can be provided in managed forest. It is tempting to think that prescriptions could be made for the various forest types that occur in managed forest that would encourage certain species of birds to colonize them. For instance, a prescription might be written for young pole-stage pine that could be expected to make it attractive to woodpeckers, but would involve too much speculation to be of value. Instead it is better to look at the habitats that have already been exploited and analyse how they have been used before recommending how similar forest types can be improved. The following section attempts to do this.

Except where stated, information is derived from the following references: Witherby et al. 1943, 1944; Peterson 1947, 1961; Cramp et al. 1977 et seq.; Newton and Moss 1977; Newton 1979, 1983, 1986b; Harris 1983 (which includes recorded details of densities in coniferous plantations); Flint 1984; De Schaunsee 1984; and Knystautas 1987. Table 12.1 shows the categories into which birds using forest and forest with open space habitats can be divided. This provides information on which conservation within productive forests can be based, and summarizes the following discussion. The value of plantations, both coniferous and broadleaf (often with conifer nurses), has been emphasized because their value to birdlife has not always been recognized compared with that of semi-natural woodland.

(a)

(b)

Plate 12.1 The rare crested tit (*Parus cristatus*) and the siskin (*Carduelis spinus*), previously confined to native pine woods have both extended their range in the second half of this century as upland coniferous plantations have provided and increasingly suitable habitat for them. (a) Crested tit; (b) siskin. (Photos: Forestry Commission)

FOREST BIRDS AND FOREST STAGES USED BY THEM IN NATURAL, COMPARED WITH PLANTED, FOREST

Thicket

Thickets and thick undergrowth under mature trees occur in the taiga, mixed and broadleaved forest zones, and are readily colonized by many species of birds. The same bird species have also been recorded using the thicket stage of coniferous, mixed and broadleaved plantations, in both newly afforested and restocked areas. Some bird species, after colonizing thicket, persist through later forest stages; others are found in mature forest with an understorey; many also find suitable habitats in gardens, parks and hedgerows.

Thicket 1 (TH 1) Birds that occur here have a northern distribution and use thicket in the natural *taiga and mixed forest zones.*

Both the redpoll and bullfinch are northern species that do not extend into the broadleaved zone in most of their range, although they do occur in broadleaves in Britain. A shrub understorey under mature trees is used as well as thicket. The number of redpolls has increased with conifer afforestation (Avery and Leslie 1990). In upland coniferous plantations in Britain, willow and birch (components of the taiga), and also alder at lower elevations, widen the choice of winter food available, insects being eaten in the summer.

Thicket 2 (TH 2) These birds have a general distribution, using thicket in *taiga, mixed and broadleaved forest.*

The chiffchaff, garden warbler and willow warbler are all summer visitors, whilst the long-tailed tit is resident. The garden warbler does extend into the taiga but not as far north as the other species (Harris 1983). All feed on insects and are found in mature forest with an understorey, as well as in thicket. Tall trees adjacent to thicket, or trees left on restocked areas, are used by chiffchaffs as song posts.

The greater and lesser whitethroat are also summer visitors but they use thicket only, on both afforested and restocked areas. The tree pipit, a summer visitor, likes scrub conditions and colonizes the early thicket stage of plantations; it is abundant on restocks in Thetford Forest (Fuller 1995). It can also be found in open forest. The tree pipit likes a tall song post and therefore prefers thicket adjacent to tall trees, or makes use of trees left on felled areas. The chaffinch, hedge sparrow, wren and robin colonize thicket after afforestation and remain as breeding birds into mature forest. All feed principally on insects in summer, whilst seeds and berries are eaten as well in the winter. The robin, wren and hedge sparrow nest near the ground, and brash left in heaps provides cover and nesting sites in pole and mature forest.

The reed bunting and sedge warbler normally frequent thickets by water but are sometimes found in the thicket stage of conifer plantations away from water.

Table 12.1 Forest zones and the forest stages used by birds

Natural range of species from north to south	Forest birds: forest stage used			Birds using forest but requiring open space		
	Thicket	Pole and mature		Small clearings and forest edge		Large clearings 2.0 ha upwards, and forest edge
		Holes not required for breeding	Holes required for breeding	Holes not required for breeding	Holes required for breeding	
Birds with a northern distribution using: Taiga and mixed forest	TH 1	PM 1	PMH 1	FE 1		FO 1
Birds with a general distribution using: Taiga, mixed and broadleaved forest	TH 2	PM 2	PMH 2	FE 2	FEH 2	FO 2
Birds with a southern distribution using Coniferous			PH 3a		FE 3a	
Mixed and broadleaved forest	TH 3	PM 3	PMH 3b	FE 3	FEH 3b	FO 3

Thicket 3 (TH 3) This group of birds has a southern distribution and uses thicket in *mixed and broadleaved forest.* They do not extend into the taiga or coniferous forest on mountains.

The blackcap and nightingale are both summer visitors to Britain that colonize thicket and shrub undergrowth in mature mixed and broadleaved forest. Both have been recorded in coniferous plantations at the thicket stage at low altitudes in Britain. The blackcap has been recorded in mature, well thinned coniferous forest with shrubby undergrowth and it also favours thickets next to pole stage or mature crops.

The nightingale is at the northern edge of its range in Britain; its decline in recent years may be the result of climatic changes rather than changes in forestry practice. Woods that held this species not long ago, and that still appear to have a suitable structure, are no longer used. Oak standards with hazel coppice was a favoured habitat in the past as this provided continual thicket in various stages with varied cover, thus allowing the birds to feed on the ground. It nests in conifer thicket with a broadleaved element (Fuller 1995). The nightingale is not a canopy feeder.

The grasshopper warbler is a summer visitor using the thicket stage. Young, open conifer plantations, including restocks, are another major habitat (Fuller 1995).

Conservation for birds favouring thicket

No special measures need to be taken for the conservation of birds using thicket. All these species eat a wide range of insects and also take berries and small seeds. The finches eat seeds, including those of conifers. The use of spot weeding rather than blanket weed control, and wide spacing with tree shelters, will provide a varied range of foods such as weed seeds and berries, particularly of bramble. Birch and willow widen the food range in upland forests as do berry-bearing trees and shrubs in lowland areas. Optimum habitats for some species are provided if pole stage or mature forest adjoins thicket, as the former provides song trees and high canopy feeding places.

Pole stage and mature forest

Birds using pole stage and mature closed-canopy forest feed mainly in the canopy on insects, seeds and fruits. They colonize at the pole stage and remain until the forest is felled. They may be divided into two categories: those that do not require holes for breeding (PM) and those that do (PMH). The latter are found mainly in mature and unmanaged forest where holes are available in old and damaged trees, but some hole nesting species excavate their own holes.

Pole stage and mature forest (PM 1) Birds in this group have a northern distribution, occurring in the *natural coniferous zone,* but they may also extend into the *mixed forest zone.* They are therefore found in Britain mainly in coniferous plantations or in mixed plantations with a high proportion of conifer.

The crossbill colonizes spruce and pine plantations in Britain when coning begins in the early pole stage. This species has a smaller bill than the Scottish race of crossbill and feeds mainly on spruce seed. Coniferous plantations have extended its range in Britain; it was an uncommon species early in the twentieth century. The Scottish race of crossbill, which has a larger bill, is confined to pine in Scotland, as it has adapted to feeding on pine in the absence of other conifers in the past. Whether it is sufficiently distinct genetically not to breed with the type race crossbill, or whether its feeding habits will alter to take advantage of the wider range of conifer seeds now available, are both unknown. The rare parrot crossbill has also bred in managed woods in Britain (Thetford Chase).

The goldcrest colonizes forest at the pole stage and is found in coniferous and mixed plantations. It also breeds in broadleaved forest in Britain, to which it has adapted in the past in the absence of conifers. All forest stages are used in the winter and insects are the main food.

Pole and mature forest (PM 2) Birds in this group have a general distribution throughout the *natural taiga, mixed and broadleaved forest zones*. They are therefore found in all types of managed forest.

The tree creeper, replaced in the broadleaved zone in parts of Europe by the short-toed tree creeper, nests in bark crevices, split trees and behind ivy. Nesting boxes can usefully be provided. Though insects form the main diet, small seeds are also eaten, including those of conifers. Flegg (1975) recorded that nesting success was better in conifers.

The siskin, once uncommon, has spread recently into coniferous plantations. Insects are eaten in summer and the presence of birch and alder widens the food supply in winter by providing seeds. It is a species that has benefited from afforestation.

The wood warbler is a summer visitor to Britain and is confined to the southern taiga and lower altitude coniferous forest on mountains elsewhere. It requires a sparse ground vegetation and no understorey of leafy shrubs; in Britain, it breeds in larch and birch where these conditions occur, in conifer plantations with a few broadleaves (Wales and Lake District pers. observation), as well as in mature oak, beech and sycamore woods. The main food is insects taken in the canopy.

Pole and mature forest (PM 3) These are birds with a southern distribution occurring (1) in *natural mountain coniferous forest* and mixed forest and (2) in the *natural mixed and broadleaved forest zones* but not in taiga.

1 The firecrest is a southern species that is not found in taiga but occurs in mountain coniferous forest in continental Europe. It is now breeding in coniferous and mixed plantations in southern England.
2 The hawfinch occurs in southern Britain where broadleaved trees fruit

sufficiently heavily to provide an adequate winter food supply. To maintain a presence it requires trees with large seeds, such as hornbeam, cherry, maple, lime, sycamore, beech, yew, hawthorn and sloe. Insects are taken in summer.

Conservation in pole stage and mature forest for birds that do not need nesting holes

No special measures need be taken for most of the species that use pole stage and mature forest. Brash left on the ground or in windrows will help some by providing cover, nesting and feeding sites. Crossbills require conifers old enough to produce cones, mainly spruce, before they will colonize. Although insects (such as the aphid-like *Adelges* spp.), buds and fruits are eaten, crossbills feed on dry seed much of the time and the provision of water will therefore assist their colonization. Hawfinches can be encouraged in mixed forest by the planting of some large-seeded broadleaves, as colonization will eventually be determined by the amount of ripe seed available for winter food.

Pole stage and mature forest with holes available for nesting (PMH 1) These are birds with a northern distribution occurring in *natural coniferous and mixed forest.* They are therefore also found in coniferous plantations and in mixed plantations with a high proportion of conifer.

The crested tit is confined to Scots pine in Scotland because of the lack of other coniferous species in the past. As with the Scottish race of the crossbill, it has yet to be seen whether the crested tit will use other conifers in the way that the crested tit does on the European mainland. Pine plantations have been colonized for several decades and this species is spreading north and west from its strongholds in the Spey valley (pers. comm.; Collier 1988). Petty and Avery (1990) speculate that in future there may be a higher proportion of crested tits living in plantations than in natural forest. Both this species and the more widespread willow tit excavate their own nesting holes in decaying timber, especially pine and birch. Nesting sites can be provided by leaving suppressed and windsnapped trees, also by ringbarking birch that has grown up in the crop; such trees do not need to be large (20 cm in diameter is adequate). Nest boxes are used by both species if these are lined with woodchips to simulate an excavated hole. Insects, pine seed and berries are eaten.

The coal tit is another coniferous species that has adapted to living in broadleaved woodland in Britain and now makes use of conifer plantations. Its stouter bill enables it to deal with a wide range of seeds. As nests are made in small holes and crevices, often below ground, the coal tit is not so dependent on tree holes as other tits; it therefore colonizes pole stage forest more readily than other tits. Insects, seeds and berries are eaten.

Pole stage and mature forest with holes available for nesting (PMH 2) These are birds with a general distribution that occur in *taiga, mixed and broadleaved forest.*

They are wide ranging and adaptable, so may be found in all types of managed forest.

The great tit colonizes pole stage and mature plantations of all kinds if nest boxes are provided. Food includes buds, insects, berries, conifer seeds and beech mast.

Although the nuthatch is considered a broadleaved species in Britain, it occurs widely in the coniferous zone in other parts of its range. The absence of conifers in the past has limited its choice of habitat in Britain. Nests are usually made in tree holes but holes in walls and sometimes old birds' nests have been used for nesting. Nest boxes can be provided for nuthatches but the entrance holes need to be larger than for tits. As a result, boxes are often taken over by starlings and sparrows. Food includes both the adult and larval stages of insects and a wide range of seeds, such as conifer seed, hazel and beech nuts and acorns.

The greater spotted woodpecker colonizes pole stage forest; quite small diameter wind-snapped pine, larch and birch are used for nesting. Holes are excavated by the birds themselves. Broken, wind-damaged, suppressed and ringbarked trees,

(a)   (b)

Plate 12.2 Woodpeckers do not depend only on large, old, dead trees. (a) 20 cm diameter birch (*Betula pendula*) ring barked to favour a crop of beech (*Fagus sylvatica*); (b) a supressed Scots pine (*Pinus sylvestris*) utilized whilst still living. (Photos: E. H. M. Harris)

particularly birch, can be left for their use and will also provide insect food. Dead trees or high dead branches provide display (drumming) sites. Damage may occur to some trees by sap sucking, the bird boring small holes in the bark to reach the sap. A wide range of food is eaten, including insects (especially bark beetle larvae and other wood-boring insects), young birds, fruits and seeds. Woodpeckers sometimes raid nest boxes which may therefore need some protection, perhaps by surrounding the entrance hole with metal.

Pole stage and mature forest with holes available for nesting (PMH 3) These are birds with a southern distribution, occurring in *natural mixed and broadleaved forest* but not in taiga.

In Britain, the blue tit nests in coniferous plantations from the pole stage onwards if nest boxes are provided. Insects, especially aphids and caterpillars, fruits and seeds such as those of birch, beech, and Scots pine are eaten. The marsh tit is found mainly in mature, mixed and broadleaved forest. It extends into the coniferous zone abroad, particularly in areas where the closely related willow tit is absent. Insects, seeds and berries are all eaten.

Conservation in pole stage and mature forest for birds requiring breeding holes

No special measures need to be taken for birds not needing old trees or holes for nesting. However, the range of such species can be encouraged by varying the structure of the forest. Management directed towards the 'normal' forest will go a long way towards achieving this aim. Some species will be attracted by the provision of an understorey in older stands. This can be achieved by shelterwood systems of regeneration or by using a coniferous shade-bearer as understorey. The presence of evergreen cover in a wood encourages more species to remain for the winter. In the past, commercial plantations were kept tidy so they lacked dead trees and older trees containing natural holes. Whatever the species being grown, birds need these components for nesting and can be encouraged in two ways. Firstly, old, suppressed and wind-damaged trees or ringbarked uneconomic species can be left when thinning or felling takes place. Secondly, artificial provision can be made for hole nesting birds by supplying nest boxes, and their populations will then soon build up. The boxes do need to be kept in good order. Old walls or buildings can also be left as these provide nesting holes.

BIRDS USING FOREST BUT REQUIRING OPEN SPACE

All the birds we have so far discussed are able to use forest as their sole habitat, although many also exploit similar habitats in gardens, parks and hedgerows. (There are now nearly half a million hectares of gardens in Britain, together providing a significant habitat for woodland birds.) Other birds using forest or woodland require situations where there is access to open areas for feeding, whilst the trees are used for breeding, roosting and cover. The forest edge is therefore exploited by a wide range of species. In addition, rocky areas, clearings (formed by wind, fire, grazing or waterlogging), streams, ponds, lakes and flood plains are all used in natural conditions. In managed forest, roads, rides and clear cut areas provide similar suitable areas. Places where trees will not grow, such as swampy and rocky areas, as well as frost hollows, can provide open space similar to that found in natural forest. The amount of open ground needed varies for different species but in general the smaller species need only small open areas to make conditions attractive to them, whilst larger species need larger areas. This section is therefore divided into three parts: small clearings of less than 2 ha, small clearings with holes available near by for nesting and larger clearings of at least 2 ha upwards. The figure of 2 ha has been selected as the dividing line because anything larger results in the increasing loss of forest conditions.

Small clearings and forest edge

Small clearings and forest edge (FE 1) Birds in this group have a northern distribution and use open space within the forest and the forest edge in *natural coniferous forest,* sometimes extending into the *mixed forest* zone. Managed coniferous forest in Scotland is used by these species.

The capercaillie has been declining in the last few decades, both in Britain and elsewhere in Europe. Various reasons have been put forward for this decline, including poultry infections from domestic stock and changes in forestry practice. It is more probable that climatic factors are mainly responsible as the decline is so widespread. Recent decades have been cooler and wetter in summer in Britain and the rest of western Europe, resulting in poorer conditions, particularly during the critical early weeks when the young chicks are most at risk from cold and wet weather, and also a reduction in the supply of insects. Until their feathers grow, the young birds live on the ground and are easily chilled. Immediately after hatching, a supply of high protein food is required, mainly in the form of insects; these are hard to find in cool, wet weather. Once the feathers have grown (3 to 4 weeks), the chicks are able to fly up into the trees and bushes where they are less vulnerable to inclement weather and to predators. Even so, feeding is still mainly on the ground, grasses and ericaceous berries being eaten, so that the presence of suitable ground vegetation is important. For these reasons, introductions of capercaillie to western Britain have so far not succeeded. By contrast, during the warmer, more continental summers of the 1950s, conditions in eastern Scotland favoured capercaillie to the extent that it reached pest proportions and it was causing damage to valuable tree crops. Successful breeding in north-east Scotland, after a series of bad years, is related to dry weather in June (Tasker 1989).

The capercaillie uses all ages of coniferous forest and is by no means confined to Scots pine. Clear felled areas and clearings are used for feeding, especially by the young chicks, and they are also used for display. Nesting occurs on the edge of thicket stage crops, and in open pole stage and mature forest where there is ground cover. Nests can also be found in quite open places, the bird relying on its colouring to be inconspicuous. Both the thicket and pole stages are used for cover and roosting, the old cocks often hiding in young, dense plantations (Harris 1960). Conifer shoots and buds, especially of pine, form an important winter food, the birds feeding high in the trees. It is sometimes suggested that as capercaillie are large they require old trees with big branches to be able to feed in. Whilst large trees are used when present, they are not essential and both sexes can be seen feeding in quite young trees in the winter.

The conservation of this species depends on 'normal' forest in which all age-classes are represented to provide the mosaic of conditions required at different times of year. Unplanted rock outcrops and wet areas, in situations where sunlight reaches them and thereby encourages insects and berry-bearing ericaceous shrubs, provide feeding places for the chicks. Heavy thinning will also encourage such ground vegetation.

This group also contains the redwing and fieldfare, which are northern thrushes that do not normally extend south into the broadleaved zone. In Britain they only nest in Scotland. Thicket, pole stage and mature trees have been used on the forest edge for nesting. The redwing nests in spruce mixed with alder and birch in natural situations, and plantations can provide similar sites. The brambling is another northern species that has nested in Scotland. Birch is a favoured species in most of its range, and nesting has occurred in Scotland in birch and on coniferous plantation edges.

Small clearings and forest edge (FE 2) This group includes birds with a general distribution using open space within the forest and forest edge in natural *coniferous, mixed and deciduous forest.* These species have colonized all types of managed forest using the open ground of the establishment stage and clear felled areas.

The stonechat and whinchat (a summer visitor to Britain) colonize the establishment stage and early thicket stage of both afforested and restocked plantations, as long as tree cover is not too dense. The whinchat prefers a ground vegetation of grass. They leave once conditions no longer resemble scrub. The primary habitat of these two species is heath and moorland but coniferous and mixed plantations are used. Partridges too nest on areas where young trees provide suitable cover. In Russia the nightjar uses forest fringes, dry conifer forest and mixed woodland (Tate 1989). In Britain nightjars make use of the scrub conditions that often occur on felled areas for nesting and are now regular breeders on such areas in coniferous plantations; the felled areas, however, need to be 2 ha or more in size (Leslie 1981). However, studies in Suffolk, England, put the minimum area required at 10 ha; areas treated with weedkiller were favoured (Ravenscroft 1989; see chapter 4 above). The British Trust for Ornithology Survey in 1992 showed that numbers were increasing. Approximately 50 per cent of the British population is now thought to live in conifer plantations (Fuller 1995). Nightjars will remain on restocked areas up to 15 years after planting, depending on the tree height.

There are a number of species that use all ages of forest adjacent to open ground. Clear felled areas of between 0.5 and 2 ha suit many of the smaller birds, whilst sites of 2 ha and upwards attract larger birds. Species include song and mistle thrushes, yellowhammer, wood pigeon, jay, pheasant, woodcock, cuckoo and turtle dove, the last two being summer visitors to Britain. The woodcock benefits from ground cover in pole stage and mature forest and brash left on the ground will provide this. Thicket stage plantations are used in Thetford Forest (Fuller 1995). Avery and Leslie (1990) state that the woodcock is abundant in plantations in south-west Scotland. Roding (courtship display) takes place over cleared areas with adjacent tall trees. The woodcock is a ground feeder and the prey taken varies with availability, earthworms being the main item on good soils. Beetles and larvae, bugs, caterpillars and other prey are taken in areas with more acid soils.

The spotted flycatcher is a summer visitor to Britain that uses pole stage and mature forest as it hawks for insects over adjacent open space. Restocks are also used (pers. obs.) Nests are made behind loose bark and in ivy but the bird can be encouraged by providing open-fronted nest boxes. These are also used by robins.

Birds of prey and owls do not frequent the thicket stage but use pole stage and mature woodland for nesting, whilst open areas, such as windblown and felled sites, are used for hunting. Brashing, high pruning and regular thinning helps the tawny owl, long-eared owl, goshawk and sparrowhawk to hunt under the trees; goshawks prefer more open stands than sparrowhawks (Petty 1989). The goshawk has increased considerably in British coniferous forests (Fuller 1995). Sparrowhawks have been shown by Newton (1979) to achieve only a low breeding success in large forests where they are at a distance from open areas for hunting. Once the second rotation starts, clear felling will diversify the forest and increase the number of small birds on which the sparrowhawk feeds. The tawny owl can be encouraged to nest by providing nest boxes (Petty 1987), but both it and the long-eared owl also use old birds' nests. The tawny owl also nests on the ground in young plantations. The two most widespread birds of prey in conifer plantations are the tawny owl and sparrowhawk (Fuller 1995).

The honey buzzard is a summer visitor confined to the warmer, southern and eastern areas of Britain where it nests in all types of managed forest. Although elsewhere in its range it extends into the taiga zone, this increased range appears to be the result of the more continental climate, with warmer summers that provide adequate insect prey on which this species mainly feeds. The red-backed shrike also requires a continental climate and has become scarce, but a pair has nested on the edge of a forest area in eastern England, which is in keeping with its usual habitat elsewhere.

Small clearings and forest edge (FE 3)
FE3a: Birds with a southern distribution using open space on the forest edge in *coniferous mountain forest*. The single British representative of any note is the ring ousel, which nests on plantation edges in Britain.

FE3b: Birds with a southern distribution using open space within the forest and forest edge in *mixed and broadleaved forest.* All have been recorded in coniferous plantations in Britain.

The goldfinch, linnet, greenfinch and blackbird colonize thicket near open space. The greenfinch and blackbird also occur in mature trees with an understorey. The finches all feed on conifer seed in winter, including Sitka spruce (Shaw and Livingstone 1991).

The woodlark is a summer visitor to Britain that was declining, but it has found a suitable habitat in pine plantations (both Scots and Corsican pine) in southern England since felling began in those areas. Pole stage and mature crops adjacent to felled and restocked areas are favoured by this species, which is found on forest edge and in scrub elsewhere in its range. Approximately 50 per cent of the British population is now thought to live in conifer plantations (Fuller 1995).

The Dartford warbler is only found in southern England and is a heath and scrub species. Cold winters sometimes severely deplete the population, but as it breeds in young Corsican pine plantations at the thicket stage, it is possible to provide suitable habitat for this species on restocked areas.

The golden oriole is a summer visitor to Britain but now breeds regularly in poplar plantations in eastern parts of the country. Breeding takes place in southern counties in hot summers.

Small clearings and forest edge with holes available nearby for nesting
No northern birds in this group are found only in taiga, so we have omitted group FEH 1. This group of birds naturally occurs mainly in mature or overmature forest where holes are available for nesting. The provision of holes in the form of nest boxes or old walls enables pole stage forest to be colonized and can increase populations in all types of woodland. Felled and windblown areas are used as open space.

Small clearings and forest edge (FEH 2) This group comprises birds with a general distribution and to be found in *taiga, mixed and broadleaved forest zones.*

The pied flycatcher is found in the pole and mature stages, especially near water where insects are plentiful. Oak woodland is a common nesting site in western Britain for this summer visitor but flycatchers may be found nesting in other tree species as well. Nesting takes place on coniferous plantation edges and in open, mature conifers, especially larch. Holes in trees are used, including those made by woodpeckers, whilst nest boxes are used readily. The pied flycatcher often nests in colonies, so boxes should be placed in groups; the entrance holes should be sealed until May to prevent the boxes being used by tits. In Wales, research has shown that oak woods that have been grazed and that have no understorey are favoured (Stowe 1987), but this has not been noted as a requirement in other parts of its range.

The redstart is also a summer visitor to Britain, using thicket on the forest edge and open mature forest for nesting. Nest boxes are used, as are holes in old walls and buildings. Boxes have encouraged nesting in high elevation spruce (Bamford 1991).

The starling needs little conservation and is more of a nuisance than an enhancement to a woodland, as it takes over breeding holes used by other species and occasionally kills trees with its droppings at roosting sites. On the credit side, a large number of insects are taken to feed the young birds. However, starlings do not like dense or extensive forest areas.

The wryneck is an uncommon summer visitor to Britain, until recently known only in southern counties. It is now colonizing eastern Scotland, where it is breeding in old, open Scots pine. These birds may have colonized from Scandinavia. It is a northern bird and why it was previously confined to southern Britain is not known but it does seem to require a continental climate as the ants and other insects it feeds on are more abundant under such conditions.

The tree sparrow nests on the forest edge in thicket. Old nests are used for breeding, as are tree holes and nest boxes. It is mainly a lowland species found near farmland.

Small clearings and forest edge (FEH 3) This group comprises birds with a southern distribution that are found in the *mixed and broadleaved forest zones.*

Both the green and lesser-spotted woodpeckers frequent open mature forest and excavate holes for nesting in dead trees. The green woodpecker has spread northwards into Scotland in recent years. It feeds on the ground in clearings, especially on ants, and old over mature trees on coniferous restocked areas have been used for nesting (Currie and Bamford 1981). Leaving overmature, wind-damaged and suppressed trees will augment the insect food resource for both these species.

The stock dove also nests in old tree holes but is not restricted to them. Nesting takes place in buildings, burrows and old birds' nests. There is little information on the numbers of stock doves using managed forest but the provision of nesting holes could augment their populations.

Conservation for birds using small clearings and forest edge

All the above birds requiring open space will find suitable habitats in 'normal' forest because there is a range of age-classes, provided this is managed on group or clear cutting systems. Selection forest reduces the open space available to roads, rides and forest edge. Species requiring holes for nesting need particular consideration in forests managed primarily for production. Woodpeckers require old or dead trees for nesting, food and display (drumming) and holes are also needed by other species. Hole-nesting species will be encouraged by leaving some old trees, wind-snapped trees and diseased trees; the ringbarking of unwanted species, especially birch whose soft wood provides good nesting sites for hole-nesting birds; and the provision of nest boxes. Many of them use quite small areas of open ground and all use regeneration coupes from 2 ha upwards. Leaving trees of poor quality or of non-economic species on clear cut areas will diversify the woodland structure. These trees are used as song posts by some species and as perches by birds of prey. They can be ringedbarked later as the new crop grows up, to provide dead wood and nesting holes.

Large clearings from two hectares upwards and forest edge

Large clearings from two hectares upwards and forest edge (FO 1) This group contains birds with a northern distribution, to be found in *taiga and mixed forest*. These species occur in upland coniferous forests.

The black grouse is a bird of the forest edge, rather than an inhabitant of the forest itself. Like the capercaillie it has been declining, but over a longer period; again various explanations for this have been put forward, including disease and destruction of habitat in southern Britain. It is probable, however, that climate is the primary cause, although the black grouse is more tolerant of high rainfall than the capercaillie. In common with the capercaillie and many other game birds, the young chicks are vulnerable in their first few weeks and a high protein insect diet is essential for rapid growth. A warm, dry continental-type climate provides optimum survival conditions in these early stages and cool wet weather causes mortality. The black grouse avoids dense closed forest but is found where trees are scattered and there is thick undergrowth. In managed coniferous forest the thicket stage is used for feeding and nesting but this needs to be adjacent to large felled areas or open moor containing ericaceous shrubs. These open areas ('leks') are used for display in the spring and for feeding later in the summer, especially by the chicks. Black grouse cause considerable damage in young plantations in the establishment and early thicket stage but pole stage forest is too dense for them. When mature forests are being regenerated more extensively in the future, the juxtaposition of open space and thicket may help numbers to increase again. In addition to conifer shoots and buds, birch buds and catkins are a favourite winter food and the encouragement of birch will add to the winter food supply.

Large clearings from two hectares upwards and forest edge (FO 2) Birds in this group have a general distribution and are found in *taiga, mixed and broadleaved forests*.

These birds are mainly scavengers and predators that range over a wide variety of habitat from tundra and natural coniferous, mixed and deciduous forests to steppe. Although they have a wide distribution, their range in Britain has contracted in many cases. The scavengers and predators have suffered extensive persecution, with the result that they now occur in the less-populated areas in the uplands where managed forest is mainly coniferous. The minimum amount of open ground required, both within the forest and outside it, by many of these species is unknown, but the proportion of open ground is important when considering land use. Unless ownership extends over land outside the forest, the opportunity to plan overall conservation may be limited. Lowland woods tend on the whole to be of a small size, so the need for a mixture of wooded and non-wooded land presents no problem there. The amount of land that can be left open in the uplands depends on other land uses, particularly farming. The relative conservation value of open ground and forest depends on land use but more knowledge is needed on the requirements of some bird species. Considering their use of forest habitats throughout their range, there must be considerable scope for creating more suitable areas within managed forest for many of these species. Most of the predators attract considerable high conservation interest, so much more needs to be known about their requirements. From continental European evidence it appears that 50 per cent forest cover is suitable for many raptor species, the numbers of predators an area can sustain depending on the food supply. Regular spacing of territories and interspecific competition between predators, will also affect the use of an area by particular species.

Rook and jackdaw both breed in pole stage and mature crops. Since the decline of elm in Britain, rooks are breeding in a variety of young trees 5 to 7 m high. Although rooks feed mainly outside woodland, caterpillars, especially defoliating *Tortix* species, are eaten in the canopy in years when they are abundant. Jackdaws nest both in trees (in holes and old nests of other species) and on cliffs in forest areas. The magpie nests in the late thicket and early pole stage crops.

Hooded and carrion crows are successful scavengers and their old nests, in pole stage and mature forest, provide nesting sites for many other species, including predators. The raven is also a scavenger and was previously more widespread in lowland areas. Outside Britain it is still found in such areas where forest and agricultural areas are mixed. Whilst these conditions exist in Wales, where the species is breeding successfully, extensive forestry in southern Scotland has taken much open ground. It has been suggested that this is the cause of the raven's decline, as it reduces the amount of sheep carrion available (Marquiss et al. 1978). As windblow and harvesting open up these forests, felled areas and restocked sites will make conditions more favourable. Deer will have to be culled in the forests in increasingly large numbers and offal could be left to provide food during the winter months, to the benefit of this and other species.

The short-eared owl and the hen harrier are both attracted to forest areas in the uplands, as these nomadic species find good nesting conditions there during the thicket stage, when the small mammal population increases rapidly in the ungrazed vegetation. The short-eared owl also breeds on large felled and restocked areas; these sites also hold good numbers of small mammal prey. There are now indications that restocked sites will also attract hen harriers, which have been recorded nesting on a restocked area of 25 ha near the forest edge (Petty and Anderson 1986). Both species normally nest on the ground. However, in 1991 and 1992, a pair nested in Sitka spruce 4.5 metres above the ground (Scott and Clarke 1993).

Kestrel and buzzard breed in pole stage and mature trees in coniferous and mixed commercial plantations. The birds use windblown, felled and restocked areas for hunting. The kestrel will use artificial nest boxes but usually nests in old birds' nests. The buzzard, which is often considered a bird of open ground, breeds and hunts within unmanaged forest in Poland (Harris and Harris 1988a).

The next three species are all falcons and their prey is caught in flight. They prefer a mosaic of open ground and trees. The hobby nests on the edge of managed forest, using old nests in pole stage and mature forest. Although the hobby occurs throughout the taiga zone, in Britain its distribution is limited to the south and east, the result of its preference for a continental climate. Its feeding on insect prey is limited by temperature and only takes place above 13°C (Milsom 1987). Small birds are taken at other times.

The merlin is now found mainly in upland areas in Britain. Its range elsewhere covers tundra and steppe as well as all three forest zones. In forest areas it prefers a mixture of trees and open areas, or open woodland with about 50 per cent tree cover. Most British merlins nest on the ground but tree nesting does occur in old nests of other birds. In upland forest in Britain merlins nest in the establishment and early thicket stages of coniferous plantations. They are also to be found nesting in older crops (25 to 30 years) on rides about 8 m wide, just within the forest and in trees on the forest edge. The decline in numbers has now been halted (Everett 1995) as merlins are increasingly using trees for nesting (Little and Davison 1992; Orchel 1992). There seems no reason why restocked areas near forest edges should not be used in the future, as well as large windblown areas and clear felled areas, as they support good populations of small birds on which the merlin preys; these sites are similar to habitats used by the merlin in the rest of its range. The length of time elapsing between felling and replanting may be a factor determining colonization. Many merlins are faithful to particular areas and in known breeding places short rotations would provide a greater proportion of open ground.

The peregrine is found over the same range of habitats as the merlin. A combination of cliff and open areas, or tall trees and open areas, is used elsewhere in its range, and in managed forest in continental Europe, nesting occurs in mature trees on the edge of clearings. Mature stands surrounded by younger stands are also used. The mosaic that will eventually be produced in upland plantations when regular regeneration is taking place will provide similar conditions. Nesting baskets in trees are occupied on sites in mainland Europe and tree nesting has been recorded recently in Britain (Ratcliffe 1984).

Eagles are the largest birds of prey and occupy large territories, though the size depends on food supply and is smaller where food is plentiful. Unlike most of the other birds of prey, the golden eagle does not extend into tundra and forest tundra but it does extend into steppe. In the rest of its range, it favours areas where cliffs and mature forests alternate with open space. In the taiga it is found where open space adjoins forest, especially in river valleys. In mountainous areas it hunts over alpine meadows and clearings, including, in Austria, felled areas of no more than 2 ha (pers. comm.). In Britain the eagle is regarded as a bird of treeless open country but this is the result of a lack of tree cover in the past; it is by no means dependent on a treeless habitat and some eyries are built in old Scots pine. In spite of extensive afforestation in Scotland, the eagle population has increased from an estimated 190 breeding pairs in 1957 to 510 occupied territories in 1987 (NCC 1987b). Persecution is, however, still a major factor, as it is with most birds of prey. How the golden eagle will interact with managed forest in Britain is not yet known as most upland plantings are still young, but from experience gained on the European mainland, they seem likely to be able to adjust to the changing situation. Clear felled areas and windthrow will provide hunting areas for the birds in the future. Carrion from deer culling left in open areas could compensate for the disappearance of the dead sheep that provided the main winter food source before afforestation. Although this has been tried with limited success (Petty 1987), carrion supply is only an important factor for winter survival. Live prey is more important in the nesting season, as it is with the red kite.

The sea eagle, or white-tailed eagle, is found over a wide range of habitat, from tundra and forest tundra, through coniferous, mixed and broadleaved forest zones, to steppe, usually near water. Nests are in trees and on cliffs through most of its range and tree nesting has already been recorded in Britain. The species was formerly widespread in Britain but because of persecution became extinct and has since been reintroduced. Forest areas, especially near to suitable fishing waters, will no doubt be colonized as numbers increase and these sites become similar to habitats used outside Britain. The white-tailed eagle does not rely on fishing alone and a wide variety of prey is taken. Nesting occurs away from water in other parts of its range. If it is not persecuted, this species could become much more widespread in the future.

Large clearings from two hectares upwards and forest edge (FO 3) This group includes birds with a southern distribution, to be found in mixed and broadleaved forest. Two rare birds of prey in this group are on the edge of their range in Britain and the barn owl too is steadily declining in numbers.

Montagu's harrier has bred in the early thicket stage in the past but breeding had ceased in Britain until recently, when a few pairs began to nest again. Restocked areas could provide suitable breeding places. The red kite was much more widespread in the past, feeding on carrion in towns, but is now restricted to Wales where it nests in trees on the edges of pole stage and mature forest. A supply of carrion is important for winter survival, live prey being taken in the nesting season. The mixture of agriculture (sheep farming supplying carrion) and forestry that exists in Wales benefits this species, as forestry is not on as large a scale as in Scotland (Newton et al. 1981). Kites have now been introduced successfully into Scotland and England.

Barn owl numbers have declined in recent years and a variety of factors seem to be involved; one of the most important is climate, as this species is on the north-west edge of its range. Changes in farm practice have meant fewer nest sites in old buildings but suitable nesting sites can be provided. The extensive use of rodenticides has also reduced food supply (Bunn et al. 1982; Shawyer 1987), which used to be plentiful in farmyards when grain was stored in stacks for threshing. Barn owls continue to use old buildings within forests and they hunt both outside the forest and over windblown and felled areas. Old trees covered with ivy are used for roosting and nesting. The barn owl is increasing in southeast Asia because of the large numbers of rats feeding on oil palm fruit (Wedgewood 1990).

Conservation for birds using large open spaces

Of the crow family, only the raven needs conservation, though old crows' nests provide breeding sites for many of the raptors. For the rest of the species requiring large open space, 'normal' forest management, based on clear cutting systems, will provide suitable conditions. In the uplands, the extensive blanket forestry that has attracted so much criticism will begin to diversify and become less even-aged when regeneration starts. The species grown in upland forest are pioneer trees and normal components of the taiga, so clear cutting systems will be favoured. Windblow too will diversify forest structure and in some cases the amount of clear ground opened up may be dictated by such factors rather than by planned management. Premature felling, whether planned or not, will bring forward the time taken to establish 'normal' forest which will in turn become more like the natural taiga forest (see chapter 6). Short rotations increase the proportion of the forest in the early stages and this will favour open-ground species. The amount of open-ground forest requirements for many species needs more detailed research but all species tolerate the presence of trees in other parts of their range. The forest manager can manipulate the size and position of felling areas to suit the species to be encouraged. Clear felled areas within the forest from 2 ha and upwards will attract many species; and larger clear cut areas, especially near to the forest edge, will attract species requiring more open space for hunting prey. In the lowlands, large areas of forest are unusual and the mosaic of farmland and forestry provides a varied habitat for those species occurring there; the provision of open space is therefore not so important. Selection systems provide little open habitat within the forest itself for birds requiring open space and that which is available is only provided by roads and forest edge. Many raptors use a few specific sites within their territories for nesting, so management must ensure that at least one of these is always available. Nesting boxes can be provided for owls and kestrels. Artificial 'crows' nests' can also be provided but success with these has been limited (Petty 1988).

OTHER HABITATS WITHIN FOREST AREAS

The following habitats (see table 12.2) are considered here because they are used in

Table 12.2 Other habitats used by birds within forests

Natural range of species from north to south	Short vegetation	Short vegetation and scattered trees	Dry meadows and short vegetation	Wet meadows and swamps	Lakes and ponds	Rivers and streams
Birds with a northern distribution using:						
Tundra	T					
Tundra and forest tundra		TF				
Tundra, forest tundra and taiga					W 1a	
Forest tundra and taiga				WM 1	W 1b	
Taiga and mixed forest					W 1c	
Birds with a general distribution using:						
Taiga, mixed and broadleaved forest			DM 2	WM 2	W 2	R 2
Birds with a southern distribution using:						
Mixed and broad leaved forest			DM 3		W 3	R 3

the rest of their range by birds that have adapted to nesting on open moorland in Britain. Some lists of birds alleged to have been displaced by forestry contain species that are to be found in forest areas in other parts of their range. As birds are very adaptable, planned management should be able to provide suitable conditions for these species to breed within forests and many already use forest situations for nesting in Britain.

Short vegetation

Tundra (T) These birds have a northern distribution and occur in the tundra zone. Dunlin and skuas are breeding birds in Britain; at present they are displaced by forestry, as fencing restricts grazing and vegetation growth soon makes conditions unsuitable for them to breed. Dunlin are communal nesters and nest near water. There is scope for more research to determine how much open space is needed round nesting sites, which are usually concentrated on wet areas. Although skuas are uncommon breeding birds, their presence is not always welcome near sea bird colonies as they are predators and scavengers.

Short vegetation with scattered trees

Tundra and forest tundra (TF) These are birds with a northern distribution and are found in tundra and forest tundra. The presence of trees is therefore tolerated in some parts of their range.

The red necked phalarope and Temminck's stint are two rare breeding birds that come into this category. Research is needed to determine how much open space and water are required near nesting sites and to assess the possibility of grading forest margins with scattered trees to reduce the impact of the forest edge.

Although the golden plover is found mainly in the tundra and forest tundra in some of its range, it also occurs in the taiga. It is therefore able to tolerate the presence of tree cover in parts of its range outside Britain. In natural conditions, raised sphagnum bog is used for nesting and short heathy vegetation is one of its requirements; this is lost when fencing excludes grazing and draining is carried out. Improved grasslands have provided feeding places in recent years near nesting sites. Leaving areas undrained and unplanted are measures that can be taken and, if deer are present, the vegetation will be grazed. The size and topography of such areas is important and needs research. Nesting near old upland, managed stands has been recorded where nesting areas are still grazed by sheep (pers. obs.).

The red grouse is a species associated with heather moorland in Britain. It is related to the willow grouse found in northern latitudes, which makes use of thickets for shelter and breeding. Red grouse remain on afforested land for a few years after planting. They have also been recorded on restocked areas, so it appears that some tree presence is tolerated.

Dry meadow and short vegetation

Taiga mixed, and broadleaved forest (DM 2) Birds in this group have a general distribution.

The wheatear and skylark remain on afforested areas for a few years in the uplands. Wheatears occur in small numbers on restocked areas and skylarks are now found frequently on restocks, especially where there is a grassy field layer.

The twite is found in alpine meadows and within mountain coniferous forest, where it nests on the forest edge; nesting occurs on plantation edges in Scotland.

Mixed and broadleaved forest (DM 3) This group contains birds with a southern distribution.

Although it usually nests in open areas, the stone curlew also nests on rides and restocked areas within managed coniferous forest in East Anglia.

Wet meadows and swamps

Forest tundra and taiga (WM 1) This group comprises birds with a northern distribution nesting in wet meadows and swampy areas in *forest tundra and taiga.*

In the taiga, the greenshank nests on the margins of forest lakes, streams and mires. Britain is the only part of its range where nesting occurs regularly without some tree cover. It is likely, therefore, that the greenshank will be able to accommodate to tree cover if sufficient wet areas are left undrained and trees are kept away from land surrounding open water, as these sites provide feeding areas for the young. The value of these areas will depend on topography, those receiving the maximum amount of sunshine being the most useful. There is scope here for research to combine conservation with forest planning. The greenshank is already nesting within afforested areas but will leave these once the trees become too large. Restocked areas will provide open ground and small trees in future.

Taiga, mixed and broadleaved forest (WM 2) Birds in this group have a general distribution, nesting in wet meadows and swampy areas in a wide range of habitats. Nesting occurs within forest areas in the *taiga, mixed and broadleaved zones.*

The size of the area needed for nesting needs investigation for various species but curlews and meadow pipits nest on restocked areas of about 2 ha in upland coniferous forest. Curlews also use restocks in Thetford Forest (Avery and Leslie 1990). Corncrakes have declined in Britain during the twentieth century and are now confined to the north and west coasts. Various reasons have been put forward for this decrease, including changes in farming practice. As this summer visitor feeds mainly on insects, other factors may also be involved, including climatic changes. In other parts of its range, the corncrake breeds on felled areas within forests with rank vegetation but only where the climate is continental.

A number of other waders breeding in Britain on wet moorland use wet meadows and swamps for nesting in natural forest. These include the wood sandpiper, which breeds by forest lakes and streams, ruff, snipe, redshank and lapwing.

Conservation for birds using meadows and swamps

Some of the birds in this section already use clear cut areas in managed forest and more are likely to do so in the future. In afforested areas in Britain in the past, it was the practice to plant all but the most boggy or rocky ground, whether the trees would succeed or not. Leaving more of these areas undrained and unplanted may attract many new species in the future, especially if attention is given to creating situations similar to those used in natural forest. Failed areas that do not justify beating up can be left unplanted and the stunted trees left to diversify the structure. There is scope here for research.

Lakes and ponds

As well as the true forest birds, there are a number of species that occur on lakes and rivers within natural forest. We discuss them here because many of them are included in census work when species lists are compiled and some of them appear misleadingly on lists of birds displaced by forestry. Their primary requirement is the presence of water and the majority are therefore found over a wide range of habitat types. Consideration of the water quality and the area in general are important if these species are to be maintained within forest; the latter is discussed more fully in chapter 6. In natural conditions, it is not the proximity of trees that limits the occurrence of these species but the size of the water body, the availability of nesting sites, and the food-producing capacity of the water resource.

Tundra, forest tundra and taiga (W 1) Not all birds with a northern distribution occur throughout the tundra, forest tundra and taiga and mixed forest zones, so the species have been subdivided as follows: birds using water in tundra, forest tundra and taiga (W 1a); birds using water in forest tundra and taiga (W 1b); birds using water in taiga and mixed forest (W 1c).

W 1a: Both black-throated and red-throated divers are found on water within coniferous forest in Eurasia and North America. As long as water is free of pollution, sedimentation and disturbance, there seems no reason why these species should not exist within managed forest. Artificial islands have been provided in some Scottish forests and are being used on a limited scale by black throated divers and ducks (pers. comm.). These islands are most useful where the water level fluctuates, as they are constructed to float and so move up and down with the changing water level.

A number of ducks breed on northern waters, such as the common scoter, pintail, scaup, wigeon and red-breasted merganser. All nest on water in the taiga and some of them already breed within managed forest in Britain. The common scoter is breeding on artificial islands provided for divers, and the merganser has used nest boxes.

W 1b: The whooper swan breeds on water in forest tundra and taiga only. A few pairs breed in Scotland. As trees are found in other parts of their range, the presence of trees should not affect them when water in forest areas is available. Trees, however, need to be kept back from breeding waters to provide resting places and to give the birds room to manoeuvre in flight. Lochans within managed forests in Britain are already used by this species in winter.

W 1c: The goosander and golden-eye already breed in managed forest in Britain. The numbers of golden-eye have increased considerably following the introduction of nest boxes into forest areas, as this species normally nests in tree hollows in natural conditions. The goosander also nests in hollow trees, nesting boxes and other suitable holes.

The conservation of northern still-water species

All the above northern species using water are now breeding in Britain, although some only in recent years. As they breed within forest in other parts of their range, there is no reason why they should not extend their range in coniferous forest in the uplands, provided water is kept free of pollution and sedimentation. The provision of nesting sites will encourage many of them and the species attracted will depend on the area of water available.

Taiga, mixed and broadleaved forest (W 2) Birds with a general distribution using water in *taiga, mixed, broadleaved forest and in other habitats.* These birds are widespread and may be found in all forest types as well as in many other habitats.

Mallard, tufted duck, teal, garganey, shoveler, pochard and grey lag goose all nest near suitable water in all natural forests. Many already nest in managed forest in Britain.

The black-necked grebe, Slavonian grebe and red-necked grebe also nest on suitable water in all forest zones. Reeds and other aquatic vegetation are needed for nesting. Gulls are a group sometimes claimed to have been displaced by forestry but colonies of common and black-headed gulls are to be found in managed forest in the British uplands. Islands are a favoured situation for colonies.

In Britain, the osprey is at present confined to Scotland. It has made a dramatic recovery since breeding resumed in 1954, after becoming extinct in the British Isles by the beginning of the twentieth century. Many pairs breed within managed forest and there is no reason why this wide-ranging species should not extend its range further south if protection continues to be successful. Artificial nesting platforms can be provided and are readily used in other parts of its range.

Other water birds include heron, coot, moorhen and water rail, all of which nest near water in managed woods of all kinds. The bittern is a northern species and ranges into taiga forest elsewhere but its presence in Britain is dependent on large reed beds.

Mixed and broadleaved forest (W 3) This group comprises birds with a southern distribution, using water in mixed and broadleaved forest and in other habitats. These birds occur in lowland forests on suitable water.

The great crested and little grebes need aquatic vegetation for nesting. Ponds need to be about 1 ha to attract the great crested grebe and a similar size to attract the mute swan.

Conservation measures for lake and pond habitats

All the species in groups W 2 and W 3 are found in areas containing trees in natural conditions. There is no reason why water should not be colonized within managed forests of any tree species, as long as the water quality is good and excessive shading is avoided.

Rivers and streams

There are no northern birds using rivers only in the taiga and mixed forest zones.

Taiga, mixed and broadleaved forest (R 2) In this group are birds with a general distribution in forest tundra, taiga, mixed and broadleaved forest.

The common sandpiper nests on the edges of rivers, streams and lakes. In managed forest in Britain it nests in thicket on the forest edge, as it does in natural situations. A favourite nesting place is on banks at the edge of flood plains. The ringed plover also nests on the edge of rivers, streams and lakes within forest, especially if sufficient river edge is left unplanted. The dipper nests near streams. Acidification of streams in Wales has affected dipper populations there, as the dipper feeds mainly on aquatic insects; this is discussed in chapter 6. Ledges under bridges are used as nesting sites and these can be provided if they are not present.

Both grey and pied wagtails feed near streams on insects and they also feed in forest clearings. Aquatic insects are not such an important food item as they are to the dipper, so wagtails are not so affected by acidification.

The common tern breeds on shingle banks in rivers or on islands in lakes. Some British colonies are near tall trees and do not appear to be affected by their presence.

Mixed and broadleaved forest (R 3) Birds in this group have a southern distribution.

The little tern is confined to the coast in Britain but elsewhere breeds in forest areas in similar situations to the common tern.

The kingfisher is found on lowland streams and rivers. Steep banks in which tunnels can be excavated are needed for nesting sites. Artificial tunnels have been used.

Conservation of water habitats in and near forests

All the species in groups W 1, W 2, W 3, R 1, R 2 and R 3 depend on water. In forests on basic soils, or where rivers and streams flow through base-rich headwaters and alkaline minerals have been taken up, satisfactory water quality will be maintained. In areas of acidic soil and rock, the pH may have to be maintained in coniferous forest by reducing acid runoff into headwaters or by liming in the catchment area. In all cases pollution, sedimentation and excessive periodic water flow should be avoided. The last is a problem in the early stages of afforestation and where clear felling occurs. In sensitive areas, felling should be in small coupes to minimize disturbance. The retention of trees beyond rotation age in headwaters is another useful option.

CONCLUSION: THE CONSERVATION OF BIRDS IN MANAGED WOODLAND

From the information in this chapter, we may conclude that both coniferous and mixed managed forest can accommodate all British woodland birds at some stage in the forest rotation. One forest cannot sustain all birds, the presence of particular species being dependent on the various factors described earlier. Many birds will colonize woodland without special measures being taken and additional species can be attracted by skilled management without incurring economic penalties (see figures 12.1, 12.2, 12.3). It is becoming appreciated that upland coniferous forests, especially in Scotland, support good breeding populations of many species, including some of high conservation value. Fifty five species breed regularly and are detailed in NCC 1986, whilst 59 are listed in Harris 1983. Scarce species include the raptors (Petty 1988), crested tit and black grouse (Bibby 1988), all of which can be increased in commercial forests by normal forest management. Raptors in particular benefit from lack of disturbance in the breeding season but forest operations near nesting sites can be suspended for this brief period. As many raptors occupy a series of traditional sites, their nesting areas can easily be taken account of in planning forest operations. There may be benefits in managing such areas on longer rotations. In coniferous areas in England, woodlark and nightjar have increased under normal forest management, making use of clear cut and restocked areas, and this use can be expected to continue.

The way in which clear cuts are managed will attract different species, depending to some extent on the vegetation under the felled crop. If grasses predominate, open-country birds are attracted; thick regeneration or an understorey will favour rapid colonization by scrub-loving birds. Factors such as burning, grazing, leaving slash, the application of pre-planting herbicides, and the interval between felling and replanting will all influence the species attracted.

The conflict between forestry and conservation interests has shifted away from the once common assumption that all coniferous plantations are of little value as bird habitats, because the facts do not support this contention. This view, however, still prevails with regard to mixed lowland forestry in which conifers are included, as in some way these are considered detrimental to birds. It takes no account of

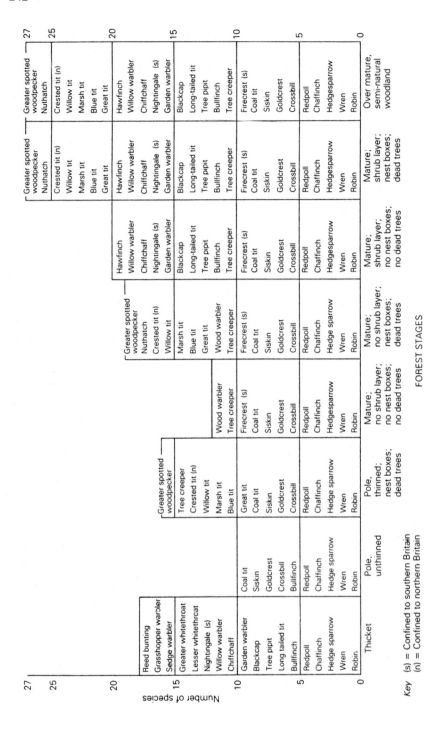

Figure 12.1 The use by birds of various stages in managed forest compared with overmature 'semi-natural' forest: (a) forest birds

many well-managed mixed woodlands with a high conservation content that hold a wide range of species.

Differences of view are now centred on the impact of upland coniferous forestry on 'open-ground' birds and, more recently, the effects on 'water birds' (NCC 1986; RSPB 1987). However, only a few species are definitely displaced by afforestation (see table 12.3).

The open-ground species usually listed as in danger from afforestation can be divided into three categories, based on the groupings given earlier in this chapter. First, group T, the true open-ground species: these are normally inhabitants of the tundra and do not tolerate a tree presence in other parts of their range, and on present evidence are displaced by afforestation. Dunlin are the most sensitive example but their colonial nesting habits and reliance on pools of water for nesting are easily identified and protected. In areas of afforestation, if this species is not to be displaced, the amount of open ground needed and the grazing required to keep the vegetation short at nesting sites both need investigation.

The second group of 'open-ground' species, TF, tolerate some trees, and includes such species as golden plover. In this case there is scope for reducing the impact of the forest edge with scattered trees where nesting is known to occur. Other factors to be considered are aspect, shading, keeping rotations short to increase the proportion of forest in the early stages and grazing levels within and outside the forest area.

The third group of 'open-country' birds, WM 1, all tolerate trees in other parts of their range, whilst open moorland is not a typical habitat for some. The green-shank, for example, in those parts of its range outside Britain does not extend into the treeless tundra. The basis of the view that these species will not survive in commercial forest is their 'unnaturalness'. One needs to consider how much this is a human assessment rather than a limitation on the use of the habitat by birds. However, there is scope here for ecologists to advise foresters how to accommodate these species within forests, rather than merely to suggest that colonization will not occur. Only time will provide the answer, as it has already in relation to woodland

---

Figures 12.1 to 12.3 summarize information in the text on the use of different forest stages by various bird species. With a few exceptions, all species colonize and use managed forest irrespective of the tree species present, structure being the most important feature. Some species using thicket are also found in mature forest with an understorey of either shrubs or young trees. In forest managed on clear cutting systems, thicket stage adjacent to mature stands with no understorey provides similar conditions and is freely used. The importance of dead trees and nest boxes is evident, as is the fact that clear felling systems attract a wider range of species than selection or group systems.

A total of 95 species use managed forest of all kinds in Britain, 87 in northern forests and 82 in southern forests; 78 species make use of managed forest in which all stages are represented without any special measures being taken, and a further 17 can be attracted with nest boxes and by providing dead trees by ringbarking.

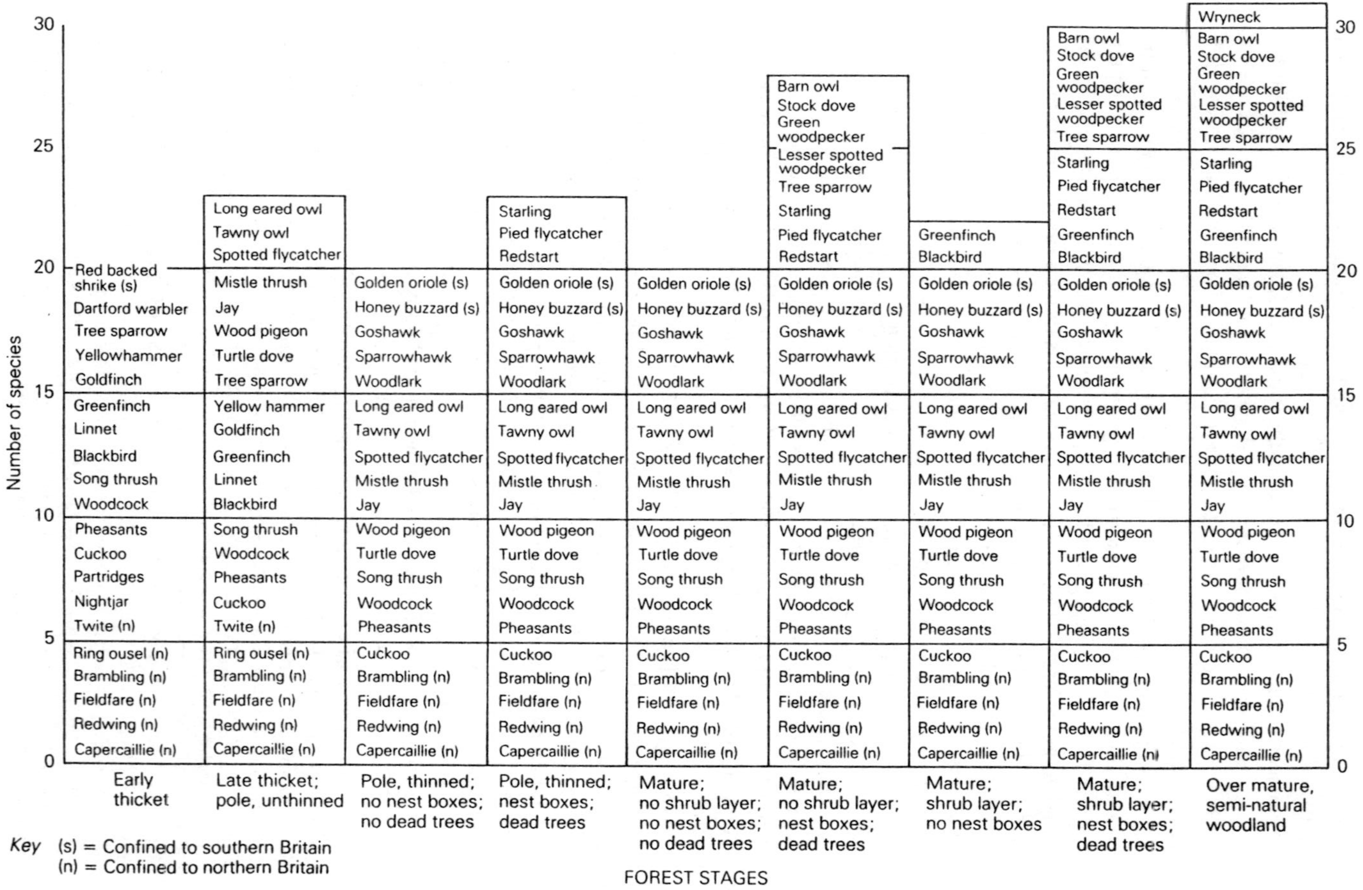
Number of species
30
25
20
15
10
5
0
Red backed shrike (s)
Dartford warbler
Tree sparrow
Yellowhammer
Goldfinch
Greenfinch
Linnet
Blackbird
Song thrush
Woodcock
Pheasants
Cuckoo
Partridges
Nightjar
Twite (n)
Ring ousel (n)
Brambling (n)
Fieldfare (n)
Redwing (n)
Capercaillie (n)
Early thicket
Long eared owl
Tawny owl
Spotted flycatcher
Mistle thrush
Jay
Wood pigeon
Turtle dove
Tree sparrow
Yellow hammer
Goldfinch
Greenfinch
Linnet
Blackbird
Song thrush
Woodcock
Pheasants
Cuckoo
Twite (n)
Ring ousel (n)
Brambling (n)
Fieldfare (n)
Redwing (n)
Capercaillie (n)
Late thicket; pole, unthinned
Golden oriole (s)
Honey buzzard (s)
Goshawk
Sparrowhawk
Woodlark
Long eared owl
Tawny owl
Spotted flycatcher
Mistle thrush
Jay
Wood pigeon
Turtle dove
Song thrush
Woodcock
Pheasants
Cuckoo
Brambling (n)
Fieldfare (n)
Redwing (n)
Capercaillie (n)
Pole, thinned; no nest boxes; no dead trees
Starling
Pied flycatcher
Redstart
Golden oriole (s)
Honey buzzard (s)
Goshawk
Sparrowhawk
Woodlark
Long eared owl
Tawny owl
Spotted flycatcher
Mistle thrush
Jay
Wood pigeon
Turtle dove
Song thrush
Woodcock
Pheasants
Cuckoo
Brambling (n)
Fieldfare (n)
Redwing (n)
Capercaillie (n)
Pole, thinned; nest boxes; dead trees
Golden oriole (s)
Honey buzzard (s)
Goshawk
Sparrowhawk
Woodlark
Long eared owl
Tawny owl
Spotted flycatcher
Mistle thrush
Jay
Wood pigeon
Turtle dove
Song thrush
Woodcock
Pheasants
Cuckoo
Brambling (n)
Fieldfare (n)
Redwing (n)
Capercaillie (n)
Mature; no shrub layer; no nest boxes; no dead trees
Barn owl
Stock dove
Green woodpecker
Lesser spotted woodpecker
Tree sparrow
Starling
Pied flycatcher
Redstart
Golden oriole (s)
Honey buzzard (s)
Goshawk
Sparrowhawk
Woodlark
Long eared owl
Tawny owl
Spotted flycatcher
Mistle thrush
Jay
Wood pigeon
Turtle dove
Song thrush
Woodcock
Pheasants
Cuckoo
Brambling (n)
Fieldfare (n)
Redwing (n)
Capercaillie (n)
Mature; no shrub layer; nest boxes; dead trees
Greenfinch
Blackbird
Golden oriole (s)
Honey buzzard (s)
Goshawk
Sparrowhawk
Woodlark
Long eared owl
Tawny owl
Spotted flycatcher
Mistle thrush
Jay
Wood pigeon
Turtle dove
Song thrush
Woodcock
Pheasants
Cuckoo
Brambling (n)
Fieldfare (n)
Redwing (n)
Capercaillie (n)
Mature; shrub layer; no nest boxes
Barn owl
Stock dove
Green woodpecker
Lesser spotted woodpecker
Tree sparrow
Starling
Pied flycatcher
Redstart
Greenfinch
Blackbird
Golden oriole (s)
Honey buzzard (s)
Goshawk
Sparrowhawk
Woodlark
Long eared owl
Tawny owl
Spotted flycatcher
Mistle thrush
Jay
Wood pigeon
Turtle dove
Song thrush
Woodcock
Pheasants
Cuckoo
Brambling (n)
Fieldfare (n)
Redwing (n)
Capercaillie (n)
Mature; shrub layer; nest boxes; dead trees
Wryneck
Barn owl
Stock dove
Green woodpecker
Lesser spotted woodpecker
Tree sparrow
Starling
Pied flycatcher
Redstart
Greenfinch
Blackbird
Golden oriole (s)
Honey buzzard (s)
Goshawk
Sparrowhawk
Woodlark
Long eared owl
Tawny owl
Spotted flycatcher
Mistle thrush
Jay
Wood pigeon
Turtle dove
Song thrush
Woodcock
Pheasants
Cuckoo
Brambling (n)
Fieldfare (n)
Redwing (n)
Capercaillie (n)
Over mature, semi-natural woodland
Key (s) = Confined to southern Britain
(n) = Confined to northern Britain
FOREST STAGES

Figure 12.2 The use by birds of stages in managed forests: (b) birds using forest but requiring open space as small clearings and forest edge

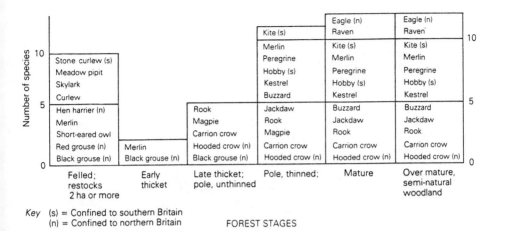

Figure 12.3 The use by birds of stages in managed forests: (c) birds using forest but requiring open space as large clearings and forest edge

birds now using commercial forest. The same principles apply to 'water birds' as most of them nest within forests in other parts of their range.

Where communities occur in habitats that are regarded as semi-natural (which is the yardstick of conservation (NNC 1986)), any land use change and not only forestry will upset the existing populations. However, land use changes have been going on for a long time, so no habitat in Britain is truly natural. There are, of course, examples that merit habitat protection. In such cases foresters can be expected to cooperate and there are many cases of them having done so. As far as the protection of rare birds is concerned, perhaps the best example is their cooperation in saving the red kite in Wales in the 1960s.

In addition to existing semi-natural communities, the potential exists for the formation of different and new communities to develop in both planted and regenerated forests by the natural exploitation of these new habitats. This process is dependent upon the adaptability of the species concerned but most birds are adaptable and there is ample evidence that they readily exploit new situations.

Managed coniferous forest in the uplands of Scotland can be expected to support an increasingly wide range of bird species as the forests mature. In the lowlands there are even more opportunities to support high numbers of birds in managed forest, as the better soil and climate means that a wider range of economical tree species, both broadleaves and conifers, can be grown with a resulting diversity of structure and increased food supply. Present indications point to intimate mixtures or small groups of broadleaves within conifer plantation sustaining more birds than pure blocks of the same tree species (Bibby 1987). Mixed forest can play a vital role, as all studies so far have shown that mixtures of conifers and

Table 12.3 Birds and forests

Species alleged to be reduced or displaced by forestry (NCC 1985, 1986)	Recorded use of managed coniferous forest in Britain
Dunlin	Remains in early stages of afforestation
Golden plover	Remains in early stages, tolerates tree presence and will nest near the outside edge of plantations
Wheatear	Remains in early stages
Stone curlew	
Dartford warbler	
Hen harrier	
Short-eared owl	Breed in early stages of afforestation and on
Curlew	restocked areas
Red grouse	
Skylark	
Twite	
Ring ousel	Breeds in early stages of afforestation but not yet recorded on restocked areas
Raven	Breeds in pole and mature forest
Black-headed gull	Breeds near suitable water
Buzzard	Breeds in pole and mature forest
Merlin	Breeds in plantations in young stages, on forest edge and on rides near forest edge
Golden eagle	Breeds in old trees in managed forest; uses younger forest for protection outside the nesting season
Common scoter	
Black-throated diver	Breed within forest near water
Red-breasted merganser	
Common sandpiper	Breeds on forest edge near water

A number of other ducks and waders listed by NCC already breed within forest areas and more will probably do so, as they breed within forests outside Britain. The goldeneye duck is one particularly successful example in recent years.

Chough and lesser-black backed gull nest mainly on the coast; any threat to breeding birds inland relates to a few specific areas. The red-necked phalarope and arctic skua, both rare as breeding birds in Britain, are limited to small areas in Caithness. The predatory nature of the latter makes it unwelcome near seabird breeding colonies.

broadleaves sustain both high total numbers and a wide range of species, as do natural mixed forests elsewhere. This is where forestry and conservation are wholly compatible; a mixed forest, especially in lowland areas, is also more attractive in economic terms than one composed solely of broadleaved species.

SCIENTIFIC NAMES AND WORLD DISTRIBUTION OF BIRDS MENTIONED IN THE TEXT

Subspecies are not included. Classification is as in tables 12.1 and 12.2.

Forest birds

Thicket

TH 1   Redpoll *(AcanthIs flammea):* Europe, Asia to Japan, North America
Bullfinch *(Pyrrhula pyrrhula)*: Europe, Asia to Japan
TH 2   Chiffchaff *(Phylloscopus collybitus)*: Europe, Asia to north-west China
Garden warbler *(Sylvia borin):* Europe, Asia to western Siberia
Willow warbler *(Phylloscopus trochilus):* Europe, Asia to east Siberia
Long-tailed tit *(Aegithalos caudatus)*: Europe, Asia toJapan
Greater whitethroat *(Sylvia communis):* Europe, Asia to north-west China
Lesser Whitethroat *(Sylvia curraca):* Europe, Asia to China
Tree pipit *(Anthus trivialis):* Europe, Asia to north-west China
Chaffinch *(Fringilla coelebs):* Europe, Asia to western China
Hedge sparrow *(Prunella modularis):* Europe, western Russia
Wren *(Troglodytes troglodytes):* Europe, Asia, North America
Robin *(Erithacus rubecula):* Europe, western Asia
Reed bunting *(Emberiza schoeniclus):* Europe, Asia to Japan
Sedge warbler *(Acrocephalus schoenobaenus):* Europe, western Siberia
TH 3   Blackcap *(Sylvia atricapilla):* Europe, south-western Siberia
Nightingale *(Erithacus megarhynchos):* Western Europe, southern Asia to western China
Grasshopper warbler *(Locustella naevia):* Europe, southern Asia to China

Pole stage and mature forest

PM 1   Crossbill *(Loxia curvirostra):* Europe, Asia, North America and Central America
Scottish crossbill *(Loxia scotica)*: Scotland
Goldcrest *(Regulus regulus):* Europe, Asia to Japan
PM 2   Treecreeper *(Certhia familiaris):* Europe, Asia, North America and Mexico
Short-toed creeper *(Certhia brachydactyla):* Central and eastern Europe
Siskin *(Carduelis spinus):* Europe, Asia to Japan
Wood warbler *(Phylloscopus sibilatrix):* Europe, west Siberia
PM 3a   Firecrest *(Regulus ignicapillus)*: Mountains in southern and central Europe
PM 3b Hawfinch *(Coccothraustes coccothraustes):* Europe, Asia to China
PMH 1 Crested tit *(Parus cristatus):* Europe
Willow tit *(Parus montanus):* Europe, Asia to Japan
Coal tit *(Parus ater):* Europe, Asia to Japan
PMH 2 Great tit *(Parus major)*: Europe, Asia to Japan

248

Nuthatch *(Sitta europaea):* Europe, Asia to Japan
Greater spotted woodpecker *(Picoides major):* Europe, Asia to Japan
PMH 3   Blue tit *(Parus caeruleus):* Europe, southern Asia
Marsh tit *(Parus palustris):* Europe, Asia to Japan

Birds requiring open space

Small clearings and forest edge: holes not required for breeding

FE 1   Capercaillie *(Tetrao urogallus):* Europe, Asia to western Siberia
Redwing *(Turdus iliacus):* Europe, Asia to Siberia
Fieldfare *(Turdus pilaris):* Europe, Asia to Siberia
Brambling *(Fringilla montifringilla):* Europe, Asia to east Siberia
FE 2   Stonechat *(Saxicola torquata):* Europe, Asia to Japan, Africa
Whinchat *(Saxicola rubetra):* Europe, Asia to west Siberia, Africa
Grey partridge *(Perdix perdix):* Europe, western and central Asia
Nightjar *(Caprimulgus europaeus):* Europe, Asia to north-west China
Song thrush *(Turdus philomelos):* Europe, Asia to western China
Mistle thrush *(Turdus viscivorus):* Europe, Asia to western China
Yellowhammer *(Emberiza citrinella):* Europe, Asia to China
Cuckoo *(Cuculus canorus):* Europe, Asia to China, India, Africa
Turtle dove *(Streptopelia turtur):* Europe, Asia to China
Wood pigeon *(Columba palumbus):* Europe, Asia to western China
Jay *(Garrulus glandarius):* Europe, Asia to Japan
Pheasant *(Phasianus colchicus):* Caucasus, Asia to China
Woodcock *(Scolopax rusticola):* Europe, Asia to China
Spotted flycatcher *(Muscicapa striata)*: Europe, Asia to western China
Tawny owl *(Strix aluco)*: Europe, Asia to China
Long-eared owl *(Asio otus):* Europe, Asia to China, North America, north Africa
Goshawk *(Accipiter gentilis):* Europe, Asia to China, North America and Mexcio
Sparrowhawk *(Accipiter nisus):* Europe, Asia to Japan, north Africa
Honey buzzard *(Pernis apivorus):* Europe, Asia to west Siberia
Red-backed shrike *(Lanius collurio):* Europe, Asia to west China
FE 3a   Ring ousel *(Turdus torquatus):* Mountains in Europe
FE 3b   Blackbird *(Turdus merula):* Europe, Asia to China
Goldfinch *(Carduelis carduelis)*: Europe, western and southern Asia
Linnet *(Acanthis cannabina):* Europe, south-west Siberia
Greenfinch *(Chloris chloris):* Europe, central Asia
Woodlark *(Lullula arborea):* Europe, Africa
Dartford warbler *(Sylvia undata):* south-west Europe, north Africa

Golden oriole *(Oriolus oriolus):* Europe, Asia to west China

Small clearings and forest edge with holes available nearby for nesting

FEH 2   Pied flycatcher *(Ficedula hypoleuca):* Europe, west Siberia
Redstart *(Phoenicurus phoenicurus):* Europe, Asia to west China, north Africa
Starling *(Sturnus vulgaris):* Europe, Asia to west Siberia
Wryneck *(Jinx torquilla):* Europe, Asia to Japan, India
Tree sparrow *(Passer montanus)*: Europe, Asia to Japan

FEH 3   Green woodpecker *(Picus viridis):* Europe, south-west Asia
Lesser spotted woodpecker *(Picoides minor):* Europe, Asia to Japan
Stock dove *(Columba oenas):* Europe, Asia to west China

Large clearings from 2 ha upwards, and forest edge

FO 1   Black grouse *(Tetrao tetrix):* Europe, Asia to China
Rook *(Corvus frugilegus):* Europe, Asia to China
Jackdaw *(Corvus monedula):* Europe, Asia to west China, Africa
Magpie *(Pica pica)*: Europe, Asia to China, India, North America
Hooded crow *(Corvus cornix):* Europe, Asia to west Siberia
Carrion crow *(Corvus corone):* Europe to Japan
Raven *(Corvus corax):* Europe, Asia to China, North America, Central America, Africa
Short-eared owl *(Asio flammeus):* Europe, Asia to China, North America, South America
Hen harrier *(Circus cyaneus):* Europe, Asia to China, North America, Central America, Africa
Kestrel *(Falco tinnunculus):* Europe, Asia to Japan, Africa, India
Buzzard *(Buteo buteo):* Europe, Asia, to Soviet Far East
Hobby *(Falco subbuteo):* Europe, Asia to Japan
Merlin *(Falco columbarius):* Europe, Asia to China and Japan, India, North America, northern South America
Peregrine *(Falco peregrinus):* world-wide
Golden eagle *(Aquila chrysaetos):* Europe, Asia to Japan, Canada, USA, north Africa
Sea eagle *(Haliaeetus albicilla):* Europe, Asia to Japan, India

FO 3   Montagu's harrier *(Circus pygargus):* Europe, Asia, to north-west China, north Africa
Kite *(Milvus milvus):* Europe, North Africa
Barn owl *(Tyto alba):* west and central Europe, Africa, India, Australia, south Canada, United States, South America

Other habitats within forests

Short vegetation

T   Dunlin *(Calidris alpina):* Europe, Asia, Canada
Great skua *(Catharacta skua):* Europe, Canada, Southern Ocean
Arctic skua *(Stercorarius parasiticus):* Europe, Asia, Canada, south
America, south Africa, India and Australia

Short vegetation and scattered trees

TF   Red-necked phalarope *(Phalaropus lobatus):* Eurasia, North America
Temminck's stint *(Calidris temminckii):* Europe, Asia to eastern Russia
Red or willow grouse *(Lagopus lagopus):* Europe, Asia to China, Canada
Golden plover *(Pluvialis apricaria):* Europe, Asia to west Siberia

Dry meadow and short vegetation

DM 2   Wheatear *(Oenanthe oenanthe):* Europe, Asia, Canada
Skylark *(Alauda arvensis):* Europe, Asia to Japan
Twite *(Acanthis flavirostris):* Europe, Asia to north China
DM 3   Stone curlew *(Burhinus oedicnemus):* Europe, south-west Asia, Africa,
India

Wet meadows and swamps

WM 1   Greenshank *(Tringa nebularia):* Europe, Asia to China
WM 2   Curlew *(Numenius arquata):* northern Europe, Asia to China
Meadow pipit *(Anthus pratensis):* Europe, western Russia
Corncrake *(Crex crex):* Europe, Asia to central Siberia, western China
Wood sandpiper *(Tringa glareola):* Europe, Asia to China
Ruff *(Philomachus pugnax):* Europe, Asia to eastern Siberia
Snipe *(Gallinago gallinago):* Europe, Asia to China, North America
Redshank *(Tringa totanus):* Europe to eastern Siberia
Lapwing *(Vanellus vanellus):* Europe, Asia to China

Lakes and ponds

W 1   Black-throated diver *(Gavia arctica):* Europe, Asia, North America
Red-throated diver *(Gavia stellata):* Europe, Asia, North America
Common scoter *(Melanitta nigra):* Europe, Asia to east Siberia, North
America

Pintail *(Anas acuta):* Europe, Asia to western China, North America
Scaup *(Aythya marila):* Europe, Asia to eastern Siberia, Canada
Wigeon *(Anas penelope):* Europe, Asia to China
Red-breasted merganser *(Mergus serrator):* Europe, Asia to China, North America

W lb Whooper swan *(Cygnus cygnus)*: Europe, Asia to China, North America
W lc Goosander *(Mergus merganser):* Europe, Asia to China, North America
Goldeneye *(Bucephala clangula)*: Europe, Asia to China, Canada
W2 Mallard *(Anas platyrynchos):* Europe, Asia, North America
Tufted duck *(Aythya fuligula):* Europe, Asia to China
Teal *(Anas crecca):* Europe, Asia to China, North America
Garganey *(Anas querquedula):* Europe, Asia to China
Shoveler *(Anas clypeata):* Europe, Asia to China, North America
Pochard *(Aythya ferina):* Europe to northern China, Africa
Grey lag goose *(Anser anser):* Europe, Asia to China
Black-necked grebe *(Podiceps nigricollis):* Europe, Asia to China, North and South America, Africa
Slavonian grebe *(Podiceps auritus)*: Europe, Asia to China, North America
Red-necked grebe *(Podiceps grisegena):* Europe, Asia to Japan, North America
Common gull *(Larus canus):* Europe, Asia to east Siberia, Alaska
Black-headed gull *(Larus ridibundus):* Europe, Asia to China, India
Osprey *(Pandion haliaetus):* Europe, Asia, North America, Australia
Heron *(Ardea cinerea):* Europe, Asia to Japan, Africa
Coot *(Fulica atra):* Europe, south-east Asia to China
Moorhen *(Gallinula chloropus):* Europe, Asia to Japan, Africa, North and South America
Water rail *(Rallus aquaticus):* Europe, Asia to Japan, India, Africa
Bittern *(Botaurus stellaris):* Europe, Asia to Japan, Africa
W 3 Great-crested grebe *(Podiceps cristatus):* Europe, southern Asia to China, India, Africa, New Zealand, Australia
Little grebe *(Tachybaptus ruficollis)*: Europe, Asia to Japan, Africa, India

Rivers and streams

R 2 Common sandpiper *(Actitis hypoleucos):* Europe, Asia to Japan
Ringed plover *(Charadrius hiaticula):* Europe, Asia to east Siberia, Canada
Dipper *(Cinclus cinclus):* Europe, Asia to east Siberia, Africa
Grey wagtail *(Motacilla cinerea):* Europe, Asia to Japan, Africa, India
Pied wagtail *(Motacilla alba):* Europe, Asia to Japan, Africa, India
Common tern *(Sterna hirundo):* Europe, Asia to Japan, North America

R 3   Little tern *(Sterna albifrons):* Europe, southern Asia, China, Japan, India, Africa, North and South America, Australia
Kingfisher *(Alcedo atthis):* Europe, southern and central Asia to Japan, India

CHAPTER 13

Amphibians
and Reptiles

As reptiles and amphibians are terrestrial, like most of the mammals, they had relatively little time in which to recolonize after the ice age. Many, such as the adder *(Vipera berus)*, did not get as far as Ireland before their spread was curtailed by the rising sea level in the late Boreal. Some, such as the edible frog *(Rana esculenta)*, did not reach Britain at all. The first to arrive were those most tolerant of the conditions prevailing after the retreat of the ice, such as the common newt *(Triturus vulgaris)*. These species still extend well into northern latitudes today. The later arrivals, the more southern species, are less tolerant of adverse climatic conditions and all these decrease in range during periods of less favourable conditions, particularly the colder conditions that have prevailed from the 1960s to the 1980s. Temperature is the most important factor for their survival; this is especially true for reptiles, who rely on external heat sources to maintain their body temperature. Amphibians are more dependent on water and damp conditions than reptiles, as they have not developed the tough skins that allow reptiles to live entirely on land.

The destruction of habitat has been another factor accelerating the decline of some species. At the same time, a number of exotic species have been introduced during the twentieth century; but only small populations of these survive in a few protected areas, so they are not important in wider conservation terms.

The amphibian species present in Britain therefore fall into two main groups, northern and southern. The northern species include the common frog *(Rana temporaria)* and common toad *(Bufo bufo)*, both of which tolerate a wide range of temperature and site conditions. Frogs are found as far north as the Arctic circle and at high elevations. Both species are present in the taiga forest zone. The natterjack toad *(Bufo calamita)* is a more southerly and continental species, requiring higher temperatures. In Britain it is now mainly confined to sand dunes, where vegetation is sparse and the ground is warm, in the west and south of England. Former habitats included grazed and burnt heath. With the reduction of grazing animals (especially sheep) on these sites, together with the dramatic reduction in rabbits through myxomatosis, such areas have been invaded by scrub and the resulting long vegetation no longer provides a warm enough environment for their needs, though a few scattered colonies still exist.

All British newts have a natural northern distribution and will tolerate a wide range of conditions. Many other species of amphibian occur further south on mainland Europe but conditions are not warm enough for most of them to survive in Britain. Some of the hardier forms, such as the edible frog, have been released artificially in a few places and are surviving.

All amphibians require a combination of terrestrial and aquatic habitats. These include suitable breeding ponds to provide sufficient food of the correct kind for the tadpoles, areas of cover for feeding once the ponds have been left after breeding and suitable conditions in which to hibernate. Apart from the natterjack toad, which only occurs in limited areas and needs open conditions, all can be usefully encouraged within commercial forest of all kinds.

Reptiles too can be divided into northern and southern species. The temperature of the environment is more important for them as they are fully terrestrial. The northern species include the adder, slow worm *(Anguis fragilis)* and common lizard *(Lacerta vivipara)*. These species have overcome the problem of obtaining enough heat to hatch their eggs by becoming viviparous (the young are born alive) or, in the case of the slow worm, ovo-viviparous (the young hatching from the eggs as they are laid). When the summer is cold, young are carried over winter inside the female and are born in the following spring.

The more southerly reptiles all lay eggs and these require sufficient warmth to hatch successfully. The grass snake *(Natrix natrix)* extends further north than the other species but not beyond 55° latitude in Britain; that is, southern Scotland. Adequate heat for the eggs to hatch is obtained by laying them in dung, heaps of grass or other rotting vegetation. The sand lizard *(Lacerta agilis)* has been decreasing in the north-west for some time and is now also declining in the south, the result of a combination of climatic influences and habitat loss. The smooth snake *(Coronella austriaca)* is confined to the south of Britain. In contrast, many more reptile species breed further north on the European mainland, where conditions are much warmer in summer. In addition to sufficient food supply, all reptiles require situations in which they can bask in safety while they warm up, especially at the beginning of the day. There is also a need for suitable cover in which to hide from predators, suitable situations for egg laying and for the development of the young, and hibernating places (where temperature fluctuations are minimal during the winter).

Amphibians and reptiles are amongst the most vulnerable of native animals to climate and habitat change. They are not normally associated with woodland but the new forests created primarily for wood production can provide new habitats for them. Such an increase in the conservation resource requires careful and sensitive management, based on proper understanding of the special requirements of amphibians and reptiles. The following accounts list factors that are important if reptiles and amphibians are to be encouraged to use woodlands. Identification details are not included. Information is mainly derived from Hvass 1978; NCC 1983b; Frazer 1983; Ballasina 1984; Spellerberg 1988a, 1988b; and personal observations.

AMPHIBIANS

Common frog *(Rana temporaria)*

Distribution Throughout Britain and Ireland. The common frog can be found breeding high on mountains.

Habitat requirements The common frog is found in all types of forest as well as in a number of other habitats providing suitable conditions. Ponds with a wide range of pH are used, though not all ponds are used for breeding and the same pond is not always used every year. The presence of suitable food for the tadpoles is a factor selected for; frogs are attracted to ponds containing algae by the smell of the glycollic acid produced by the latter. Algae are important for the young tadpoles but animal food becomes more important as they grow; tadpoles also prefer ponds with high levels of potassium and low levels of phosphate (Frazer 1983).

Frogs lay their eggs in shallow parts of the pond where it is warmer and graded edges can provide suitable conditions. In shaded ponds, spawning takes place in the areas that receive the maximum sunshine, so removing trees from the south side will encourage breeding. Spawning in shaded ponds is later than in those receiving more sunlight. Ponds on the edge of woodland are particularly favoured and are colonized soon after they are constructed.

Outside the breeding season, adult frogs feed in long grass or other damp vegetation, often at considerable distances from any pond. Young frogs live in such situations for 2 to 3 years until ready to breed.

During the winter, sites for hibernation need to provide conditions that do not fluctuate from one extreme to another. Thus many frogs hibernate in ponds protected by trees, with a layer of mud and leaves on the bottom. Others use underground burrows, stone walls, heaps of stones or vegetation and compost heaps, where the temperature range is buffered.

Food A wide range of animal food is eaten, including worms, slugs, snails, flies and beetles.

Conservation There has been concern in recent years that the common frog has been declining in Britain. The fall in numbers has been attributed to various factors. Although the collection of frogs for teaching purposes was implicated, it is not now considered to have had a significant effect (Frazer 1983). Far more important has been the decline in the maintenance of farm ponds as a result of improved agricultural water supplies. This has led to neglect, with silting up and contamination (by petrol and oil washed off roads, by fertilizer, slurry, pesticides and other contaminants) and filling in. A large number of ponds have also been lost where land has been built on. Many frogs are road casualties on their way to breeding ponds.

New ponds are now being created on agricultural land and in gardens. Within commercial forests, frogs can be encouraged by providing ponds, either by excavation or by damming streams, the latter preferably in a number of places. Several small ponds provide a wider range of choice than a single large pond and may be cheaper to construct and maintain. Ponds near the edge of plantations in sunny situations are likely to be used for breeding, whereas the more shady and protected ones will tend to be used for hibernation. Where fire dams are needed in high-risk fire areas, some thought to the design will enhance their secondary conservation value. Although these may need to be deep to hold sufficient water, grading out the side receiving most light to provide warm, shallow water will benefit not only frogs but a wide range of wildlife. Apart from the provision of water, conditions within forests can offer the other habitats required. Grass rides, woodland edges, felled coupes and clear cut areas provide feeding sites, whilst conditions in the plantations themselves provide protected hibernating sites if ponds are not used. Frogs travel long distances in relation to their size so suitable conditions do not need to be near a pond. Young frogs that have just left the breeding ponds and are dispersing can be found well into forest stands.

Common toad *(Bufo bufo)*

Distribution Throughout mainland Britain but not in Ireland. Widely distributed in Europe to 65° N.

Habitat requirements The common toad is found in all forest types as well as in other habitats, if suitable conditions are present. The toad, unlike the frog, is mainly nocturnal and spends the day in holes, crevices or burrows, under stones or tree roots, in piles of wood, or in thick vegetation. It can be seen in daytime when it travels to its breeding pond, usually in damp weather. A wide range of ponds are used for breeding, with pH varying from 4.2 to 8.2. Many toads return to the same pond to breed but new ponds are also quickly colonized. If conditions are adverse during the breeding season, egg laying may be delayed until the following season (Ballasina 1984). After breeding the adults settle down in a restricted feeding site. Young toads remain feeding on land until ready to breed at 3 to 4 years old. Hibernation has been recorded in ponds but usually occurs out of water. Toads excavate burrows in suitable soil, use burrows made by other animals and also use holes and spaces under timber to hibernate.

Food A wide range of insect food is eaten, as well as slugs, snails, woodlice and other invertebrates.

Conservation Toads, like frogs, have suffered from the loss of many breeding ponds but they are quick to colonize new ponds; the use of ponds only a year old, situated on the forest edge, has been recorded (Frazer 1984). Apart from the provision of water for breeding, conditions in all types of normal forest, including conifer forest, will provide suitable habitats for this wide-ranging species.

Natterjack toad *(Bufo calamita)*

Distribution This species is on the edge of its range in Britain, as it is a southerly and western European species, occurring in France, Spain and West Germany. It is only able to survive where conditions are sufficiently warm. In Britain it is now mainly present in sand dune areas in the south and north-west of England. Former breeding areas included heathland, where the short vegetation and light soils provided a suitably dry, warm habitat, but how many of these areas now hold natterjack populations is unknown.

Habitat requirements The natterjack toad requires shallow water to breed and tolerates quite a high degree of salinity. Depths of a few centimetres are sufficient but ideally the water should be 8 to 15 cm deep. Deeper water will prevent the temperature of the water from being high enough for tadpole development, although sufficient depth is needed so that the water does not dry out before metamorphosis is complete. This normally takes from 4 to 8 weeks, depending on the water temperature. A temperature of about 25°C is required for metamorphosis to take place (Frazer 1983).

In mainland Europe the natterjack is a pioneer species on newly created sandy sites (Ballasina 1984). It prefers dry conditions and does not appear to be able to compete with the common toad in wet vegetation. After breeding, active feeding does not occur on land until the temperature reaches 11-12°C. Toads live underground during the day, in holes they have excavated, in animal burrows or under stones. They travel away to feed at night. Hibernation is also underground. Soil which is easily burrowed into is therefore an important component of the habitat.

Food A wide range of invertebrates are eaten, including insects, spiders, woodlice and small molluscs.

Conservation The decline of this species is the result of a combination of factors. These include the cooler climate, the destruction of suitable habitat for buildings and golf courses, pond destruction and vegetation changes. The reduction of grazing by domestic stock and rabbits, mentioned at the beginning of this chapter, has provided conditions more suitable for the common toad, which competes with and displaces the natterjack. Many areas of sandy heath that formerly held toads are now under trees, both planted and regenerated naturally. If the natterjack is to be encouraged to recolonize in commercial pine forest on former heathland, several factors need to be taken account of. The natterjack is a pioneer species on disturbed, light, mainly sandy soils, hence its presence on sand dunes; it freely colonizes new areas. Thus clear felled areas with their disturbed soil, large enough for the surface to dry out, will provide conditions for a few years for natterjack survival after breeding. Breeding ponds that are used are also transient, the shallow depth favoured means that such ponds soon become silted up and unsuitable. Such shallow ponds could be formed near or on clear felled areas in sunny situations. Ruts created by extraction machinery can collect sufficient water for breeding to be successful; the next clear fall a few years later will then provide suitable conditions to maintain the population. As natterjack toads have disappeared from many places, artificial reintroduction is necessary to conserve this species and is being carried out successfully under the Endangered Species Programme. Advice on obtaining suitable stock can be obtained from the British Herpetological Society.

Common newt *(Triturus vulgaris)*

Distribution Throughout Britain, including Ireland. This species is the most widely distributed newt in Europe.

Habitat requirements The common or smooth newt uses a wide variety of habitats, including coniferous and deciduous forest. It tolerates a wide range of pH and pond types, even polluted ponds. New ponds are quickly colonized. Ditches with slow-moving water are also used for breeding. The tadpoles may fail to complete their development into adults in adverse conditions, such as low temperatures or lack of sufficient food. They then overwinter in ponds and breed the following year before they are fully adult; development proceeds and the fully adult newts leave the ponds the same summer. On the European continent, hibernating newts emerge from deciduous forest areas first, where the ground warms up earlier than under conifers. Hibernating sites in denser conifer forest, however, provide more sheltered places to buffer winter temperatures. After breeding, newts feed on land. During the day they hide in small holes or under stones and logs. The young stay on land for 2 to 4 years until they mature. Hibernation takes place deep underground and under logs, stones or moss. Some newts may return to ponds in October or November and stay there for the winter.

Food Various invertebrates are eaten, including slugs, snails, worms and insects.

Conservation The presence of ponds for breeding is the most important factor. In other respects all woodlands are suitable, as they are damp and have an adequate food supply.

Palmate newt *(Triturus helveticus)*

Distribution Throughout mainland Britain but not in Ireland. The palmate newt does not have a wide distribution in mainland Europe, being found mainly in West Germany and France.

Habitat requirements The palmate newt uses a wide range of habitats and pond conditions. It has a higher tolerance of low potassium levels than the various species mentioned above and tolerates brackish water; new ponds are colonized rapidly. In adverse seasons, development is delayed, as with the common newt. After normal breeding, newts leave the ponds to feed on land. Hibernation is usually on land but some adults and undeveloped young overwinter in ponds, entering them in October and November.

Food Various invertebrates are eaten, including slugs, snails, worms and insect larvae and adults.

Conservation The presence of ponds is the most important factor; otherwise requirements are similar to the common newt's.

Great crested newt *(Triturus cristatus)*

Distribution Throughout mainland Britain but not in Ireland. Its range in mainland Europe is extensive.

Habitat requirements The great crested newt prefers deeper water than the other species mentioned; it occurs in a wide range of pond types and habitats and is not confined to lowland areas. New ponds are rapidly colonized. Great crested newts can also be found in lakes, slow-moving water and ditches. After leaving the water when breeding is finished, adults can be found under stones, in the soil and amongst vegetation. The young are found in similar situations after leaving the water and they remain on land for 2 to 3 years before returning to breed. Hibernation is underground in burrows where the soil temperature is fairly constant, as well as under tree roots, in moss or in ponds.

Food A wide variety of invertebrates as well as tadpoles of other species are eaten.

Conservation The great crested newt is in no danger of extinction in Britain but because it is one of the species protected under the Berne Convention it is included in Schedule 5 of the Wildlife and Countryside Act, 1981. As with the other newts, the provision of ponds, which need to be deeper and unpolluted, is the means of encouraging this species.

Recommendations for the conservation of amphibians within commercial forest

The common frog and the common toad are widespread species and occur throughout the taiga, mixed and deciduous forest zones. Conservation of these two species therefore presents no difficulties. The more southern natterjack toad needs more careful consideration, as mentioned under the reference to that species above. Both the common and great crested newt have wide ranges; the status of the palmate newt is less clear but it reaches the north of Scotland, so if its habitat requirements are met, its conservation in Britain does not present a problem.

Ponds for breeding and hibernation are the most important elements of the habitats of all amphibians. Suitable water, especially for the great crested newt, can be provided by the damming of small streams in several places to form pools; or the stream can be diverted into small ditches or channels. A number of smaller ponds of various depths, situated both within the forest and on sunny woodland edges, provide a wider range of choice for breeding and hibernation than a single large pond. If the latter is chosen, the pond should be graded on the south side to provide shallow water and any trees shading this area removed to allow the sun to warm the water. The south-facing bank should be steeper to make full use of the sunshine. It is important that leaves, particularly of conifers, do not fall into the water in large quantities as their slow decomposition deoxygenates it. Pollution of the water by pesticides, some herbicides, waste fuel, or the washing out of containers should be strictly avoided. Any work done with forest machinery should avoid disturbing ponds and must not cause them to be silted up.

When ponds need to be cleaned out, the time of year should be chosen carefully. From the foregoing life histories, it can be seen that some ponds may be in use for most of the year. The period which is likely to have the least impact is from August to September, when, in normal years, the larval stages will have been completed and the young adults will have left the ponds; the mature adults will not return to hibernate until October or November. Any animals still using the ponds in late summer are active at that time and if removed with the debris will either find their way back to the water or move away.

Ponds are the most important, but not the only, significant part of the amphibian habitat. Clear felled areas, early thicket stage forest, forest edge, forest roadsides, rides, clearings and failed areas all provide habitat with sufficient light, damp vegetation and food when amphibians are not breeding.

REPTILES

Habitat factors that are important to reptiles are somewhat different to those of amphibians, because of differences in their mode of life.

Common lizard (*Lacerta vivipara*)

Distribution Throughout Britain and Ireland, from dry heath to bogs. Widespread in taiga, mixed and deciduous forest areas in Eurasia, extending east to Manchuria.

Habitat requirements A wide variety of habitats are used, ranging through heath, moor, banks and woodland edge. These need to contain open places for basking, as well as protective cover. In all types of woodland such situations occur on the forest edge, ridesides and roadsides (especially those with south-facing banks), and in clearings. Hibernation takes place in underground holes, under tree stumps and in heaps of stones or in walls.

Food A wide range of insects and their larvae are eaten, as well as worms.

Conservation No special measures need to be taken. Suitable habitat is created by clear felled areas, clearings, rides, roads and forest edge. This species is a normal inhabitant of the taiga zone and readily makes use of habitats within and on the edge of coniferous plantations as well as other forest types.

Sand lizard (*Lacerta agilis*)

Distribution This species has a wide distribution in Europe, extending to southern Sweden and into Asia. In Britain it is restricted to a few coastal counties, mainly in England.

Habitat requirements In the main part of its range it is found in a number of different habitats, including hedgerow and woodland edge, as well as sand dunes, heath and steppe. This lizard requires much warmer conditions than the common lizard and is at the limit of its range in Britain where, once widespread, it is now only found in areas where summer temperatures are high, hence its occurrence on dry sandy heath and sand dunes. The conditions required are loose soil for burrowing, and open areas (south-facing for maximum sun) for egg laying and development. The young hatch in August and September, so temperature in the summer months is important. Vegetation at least 30 cm high is needed for cover and the lizards also bask on this so that a rapid retreat into cover to escape predators is possible. Hibernation takes place underground in moist, well-drained sites, including mammal burrows.

(a)

(b)

Plate 13.1 (a) Smooth snake (*Coronella austriaca*) basking in the lower branches of a young Corsican pine; (b) juvenile sand lizards (*Lacerta agilis*) basking on the stem of a conifer. (Photos: Dr I. F. Spellerberg)

Food A wide range of invertebrates are eaten. These include woodlice and weevils in the spring. Beetles also form an important food source, as do spiders and hymenoptera (bees, wasps and ants).

Conservation Numbers have declined since the early 1960s for various reasons: the climate has become cooler in summer; sand dunes have been built on and are subject to heavy recreational use; burning of heathland renders conditions temporarily unsuitable; and the reduction of grazing has led to shrub succession causing shading. Maturing plantations have a similar effect. Against predictions, however, the sand lizard has adapted to the presence of planted woodland, particularly pine, in southern England. In the forest, home ranges are larger than on open heath, depending on the amount of open areas in the habitat. Linear territories are taken up along road verges and forest edge. The roads and rides that are favoured face east or south and are usually at least 5 m wide so that sufficient sunlight is available. Failed areas and open areas are also used. The presence of vegetation is important for protection and basking, lizards often using the lower branches of pine trees for this. For egg laying and development, bare areas are selected in soft clay or sandy soils where the chosen spot is shaded by trees no more than 40 per cent of the time; in forest areas these conditions occur amongst the smaller trees. Eggs are often laid near the tips of tree roots, which may help to maintain humidity (Spellerberg 1988b).

The suitable range of habitats in mature forest can be increased by felling coupes on the edge of the forest, especially on south- and east-facing slopes, or by making larger coupes within the forest to let plenty of sunlight reach the forest floor. Suitable forest habitats are mainly pine on sandy soils. Disturbance of the soil by extraction machinery will bare the soil, and the early years of planting (up to about 10 years (NCC 1983b)) will provide the optimum height of vegetation and cover. 'The potential value of afforested areas for the conservation of *L. agilis* (sand lizard) and other reptiles is supported by a recent document which suggested that 25% of the populations of that species were to be found within afforested areas (Nature Conservancy Council 1983)' (Dent and Spellerberg 1987). In newly planted forest the areas favoured by sand lizards are those in which there is a predominance of young age-classes. Premature felling will ensure a continuation of suitable habitat, especially on the forest edge. To minimize disturbance, forest operations should be concentrated in the winter when lizards are hibernating (from August to October through to mid-March and April). Winter gassing of rabbit burrows should be avoided, as these provide hibernation sites for sand lizards and a number of other reptiles and amphibians (NCC 1983b).

Slow worm *(Anguis fragilis)*

Distribution Found throughout Britain in a wide range of habitats but not in Ireland. In mainland Europe this species is the most widely distributed of all reptiles (Hvass 1978).

Habitat requirements The slow worm hides in damp, shady places during the day, such as in vegetation, under stones, slates and pieces of wood, emerging to feed in the evening and early morning. Daily basking is not usually as important as it is for other reptiles but is essential in the spring after emergence from hibernation. Adults can be found during the summer in drier areas but the young stay near the damp hibernation sites. Hibernating takes place in the burrows of mice and voles and in other underground holes.

Food Slugs, worms, snails and other invertebrates are taken. Worms are more often taken in the southern part of the range (Hvass 1978).

Conservation Little is known about this species and it is often overlooked. Commercial forests appear to provide habitats suitable for it without special measures being taken. Forest edge, ridesides, roadsides and restocked and failed areas would seem to provide suitable habitats.

Adder *(Vipera berus)*

Distribution Throughout Britain in a wide variety of habitats but not in Ireland. The adder is more widely distributed than other European snakes (Hvass 1978), extending right across Asia to the Pacific coast.

Habitat requirements Occurs in a wide variety of habitats, ranging from moorland to heaths, dunes, bog edges, woodland edges, woodland rides and clearings. The habitat needs to include places for hibernation, feeding areas and cover. Hibernation does not take place on open heath, moor or grassland and adders travel long distances to seek sheltered places for the winter. Mammal burrows are used as well as holes in banks, walls and under tree stumps. In the more exposed hibernating areas, banks with vegetation cover facing south are favoured.

Food Small mammals, young birds, eggs, frogs and lizards.

Conservation The usual reaction of human beings is to kill this species; its decline in many areas is the result of this cause alone (NCC 1988). The adder is a normal inhabitant of the coniferous and mixed forest zone and conditions in managed forest of all kinds produce suitable habitat for it. Rides, roads, clear cut and failed areas, as well as forest edge, combined with cover, are habitats used. Mature, managed larch woodland in South Wales held a good population of this species (personal observation).

Grass snake *(Natrix natrix)*

Distribution In Britain only up to the Scottish border; it is not found in Ireland. The range of this species extends much further north in continental conditions where the summer temperatures are higher.

Habitat requirements The grass snake is more aquatic than the other snakes, being found near ponds, lakes, ditches and streams where there is some direct sunlight. Vegetation cover is needed and woodland edge is used. Hibernation takes place in underground burrows, holes, stone walls, heaps of stone or brushwood, tree stumps and in compost heaps or other rotting vegetation. Basking takes place on low shrubs in spring or in the lower branches of trees. Eggs are laid in rotting vegetation, grass cuttings, compost heaps, old sawdust, forest litter, heaps of straw and, commonly in the past, in dung heaps.

Food Frogs and newts form the main prey. Young snakes feed on slugs, tadpoles and worms (Hvass 1978).

Conservation The provision of ponds, and other water where frogs and newts will breed, provides a food source, as will damp areas away from water suitable for amphibians. Ponds and their immediate surroundings should be kept free of shading trees. Leaving rotting vegetation in heaps (for example, if ponds are cleaned out) will provide sites for egg laying. One of the most positive conservation measures that can be taken is to prevent unnecessary killing.

Smooth snake *(Coronella austriaca)*

Distribution In Britain, this snake only occurs in the south of England and seems to require a higher summer temperature than other reptile species. It appears to be in decline in the north and west of mainland Europe as well as in Britain. This decline may be the result of climatic changes, although in Britain habitat loss is also involved. Elsewhere in Europe it can be found in hedgerows, open woods, woodland edges and on scrubby slopes (NCC 1983b).

Habitat requirements In Britain the smooth snake occurs on heathland with water near-by. It is found on heath and also at the interface of heath and woodland, so it is therefore to be found on the edge of plantations and also within managed forest. Home ranges in forest are larger than on open heath. South-facing embankments are favoured. Basking takes place in spring on vegetation and on the lower branches of trees, especially if south facing. However, this species spends less time basking than the adder. The young are born between August and October, hatching almost immediately from the eggs. Breeding is irregular, probably depending on environmental conditions, as regular annual breeding seems more usual in the warmer summers further south in France (NCC 1983b).

Food The prey taken includes lizards, mice, fledglings and other small mammals and their young.

Conservation Only a few woodlands on sandy heaths are suitable for this species. Woodland edge adjoining heath and other open areas provide suitable habitat and the provision of south-facing basking sites is important. As with the other snakes, the prevention of indiscriminate killing will help to improve its status.

The conservation of reptiles

The common lizard, slow worm and adder are northern species and conditions within most types of forest can be made suitable for them. In Britain, the grass snake is confined to more lowland areas in England and the provision of ponds, with their food supply of frogs and newts, is important. Otherwise, for the snakes, perhaps the most important preservation measure is public education to prevent indiscriminate killing. Measures to achieve this can be combined with public access to forest areas. For both snakes and lizards heat regulation is important, so basking sites need to be created that receive sufficient sunlight. Open space provided by clear cut areas will fulfil this need, as will wide rides and forest edge with some low vegetation cover. The complex set of requirements of most species will be best provided on south-facing sites. The southern species, the smooth snake and the sand lizard, can both live in commercial pine forest and young plantations, where a combination of vegetation and open ground provides a habitat. Felling coupes adjacent to colonized areas will provide continuity of habitat, and shortened rotations will ensure that more of the forest is in the younger stages.

FISH

Fish cannot here be disregarded, as water sources within forests and woodlands provide a habitat. The type of fish present will depend on whether the water is still, slow-moving or well-oxygenated. Upland, acidic waters contain the least wildlife, both in invertebrate numbers and fish depending upon them, but productivity in such waters has successfully been increased by liming. There should be both deep and shallow water in both sunshine and shade. To maintain fish populations, water quality is important. Pollution by fuels, chemicals, insecticides and herbicides should be avoided. Herbicides should be applied at the recommended time, in the right weather conditions (not in wind or rain) and well away from vulnerable water. Containers should be safely disposed of and spillage avoided. Fertilizers, except lime, should not be applied too near water. Scouring, sedimentation and blockage by branches or felled trees should also be avoided: scouring can be reduced by careful draining, and sediments can be caught in silt traps. In sensitive head waters, rotations can be extended or a selection system used. *Forests and Water* (Forestry Commission 1988a) provides full and useful guidelines. Although conifers are usually assumed to have an adverse effect on streams and fish, it must be appreciated that where this occurs, it is man-made pollution that causes toxicity (see chapter 6). Many rivers in the taiga support abundant populations of fish, which can tolerate quite low pH values in the absence of toxic chemicals.

CHAPTER 14

Insects and other Invertebrates

The formation of the English Channel after the ice age did not present a barrier to further colonization by many insects, as most species are winged and many are migratory. A small number, however, have poor methods of dispersal, especially in wingless species such as the female glow worm *(Lampyris noctiluca);* but ground-living, running beetles can cover considerable distances. In common with other forms of plant and animal life, the first insects to colonize Britain after the ice age were those found in the tundra today. As the ice retreated, wet conditions in the summer created a suitable habitat for species such as midges, which are so successful in such conditions.

Insects must have been quick to exploit available habitats as vegetation spread, especially as many spend the winter underground or near the surface, usually as eggs or pupae, where they are protected from extreme conditions; the adults emerge and fly in the warmer months, when conditions are suitable. The arrival of trees provided a wider range of habitats, together with shade and protection from wind, making the conditions suitable for the more sensitive species. The warm weather of the late Boreal and early Atlantic periods, together with the wide spectrum of tree species available in the Atlantic mixed forest, must have seen the optimum phase of insects in the southern parts of Britain and the furthest range northwards for many species.

As with other life forms, insects have physiological limits within which they must live and breed. Under optimum conditions breeding and dispersal can be rapid. The spread of Dutch elm disease illustrates this point, the infection slowing down northwards. Near the edge of an insect's range, breeding and dispersal are limited by temperature and other factors such as humidity. Breeding will take place but the conditions suitable for dispersal flights may not occur very often and, in the case of relict species, hardly at all, so that the population remains static and does not spread. Climatic changes, even on a small scale, may then allow expansion or, alternatively, hasten extinction.

Plate 14.1 The caterpillar of the native British moth, pine beauty (*Panolis flammea*), endemic on Scots pine (*Pinus sylvestris*) has caused severe damage recently to lodgepole pine (*Pinus contorta*) plantations on infertile sites in Scotland. (Photo: Forestry Commission)

Insects exploit a very wide range of foods and habitats. Because they are mostly small and breed rapidly, large numbers can occur in a small area. Insects can be grouped according to the food they eat: omnivores (eating almost anything), carnivores (predators, blood suckers, parasites and scavengers) and vegetarians (feeding on plants and fungi). The first two groups are not dependent on the plant or tree species forming woodland cover but on the habitats and food sources provided by it. Many of the plant-feeding insects too, although they do not depend directly on the trees forming the woodland, are indirectly dependent upon the tree cover as it allows the growth of the plants on which these insects feed. Many species of plant-feeding insects breed very rapidly and there have therefore evolved a number of species that are specific to their host plants. In some cases, such as aphids, races can be found which are restricted to a particular clone of a single tree species. The survival of the more specialized species is therefore dependent on the presence of the host and, if this is a tree or shrub, the type of woodland and its species composition become important. At the other end of the scale, there are vegetarian tree-feeding species that exploit a very wide range of host plants, both broadleaved and conifer. For these species the chances of survival are higher.

Vegetarian insects in particular come into conflict with human beings as they feed upon products required for human use. Others have a useful role in nature, for example the wood-boring insects that break down dead wood but are unwelcome when they destroy building timbers and furniture. For many species of insect, therefore, the problem is not conservation but control. Methods of control include chemicals (pesticides and attractants); biological control (using natural predators, parasites, viruses and the release of sterile males so that females lay infertile eggs); and physical control (removing infected elm trees killed by Dutch elm disease to curtail the numbers of beetles carrying the fungus) and trapping.

It would be impossible in a book of this kind to discuss all insect groups fully. A great deal is known about the insects of economic importance and detailed accounts of their life histories and control should be sought elsewhere; Bevan 1987 is particularly useful. For many of the harmless or useful insects, detailed knowledge of their life histories is lacking and only a few of the more attractive groups, such as Lepidoptera (moths and butterflies) and Odonata (dragonflies), have been intensively studied. Fry and Lonsdale (1991) provide a useful account of insects in woodland. As with other life forms and perhaps more so, the distribution of insects in Britain is primarily dictated by climate. Insects prefer warm conditions so more species are found in the southern parts of the country and even there many of them are at the northerly edge of their range. As we noted above, most species must have been at their most abundant in the Atlantic period but the gradual decline in temperatures since then has contracted the range of many with the result that the most diverse insect fauna is confined to southern Britain.

Woodlands have their own range of insects depending on the vegetation and tree species present. It would be impossible to provide habitats for all woodland insects in one particular area but woodland management can increase the range of habitats by maintaining dynamic and changing conditions. A succession of insect species takes place as woods and forests mature (NCC 1987d), and succession also occurs in the different stages of dead and dying wood (Evans 1977). Species feeding on broadleaves are different to those feeding on conifers, and tree species too have their own specialized insects (Winter 1983; Bevan 1987). The actual tree, shrub and plant species found or planted within the woodland will therefore dictate the presence or absence of many species. Southwood (1961) first drew attention to the relationship between the number of insects on trees and the time the tree species has been present in Britain (see table 14.1). More recent work has related the number of insects on trees and plants to the geographical range of the tree species and the colonization of species has been shown to be more rapid if taxa are related (Strong 1974a, 1974b, 1979; Strong et al. 1977). Studies of exotic trees bear this out, colonization of these trees, where they are related to established species, by insects being more rapid than expected (Welch 1981). Later, Welch (1995) goes on to say 'there is some evidence the more commonly and widely planted oaks have acquired the largest gall fauna, irrespective of their country of origin or length of time since their introduction to Britain, thus conforming to the theory of species-area relationship'. Strong and Levin (1979) also showed that trees have a wider range of insects feeding on them than feed on shrubs, which in turn have more than herbs. The more complex structure of a tree is thought to provide a greater variety of niches for insects to exploit. However, all these factors only partially explain the differences in species richness that occur in woodlands (May 1979a).

Many insects are not confined to one host but specialized species can only be present if their specific host is also present. Although oak is often quoted as having the most insect species associated with it (and it is not always appreciated that all species cannot be found on one oak), many of them are not specific to oak alone and, as records accumulate, oak stands out less as the only valuable host. To plant only oak provides no habitat for species dependent on other trees, both conifer or broadleaved. Mixed plantations of conifers and broadleaved trees will support the widest range of insect species feeding on trees.

Commercial woods have often been dismissed as a habitat for insects in much the same way as they have been for birds. However, the research that has been carried out shows rapid colonization of afforested areas (Young 1986) and the potential for the transfer of more species to exotics is significant (though not necessarily welcomed by the forester). Of 129 beetles found in semi-natural Scots pine, 77 are also found on spruce in Finland and some have already colonized conifer plantations (Hunter 1977; Welch 1986). A large number of insect species are not dependent on the tree species forming a particular wood but depend on the range of microclimates available within it; and this in turn depends on management. Hambler and Speight (1995) point out that the canopy of dense woodlands can sustain a high biomass of insects. Recent research by Ozanne (1996) in commercial conifer woods (Corsican

Table 14.1 Numbers of insects associated with trees and shrubs

Tree species	Number of insect species
Willows	450
Oaks	423
Birches	334
Pines	263
Hawthorns	209
Poplars	189
Spruce	178
Blackthorn	153
Alder	141
Elms	124
Silver fir	123
Apple	118
Hazel	106
Larch	100
Beech	98
Southern beech	93
Ash	68
Rowan	58
Limes	57
Hornbeam	51
Field maple	51
Sycamore	43
Juniper	32
Sweet chestnut	11
Holly	10
Horse chestnut	9
Douglas fir	9
Walnut	7
Yew	6
Evergreen oak	5
Elder	5
Bird cherry	3
Buckthorn	3
Box	2
Whitebeam	2
Alder buckthorn	1
Dogwood	1

Sources: Southwood 1961; Welch 1981, 1986; Winter 1983; Kennedy and Southwood 1984.

This table gives no indication of host specificity or the distribution of the insects concerned, but for many tree and shrub species the majority of insects only occur when the trees are growing in the south of Britain. These figures will continue to change as more information comes forward, but they give a relative indication of the insects supported by various trees and shrubs. Further data on insects are being collected by the ITE's Phytophagus Insects Data Bank. Information is lacking on the colonization of exotics; for example, some leafhoppers previously thought to be specific to oak have recently been recorded on southern beech (Welch 1981).

and Scots pine in the New Forest, Norway and Sitka spruce in South Wales) has revealed 'a very high density of insects and species too. For some types of insects, the number of species is comparable to that of broadleaf woodlands' and 'there are fewer insects on the edge trees; the forest core has a far higher density'. This research also revealed that springtails, normally regarded as a ground living species, occur 'in large quantities in the conifer canopy because it is possible that the conifer canopy provides a stable, humid environment with a constant food supply'.

About 20 000 species of insect occur in the British Isles and these are divided into two main groups. The first group consists of those insects that are wingless, including the springtails and bristletails that form an important constituent of forest soil fauna. They have not been studied in detail but they occur in a wide range of soils and tolerate pH levels as low as 3.6. Activity is greatest when soil temperatures are between 7 and 13°C. Numbers are least in more acid soils and those lacking decomposing humus but even then are numbered in millions per hectare. They form an important food source for other species.

The second group contains those having wings as adults. In some species wings are vestigial so that dispersal is limited. The winged insects are further divided into those whose larvae do not look like the adult and have a pupal stage in their development (such as butterflies) and those whose larvae are like wingless adults and do not have a pupal stage (such as aphids). Many species of both kinds have larvae that live in water, thus water sources in forests are particularly valuable.

The main insect orders of importance in woodland situations and that have a pupal stage include butterflies and moths (Lepidoptera); ants, bees and wasps (Hymenoptera); lacewings and snake flies (Neuroptera); flies (Diptera); and beetles (Coleoptera). Caddis flies (Trichoptera) have aquatic larvae and are therefore dependent on a water source.

The most important insects in woodland that do not have a pupal stage are the bugs (Hemiptera); grasshoppers and crickets (Orthoptera); earwigs (Dermaptera); thrips (Thysanoptera); and plant lice (Psocoptera). Water sustains some useful and harmless species such as dragonflies (Odonata), mayflies (Ephemeroptera) and stone flies (Plecoptera).

AQUATIC INSECTS

The presence of aquatic insects depends on a water source within the wood. Caddis fly larvae are found in most types of water and form an important food source for fish and water birds. The short-lived adults hatch in large numbers and provide food for fish and other animals, such as bats and birds. Caddis flies fly mainly at dusk, resting during the day on vegetation, trees and bushes near water. Some species lay their eggs on overhanging vegetation so the often recommended practice of removing all trees from stream and pond edges is not always beneficial.

The input to streams and other water bodies of terrestrial insects falling from trees and vegetation forms an important food source for fish and other aquatic animals, and so do falling leaves, including conifer needles once they have decomposed (Ormerod et al. 1987). Alder flies (Neuroptera) have aquatic larvae, usually found in ponds and slow-moving water; these are predatory on other aquatic organisms. The snake flies rest on waterside vegetation, bushes and trees.

A number of beetles (Coleoptera) are also aquatic, the adult water beetles as well as the larvae living in the water. Both the larvae and adults of diving beetles are predators and take a wide range of aquatic animals, including small fish. Other water beetles are vegetarian in the adult phase but the larvae are predatory. The largest water beetle, the silver diving beetle *(Hydrous piceus),* is on the edge of its range in Britain and is only found in the south and east of the country. Other water beetles are found in a wide range of water types, including quite stagnant ponds. They are able to tolerate pollution because both adult and larvae come to the surface to take in air as they have no gills, unlike the larvae of species such as dragonflies. As the adults are strong fliers, new ponds are quickly colonized. The larvae of the much less welcome midges and mosquitoes can be found in most still waters, even quite small puddles. Like those of water beetles, the larvae need to come to the surface to breathe. Water beetles, midges and mosquitoes are rapid colonizers and water quality is not important to these and other surface breathers.

Of the aquatic insects that have no pupal stage, mayflies and stoneflies are the best known, especially to fishermen, as they are an important fish food. Both tend to be found in well oxygenated lakes, rivers and streams. Stone flies on the whole prefer clear, flowing streams so more species are found in hilly districts. Sympathetic management is therefore necessary near such streams to maintain water quality. Operations, such as draining and clear felling, that cause excessive release of sediments or water surges need to be avoided. Scouring destroys stream structure and underwater habitats and these take some time to reform. Different habitats provided in streams are occupied by different species (Ormerod et al. 1987; Leeks and Roberts 1987). (The more complex problem of acidification due to pollution is discussed in chapter 6).

Dragonflies have received wider attention than other aquatic species because of their striking appearance and harmless nature. Britain has only 40 species and none of these is confined to Britain. The smaller, delicate species are known as damsel flies and the larger, heavier-bodied ones as hawkers or darters, depending on the way they catch their prey. They form two geographical groups, northern species and southern species.

Most of the northern species colonized Britain soon after the ice retreated because many of them are strong fliers. A few are thought to be relict species that survived the ice age in the west, as some tree species are now thought to have done. *Sympetrum nigrescens* is a rare example which breeds in brackish water on the west coast of Scotland. Most northern species have a wide distribution today in the northern hemisphere and some migrate over long distances. The northern species are adapted to living in more acid waters and harsher conditions than their southern relatives and many occur in the taiga zone. Eggs and larvae are the stages that over-winter. The eggs of some species, even if laid early in the summer, do not hatch immediately but enter a resting stage which is broken by certain conditions of temperature and daylight. Depending on temperature and other factors, larvae vary in the time taken to mature. Many species mature much more quickly in the southern part of their range than further north. For example, the normal time for the larvae of *Libellula depressa* to mature is more than two years. In a small and shallow southern pond, after a mild winter, adult dragonflies emerged the spring after egg-laying the previous summer (personal observation). The hatching of adults from larvae is often synchronized in many species and this increases the chance of mating, especially when adult flying periods are short. Some species emerge mainly in spring, whilst others fly during the summer months. Most species of damsel fly normally complete the life cycle in one year; the larger dragonflies require one to five years depending on temperature and feeding conditions.

The southern species of dragonfly are less adapted to adverse conditions and many of them are at the edge of their range in Britain. Indeed, of the British dragonflies lost since 1950, three were only found in the slightly more continental climate of the eastern counties and the others only in the south-west (NCC 1984). Although the loss of suitable breeding waters from various causes, including pollution and, in one case, by inundation from the sea, has caused a reduction in habitat, in view of the slender hold these species had in Britain climatic factors may also be involved. Climatic changes may also be a factor in the decline of a further four species, though afforestation is given as a cause (NCC 1986).

The great majority of dragonfly species will colonize ponds within forest areas, provided the ponds get enough sunlight. Forest rides, restocked areas and clearings in any type of managed forest are used by many species to feed in after hatching. During this time they can be found hunting a long way from water while they mature: at maturity, they find their way back to water to mate and lay eggs. As new bodies of water are quickly found, dragonflies must have the ability to detect water from some distance.

In the following subsection we describe the range and habitats of dragonflies in Britain. The information is mainly derived from Corbet et al. 1960 and Chelmick et al. 1980.

Dragonflies

Northern species using small and large bodies of still or slow-moving water

Members of this group tolerate a wide pH range from acid to basic. Most of these species have a general distribution.

Damsel flies Emerging in spring: the blue damsel fly *(Enallagma cyathigerum)*. Emerging in summer: the green lestes *(Lestes sponsa)*; the common ischnura

(Ischnura elegans); the common coenagrion *(Coenagrion puella);* the variable coenagrion *(Coenagrion pulchellum).*

Hawkers Emerging in spring: the downy emerald *(Cordulia aenea);* the brilliant emerald *(Somatochlora metallica);* the restriction of this latter species to two main areas, Scotland and southern Britain, is thought to be the result of two separate colonizations after the ice age, one from the west and the other a reinvasion from the south. Emerging in summer: the brown aeshna *(Aeshna grandis);* this species is not found in Scotland and appears to require a more continental climate.

Darters Emerging in spring: the four spotted libellula *(Libellula quadrimaculata)* .

Northern species using small acid pools or peat bogs

Damsel flies Emerging in summer: the northern coenagrion *(Coenagrion hastulatum),* confined to Scotland; the scarce green lestes *(Lestes dryas),* which breeds in low fens so is restricted to the south-eastern counties and may now even be extinct in Britain (NCC 1984).

Hawkers Emerging in spring: the blue aeshna *(Aeshna caerulea),* confined to Scotland; the northern emerald *(Somatochlora arctica),* confined to Scotland and south-west Ireland. Emerging in summer: the common aeshna *(Aeshna juncea);* and the black sympetrum *(Sympetrum scoticum* and *Sympetrum nigrescens),* the latter confined to north-west Scotland.

Darters Emerging in spring: the white-faced dragonfly *(Leucorrhinia dubia);* the keeled orthetrum *(Orthetrum coerulescens).*

Northern species using streams and aerated water

Damsel flies Emerging in spring: the banded agrion *(Agrion splendens),* not found in Scotland and requiring a muddy bottom; the demoiselle agrion *(Agrion virgo);* the large red damselfly *(Pyrrhosoma nymphula).*

Hawkers Emerging in spring: the club-tailed dragonfly *(Gomphus vulgatissimus),* found only in southern England and requiring a silty bottom; the golden-ringed dragonfly *(Cordulegaster boltoni),* requiring a muddy bottom.

Southern species using a wide range of stagnant waters
These dragonflies are found mainly in southern Britain.

Damsel flies Emerging in summer: the small red damsel fly *(Ceriagrion tenellum),* not found in Ireland; the scarce ischnura *(Ischnura pumilo).*

Hawkers Emerging in spring: the hairy dragonfly *(Brachytron pratense)*, not found in Scotland.

Darters Emerging in spring: the emperor dragonfly *(Anax imperator);* the back-lined orthetrum *(Orthetrum cancellatum)*. Emerging in summer: the common sympetrum *(Sympetrum striolatum)*, extending into the Scottish lowlands; the ruddy sympetrum *(Sympetrum sanguineum)*.

Hawkers Emerging in summer: the scarce aeshna *(Aeshna mixta)*, not found in Ireland; the southern aeshna *(Aeshna cyanea)*, not found in Ireland.

Southern species using more oxygenated waters, either still or slow-flowing

Damsel flies Emerging in spring: the red-eyed damsel fly *(Erythromma najas)*; the white-legged damsel fly *(Platycnemis pennipes)*.

Hawkers Emerging in spring: the Norfolk aeshna *(Aeshna isosceles)*, found in Norfolk only.

Darters Emerging in spring: the broad-bodied libellula *(Libellula depressa);* the scarce libellula *(Libellula fulva)*.

Southern species using streams

Damsel flies Emerging in summer: the southern coenagrion *(Coenagrion mercuriale)*.

Conservation measures for aquatic insects

The provision of water in forests benefits all aquatic insects. The creation of ponds is a positive measure, as they are quickly colonized. Some insects will colonize any body of water and tolerate a wide range of pH and, in some cases, pollution. Sunlight on the water is important for dragonflies, as also is vegetation that gets sun early in the day and on which newly hatched dragonflies can rest. Many species hatch in the early morning and require sun to warm them up for flight. The sooner they are able to fly the less vulnerable they are to predators. If ponds require cleaning out the timing is made critical by different emergence times and lengths of life cycle, and so depends on the species present. If a wide range of species with varying requirements is present, cleaning should only be done infrequently and when really necessary. Even then it is wise to clean half or less at a time as many species have a larval life of more than a year. The optimum cleaning time for most species is late summer, as the summer species are still laying eggs and some of the spring species have well grown larvae that can find their way back into the water if mud and vegetation taken from the pond is left near the edge. Eggs in a resting stage are vulnerable.

At adult emergence, which often takes place overnight, many species require upright water plants to climb for ecdysys (change), which has to take place out of the water; others use gently sloping banks. Early morning sun helps newly emerging dragonflies to fly. Some species require clear water on which to drop their eggs, so it is important that such water is available at egg-laying time if certain species are to be attracted. In upland areas where trees are planted, wet peat pools and sphagnum bogs need to be left for some species, again with sufficient penetration of sunlight from the south and east, so the ultimate effect of the height of mature trees needs to be taken into consideration at the time of planting. Alternatively, any shading trees can be removed earlier as small produce in order to get some return and to keep the pools in a suitable, unshaded state. A few trees and bushes near the edge of the water are important for resting places.

BUTTERFLIES

Butterflies have received considerable attention and much is known about their life histories. The majority of the 57 British species occur in southern Britain where the climate is warm, and include southern species at the northern edge of their range and species with a northern distribution elsewhere but requiring a continental climate. Butterflies usually prefer open, sunny situations and many of them are found in meadows where their food plants grow. The increasing use of leys (selected grasses grown for a few years and then reseeded) in place of the more traditional permanent grassland has meant a reduction in the areas providing suitable habitat and this, together with the use of pesticides, has meant that woodland refuges have become important for many species. Although very few butterflies are dependent upon a particular species of tree as a food source for their caterpillars, many do depend on the sheltered conditions and microclimates provided by woodland (Young 1986). A wide range of food plants other than trees and shrubs are used and closely related butterflies tend to use the same food plants.

Butterflies are dependent on the presence of food plants in sufficient quantity for their larvae to feed upon. The ability to find food plants is the result of pheromones, chemical substances produced by plants and insects that act as attractants. The swallowtail, for instance, is able to distinguish between closely related Umbellifera; these plants contain a chemical in one of two forms. Eggs are only laid on plants containing the chemical in the appropriate form. Experiments have shown that the development of the larvae is superior on plants chosen naturally by the butterflies.

The suitability of a woodland for butterflies depends on (1) conditions providing protection and shade, (2) the presence of food plants for the larvae and (3) the presence of plants in flower providing a nectar source at the time the adults are flying.

The vulnerability of these delicate insects explains why many butterflies occur in woodland clearings or open scrub, especially in the more exposed parts of their range. Here both protection from wind and rain can be found, together with varying conditions of light and shade. Sun is important to butterflies, as it is needed to provide warmth for flight. Basking achieves this: butterflies either spread their wings wide or close them completely, when dark colouring near the body absorbs heat. Many northern butterflies are dark for this reason. White butterflies, on the other hand, reflect light from partially closed wings onto the body (Kingsolver 1982). Bare ground on rides provides ideal basking space.

Nectar from flowers provides energy for flight, so a suitable source for the adult butterflies is important. Willow and blackthorn are important sources in early spring and encouragement of these species on forest edges also benefits other species. The adult life of butterflies is short, so flight is vital to find mates, to find food plants on which to lay eggs and to disperse the species.

Butterflies do best in regions with hot summers and the fluctuations in numbers in Britain are largely the result of variations in the climate (Pollard 1981). The great majority of butterflies are limited in their range in Britain, many of them confined to the southernmost counties of England. Others only reach as far north as the Scottish border. Relatively few species occur in northern Britain because of the cooler summers there. Butterflies mainly overwinter as eggs or larvae but a few species overwinter as adults. The dependence of the large blue (*Maculinea arion*) caterpillar on a particular ant species for overwintering was the main reason for its extinction in England; lack of management led to the open, short grass that the ant required becoming rank and long (Thomas 1980). Climatic influences may also have operated, as the large blue is on the edge of its range in Britain and from this point of view it remains to be seen how successful recent reintroductions will be.

The type of food plant required by butterfly larvae gives an indication of the type of habitat required. Butterflies feeding on flowers and some grasses are to be found in open areas, whilst those feeding on plants colonizing bare land are found on woodland edges and clearings where the ground is disturbed. Only a few species feed on trees and shrubs. These still need clearings in which to fly, although some adults spend most of their time in the tree tops feeding on aphid honey dew. For the majority of species, the woodland structure and tree spacing matter as much as the tree species (Peachey 1986). For butterflies using open space within the forest, the microclimate determines their presence, not the tree species (Young 1986). Butterflies can be grouped according to their distribution and food plants. Except where stated, the following information is derived from Ford 1945; Robinson 1968; Whalley 1980; Carter 1982; Higgins and Riley 1983; and Sawford 1987.

Grass-dependent butterflies

Butterflies whose larvae feed on grasses and who have a northern and general distribution

Species	Food plants
Meadow brown *(Maniola jurtina)*	*Poa* spp., especially *P. annua;* also other grasses
Small heath *(Coenonympha pamphilus)*	*Poa* spp., especially *P. annua;* also *Nardus* and other grasses
Ringlet *(Aphantopus hyperantus)*	*Poa* spp., *Dactylis, Milium* and *Carex* spp.
Grayling *(Hipparchia semele)*	*Aira* and other grasses
Scotch argus *(Erebia aethiops)*	*Dactylis, Molinia*
Large heath *(Coenonympha tullia)*	*Eriophorum, Carex* spp.
Chequered skipper *(Carterocephalus palaemon)*[a]	*Bromus, Molinia*

[a]In the late 1970s this butterfly was thought to be extinct in Britain. However, it has established itself in the rides of conifer plantations in Scotland (Young 1986) and is no longer protected by Schedule 5 of the Wildlife and Countryside Act 1981 (NCC report number 13).

Butterflies whose larvae feed on grasses and have a southern distribution

Species	Food plants
Gatekeeper *(Pyronia tithonus)*	*Poa annua*
Wall *(Lasiommata megera)*	*Poa annua, Dactylis glomerata*
Large skipper *(Ochlodes venatus)*	*Brachypodium* spp., *Dactylis glomerata*
Marbled white *(Melanargia galathea)*	*Dactylis, Phleum* spp.

Species	Food plants
Small skipper *(Thymelicus sylvestris)*	*Brachypodium, Holcus* spp.
Essex skipper *(Thymelicus lineola)*	*Brachypodium, Agropyron* spp.
Speckled wood *(Pararge aegeria)*	*Agropyron, Dactylis, Triticum* spp.

Conservation measures for butterflies dependent on grasses

All the species listed in the two tables above depend on grasses, apart from the large heath, which could be encouraged in forest and woodland by leaving undrained areas because the plants it uses grow in damp places. Most of the grasses require strong light, so clearings or wide rides are needed to ensure their growth. Clear cut areas will also allow colonization by suitable grasses. *Poa annua,* used by many species, colonizes bare ground, so disturbance or trampling (for example on footpaths) will provide conditions for this grass to colonize, as

Plate 14.2   The chequered skipper (*Carterocephalus palaemon*), a butterfly of woodland glades which had become rare and was thought to be extinct. It has recently found suitable habitat in coniferous plantations in Scotland. As a result of populations increasing, the butterfly was removed in 1987 from the list of specially Protected Animals (Schedule 5 of the Wildlife and Countryside Act). (Photo: Forestry Commission)

will timber extraction. With increased shading, grasses such as *Holcus, Deschampsia* and *Milium* spp. increases, so narrower rides will sustain them but the amount of sunlight received is still important. Thinning will ensure grass growth in plantations, the grass species being dependent on the amount of light available. Sunlight is required by butterflies to bask and fly, so rides need to be planned or adapted to receive the maximum amount of sunshine (Anderson and Carter 1987, 1988).

Flower-dependent butterflies

Butterflies whose larvae feed on plants mainly colonizing bare ground and have a northern and general distribution

Species	Food plants
Dark green fritillary *(Argynnis aglaja)*	Violets and other *Viola* spp.
Pearl bordered fritillary *(Boloria euphrosyne)*	Violets
Small pearl-bordered fritillary *(Boloria selene)*	Violets
Green-veined white *(Pieris napi)*	Cruciferae; *Draba, Sisymbrium Arabis, Alliaria.*
Small copper *(Lycaena phlaeas)*	*Rumex, Polygonum* spp.
Common blue *(Polyommatus icarus)*	*Trifolium, Vicia* spp. *Lotus corniculatus, Ononis spinosa*
Northern argus *(Aricia artaxerxes)*	*Helianthemum, Erodium, Geranium* spp.

Butterflies whose larvae feed on plants mainly colonizing bare ground and have a southern distribution

Species	Food plants
High brown fritillary *(Argynnis adippe)*	Violets

Species	Food plants
Silver washed fritillary *(Argynnis paphia)*	Violets
Duke of Burgundy fritillary *(Hamearis lucina)*	*Primula* spp. (cowslip, primrose)
Orange tip *(Anthocharis cardamines)*	Cruciferae (Ladies' smock, hedge mustard, garlic mustard)
Large and small white *(Pieris spp)*[a]	Cruciferae
Wood white *(Leptidea sinapsis)*[b]	Tuberous pea *(Lathyrus tuberosus)*, bird's-foot trefoil *(Lotus corniculatus)*
Argus *(Aricia agestis)*	Rock rose *(Helianthemum)*, Stork's bill *(Erodium)*
Grizzled skipper *(Pyrgus malvae)*	Bird's-foot trefoil and other legumes, *Fragaria* spp. (strawberry), *Potentilla* spp. (cinquefoil and silverweed)
Dingy skipper *(Erynnis tages)*	Bird's-foot trefoil, *Hippocrepis* and *Coronella* spp.
Marsh fritillary *(Eurodryas aurinia)*	Devil's-bit scabious *(Scabiosa succisa)*, plantain *(Plantago)*
Heath fritillary *(Mellicta athalia)*	Cow-wheat *(Melampyrum)*, plantain, foxglove *(Digitalis)*, wood sage *(Teucrium scorodonia)*
Green hairstreak *(Callophrys rubi)*	Gorse *(Ulex)*, broom *(Sarothamnus)*, and a wide range of other plants

[a]Migrants augment overwintering populations
[b]Research work on this species showed that rides with about 30 per cent shade were favoured. Many of the remaining colonies of this species are in managed, often mainly coniferous, woodland where rides are managed (Warren 1980).

Conservation measures for butterflies dependent on flowers

Most of the plants listed above colonize exposed ground, so disturbance is essential if they are to increase and maintain themselves against competition from other vegetation. Many are woodland edge species and thus colonize areas where competition is reduced. The decline of fritillaries has been attributed by Young (1986) to lack of coppicing, because the ground is no longer being periodically exposed, but this explanation does not hold good throughout their range. Neglect of woodland since the Second World War has also resulted in a reduction of ground disturbance, as a lack of management in many woods has led to the shading out of ground vegetation. Felling and extracting timber creates bare ground and in many woods it is noticeable that violets flower on path edges where bare soil has been exposed. In the totally protected forest of Bialowieza in Poland, violets colonize where wild pigs turn over the soil seeking food. Indeed, one of the old names for the heath fritillary, 'the woodland follower', indicates the need for periodic clearance. Where violets are established, destruction should be avoided as the eggs of most fritillaries remain on them over winter. Some of the food plants listed above are limited in their distribution; for example, rock rose grows only on calcareous soil. The use of limestone for forest road-making encourages such plants, so the artificial introduction of them is an option to be considered.

Nettle-dependent butterflies

Butterflies whose larvae feed on nettles (Urtica spp.)

Species	Notes
Those with a northern distribution:	
Small tortoiseshell *(Aglais urticae)*	Hibernates as an adult. Woodsheds were a favoured place in the past; woodpiles in the forest could provide wintering places
Peacock *(Inachis io)*	Adult also hibernates, like the tortoiseshell
Those with a southern distribution:	
Comma *(Polygonia c-album)*	The caterpillar also feeds on elm and willow
Red admiral *(Vanessa atalanta)*	Migrant: broods are reared here and there is evidence of a return migration (Williams 1958); also found on thistles
Painted lady *(Cynthia cardui)*	As the red admiral

Conservation for butterflies needing nettles
Many of the most attractive British butterflies depend on nettles. Clumps of nettles need to be left in sunny and semi-shaded situations; this is often particularly appropriate at ride junctions and around the sites of old buildings.

Woodland butterflies

Butterflies dependent on woodland shrubs and trees
All are southern species. Many of these butterflies spend most of their time in the tree canopy and are rarely seen at ground level; consequently they often escape detection.

Species	Food plants and other details
White admiral *(Ladoga camilla)*	Honeysuckle (*Lonicera* spp.) Eggs are laid on plants in light shade; plants near the woodland edge or in light or mature woodland are therefore the most important. This species is said to have benefited from the increase in neglected coppice (Pollard 1981,1979) but has probably also gained from general woodland neglect since the Second World War. Its expansion in range between 1920 and 1961 has also been attributed to the rise in mean summer temperatures during that period (Joyce 1988)
Holly blue *(Celastrina argiolus)*	Ivy (*Hedera* spp.), holly (*Ilex* spp.), buckthorn (*Rhamnus* spp.)
Brimstone *(Gonepteryx rhamni)*	Buckthorn, alder buckthorn (*Frangula* spp.)
Black hairstreak *(Strymonidia pruni)*	Plum and blackthorn (*Prunus* spp); clumps facing south-west and south-east are favoured but these need to be in sheltered, unshaded situations (Thomas 1980). Some of the remaining British colonies are in managed woodland, including coniferous plantations in the Midlands (Forestry Commission, pers. comm.)
Brown hairstreak *(Thecla betulae)*	Blackthorn, plum, damson (*Prunus* spp.), birch (*Betula* spp.)

Species	Food plants and other details
Purple emperor *(Apatura iris)*	Willow, sallow. Found in beech, oak and mixed woods outside Britain so is not dependent on oak as is often stated. Certain trees are used as 'meeting' places. On the West Dean estate in Sussex, where coppicing was undertaken leaving a 35 per cent oak canopy cover, the population of this species suffered a severe decline. It was suggested that the adults required a more continuous canopy but it is also likely that, as the butterfly overwinters as a caterpillar high up on its food plant and not near ground level, the open conditions were too extreme for winter survival (Williamson, pers. comm.)
Purple hairstreak *(Quercusia quercus)*	Oak; sometimes ash, willow and sweet chestnut
White-letter hairstreak *(Strymonidia w-album)*	Lime, elm (all species) and other trees
Large tortoiseshell *(Nymphalis polychloros)*	Elm, also willow, poplar, sallow, apple and cherry. This species has declined, the disappearance of elms drastically reducing available food supply

Conservation measures for butterflies dependent on trees and shrubs

The presence of the food plant is the key factor in the encouragement of these butterflies but the degree of shelter and sunshine is also important. Thomas (1974) has claimed that the black hairstreak requires old bushes of blackthorn to survive and that other apparently suitable blackthorn bushes are not colonized, so other factors of more importance to the butterflies may be operating. On the whole, all insects mature more quickly when they are feeding on young leaves and the time when these are available varies among different plants of the same species (Hodkinson and Hughes 1982). It has been observed that hairstreaks will colonize young plants of blackthorn and on such plants leaves will be more nutritious for longer than on older ones (Pollard 1981). The planting of new clumps therefore helps the survival of the species but it would seem that the planting needs to be on warm and sheltered, preferably south-facing, sites.

General conservation measures for all butterfly species

The presence of the food plant is most important. For butterflies laying single eggs the density of plants needed to sustain the caterpillar is less than for species laying their eggs in groups. Sunny, sheltered situations provide the best areas for food plants. Shelter and protection are important and curving rides form very suitable conditions as they prevent wind blowing straight along them; long straight rides need modification for this reason. All above butterflies need nectar (or 'honey dew' for some canopy species) to supply energy when flying, so that a succession of flowers in a woodland is important. Many of the summer-flying butterflies feed on bramble flowers. Treating rides as meadows and mowing them in rotation has successfully maintained flowering plants in the Forestry Commission's Bernewood Forest near Oxford. Rides form one of the most favoured habitats for many species, so in existing woodlands shaded rides may need widening and trees should be removed and thinned selectively to allow the maximum amount of sunlight to penetrate. Wide rides and clear felled areas have been shown to have greater numbers of butterflies than neglected woods (Steel 1989). Widening ride junctions by removing corners creates small glades. Timber loading bays sited in sunny situations will provide areas that are periodically disturbed, thereby encouraging the ground flora. Periodic cutting of shrubs and mowing of rides is necessary to keep them in a suitable condition but this should be undertaken in the autumn (Carter and Anderson 1987; Roberts and Warren 1994). In existing woodland, restocked areas will provide disturbed and open habitats, the butterflies favoured depending on the size of felling coupe and the vegetation cover present. Coppicing in the past provided open habitats with different insects using different ages (Waring and Haggett 1991).

A wide range of butterfly species can be encouraged in coniferous, mixed and deciduous woodland, depending on geographical situation, broadleaved trees and shrubs only being of importance for a small number of species. The butterflies that use broadleaves as their food plant are all southern species, confined in Britain to southern England where mixed and broadleaved forestry is a more economic proposition. The retention of food shrubs near ridesides, either by planting or by allowing natural colonization, will enhance the conservation value of these woods. Woodland shrubs grow and flower best on south-facing woodland edges. Group regeneration will provide small clearings favoured by the speckled wood, and crown thinnings will create diverse light patterns. If conifers are to be grown, larch allows more light to penetrate the canopy than other species, though the amount of light penetrating in plantations can be increased by brashing, high pruning and early thinning.

MOTHS

There are many more British moths than butterflies, so we cannot consider all species in this chapter, but the general principles concerning their conservation are similar. The distribution and life histories of moths, especially the smaller forms, are not as well documented as those of the butterflies.

Moths mainly fly at dusk, spending the day resting, though there are some day-flying species. Many of them are camouflaged to blend with their surroundings and are not easily detected by daytime predators. There are over 2000 species of moth in

Britain and, as with other insects, the majority of species have a southern distribution. The range of food plants used by moth larvae is much wider than those exploited by butterflies, and many moths are pests. Some adult moths do not feed; others feed on nectar, both during the day, and at dusk when flowers such as honeysuckle produce scents that attract moths and other pollinating insects. There are several moths whose caterpillars are severe defoliators of trees, occasionally reaching pest proportions and that consequently do not need conserving.

Many moths are not dependent on trees for their caterpillar food and cause no damage. Where they occur in woodlands they benefit from similar conditions to those that favour butterflies. Species whose larval food plants normally grow in meadows and hedges make use of woodland rides and these too will be assisted by measures taken to conserve butterflies.

Non-woodland moths
A few of the better-known non-woodland species are listed below.

Moths using grasses

Species	Food plants and other details
Drinker *(Philudoria potatoria)*	*Dactylis* (cocksfoot), *Agropyron* (couch) and other grass species are used; this moth is a wide-ranging northern species, to Siberia
Common wainscot *(Mythimna pallens)*	Cocksfoot, *Poa* and other grasses
Dark arches *(Apamea monoglypha)*	*Poa* and other grasses; the moth is a wide-ranging northern species, to Siberia
Grass moth *(Crambus pascuella)*	*Poa* and other grasses

Moths using plants that colonize bare ground
A selection of commonly used food plants is given below, but most species will feed on a wide range of herbaceous plants.

Species	Food plants and other details
Elephant hawk moth *(Deilephila elpenor)*	Willow herb (*Epilobium* spp.)

Species	Food plants and other details
Small elephant hawk moth *(Deilephila porcellus)*	Willow herb
Yellow underwing *(Noctua pronuba)*	Sorrel; dock *(Rumex* spp.)
Setaceous hebrew character *(Xestia c-nigrum)*	Sorrel, dock; this moth is widespread in Eurasia and North America
Wood tiger *(Paresemia plantaginis)*	Plantain *(Plantago* spp.); this is a northern species ranging to Siberia
Buff ermine *(Spilosoma lutea)*	Plantain
Cream spot tiger *(Arctia villica)*	Chickweed *(Cerastium* spp.); this is a central and southern European species, confined to the south in Britain
Netted pug *(Eupithecia venosata)*	Red campion *(Melandrium rubrum)*
Convolvulus hawk moth *(Agrius convolvuli)*	*Convolvulus* spp.; southern Europe, migrating to Britain
Peach blossom *(Thyatira batis)*	Bramble *(Rubus* spp.)

Moths using nettles (Urtica spp.):

Plain golden Y *(Autographa jota)*

Dark spectacle *(Abrostola trigemina)*

Small magpie *(Eurrhypara hortulata)*

Moths using other ground vegetation

Species	Food plants and other details
Small angle shades *(Euplexia lucipara)*	Bracken *(Pteridium aquilinum)*

Species	Food plants and other details
Great brocade *(Eurois occulta)*	Bilberry *(Vaccinium* spp.); found in northern Britain, this is a northern species ranging over Eurasia and North America
Grass emerald *(Pseudoterpna pruinata)*	Gorse *(Ulex* spp.) and broom *(Sarothamnus* spp.)

Conservation measures for moths not dependent on trees

Conservation directed towards butterflies will also provide feeding plants for a wide range of moths. The presence of non-woodland moths, as is the case for butterflies, will depend more on the microclimates provided by the woodland, together with the presence of the right food plants, rather than on the actual tree species forming the woodland.

Woodland moths

Moths dependent on woodland shrubs

Species	Food plants and other details
Broad-bordered bee hawk moth *(Hemaris fuciformis)*	Honeysuckle; found in southern counties only
Chinese character *(Cilix glaucata)*	Blackthorn *(Prunus* spp.) and hawthorn *(Crateagus* spp.); a southern species found mainly in England and Wales
Brimstone moth *(Opisthograptis luteolata)*	Blackthorn and hawthorn
Pebble prominent *(Eligmondonta ziczac)*	Sallow and willow

Moths dependent on trees

In contrast to butterflies, many species of moth feed on trees. Some are foliage feeders and some are leaf miners, whilst others attack seeds; a few live on wood. Some species feed on a wide variety of trees and are found on both broadleaves and conifers; others are confined to feeding either on broadleaves or conifers. Some are restricted to one tree species, whilst others are more specialized still and fill a particular niche on a specific tree species. The leaf miners in particular show this

292

specialization; for example, the various species of midget moths of the genus *Phyllonorycter*, which form blotches where they have mined leaf tissue. *Phyllonorycter* spp. attack a wide range of broadleaved trees but each species is restricted to a single species of tree host. For example, out of the 26 species listed in Winter 1983, eight are found only on oak, one is found on oak and on the closely related sweet chestnut, five on alder, three on birch, three on willow, two on elm, one on poplar, one on beech and one on sycamore. Only one species is recorded on a number of other broadleaved trees besides oak. Other moth genera are restricted to particular tree species in the same way (Roberts 1996). Amongst the insects, this pattern of distribution is not confined to the moths but it illustrates that both conifers and broadleaves each support their own range of insects as well as sharing others.

As with other plant and animal species, many moths are limited in their range by climatic conditions. In Britain, some only occur in the south of the country (these include southern species at the edge of their range and northern species requiring a more continental type of climate); others are more widespread and some are found mainly in the north, becoming rare in the south. The insects that occur in Scotland, including moths, are a much impoverished fauna compared with southern Britain (Welch 1986). The following examples of moths feeding on trees illustrate the above points. Similar range restrictions have been mentioned earlier in the list of moths not dependent on tree and shrub species.

Species	Food plants and other details

Some species feed on only a few broadleaves; some are restricted to a single species

Great prominent *(Peridea anceps)*	Oak; mainly southern Britain, rare in the north
Oak hook tip *(Drepana binaria)*	Oak; England and Wales
Pebble hook tip *(Drepana falcataria)*	Birch and alder; general distribution
Lesser swallow prominent *(Pheosia gnoma)*	Birch; a northern species, ranging to Siberia but with a general distribution in Britain
Poplar hawk moth *(Laothoe populi)*	Aspen and willow; general distribution but in Britain is commoner in the north; ranges up to the Arctic circle
Poplar clearwing *(Sesia apiformis)*	Poplar; larvae live in the wood; England and Wales, extending into southern Scotland

Species	Food plants and other details
Puss moth *(Cerura vinula)*	Poplar and willow; commoner in south of Britain

Some species feed on a wide range of broadleaves; these include:

Brown oak tortix *(Archips xylosteana)*
Lackey moth *(Malacosoma neustria)*
November moth *(Epirrita dilutata)*
Peppered moth *(Biston betularia)*

Other species are general feeders and feed on both conifers and broadleaves; these include:

Rose tortrix *(Achips rosana)*
Apple leaf roller *(Epiphyas postvittana)*
Autumn moth *(Epirrita autumnata)*
Scalloped hazel moth *(Odontoptera bidentata)*
Hebrew character *(Orthosia gothica)*

Conifers accommodate a range of species that do not occur on broadleaves. Many of the species found on Scots pine are now also to be found on exotic species of pine, some of them in pest proportions, such as the pine beauty *(Panolis flammea)*. The pine hawk moth, once uncommon, has become much more widespread with the increase in coniferous forestry (NCC 1986), which has also provided habitats for a number of other moths that are not restricted to a particular tree species.

Conservation measures for moths feeding on trees
Neither the conservation nor the control of moths is straightforward as so many of them cause damage. Of the species that feed on trees, only those that are rare and do not cause damage justify conservation. Measures taken to conserve butterflies will help many of the harmless moth species; otherwise the presence of the trees themselves, as a food source, encourages a wide range of both harmless and harmful species.

BEETLES
Nearly 4000 species of beetle are found in Britain but a third of them are absent from Ireland. Over half are confined to the south, as they favour a more continental climate and are on the western or northern edge of their range in Britain. About 200 species have a northern distribution and in Britain do not extend southwards further than the Lake District and Welsh hills (Harde 1984). The remainder have a general

distribution throughout Britain. It is reasonable to suppose that the more northern species found today were also the earliest to colonize after the ice age, or were relict species that had survived in refuges. Woodland species are found in the Caledonian pine forests but many have been restricted to this habitat in the past because of the absence of other coniferous species. In other parts of their range, these species use other conifers as hosts; out of 53 species recorded on Scots pine in the Phytophagous insects data bank, only 14 are specific (Ward and Spalding 1981). Many species have already colonized plantations of Scots pine and other conifers. By contrast, many of the southern species, associated with broadleaves, colonized at the climatic optimum during the Atlantic period and at that time extended further north than they do now. The deterioration of the climate since then means that many are now at the northern edge of their range (Evans 1977) and others have been lost to Britain altogether. Some species are strong fliers and are able to colonize new situations readily, whilst others are more sedentary.

Because of their tough cuticles (outer skins), beetles are able to live in a wide variety of habitats but they are sensitive to humidity so many of them are either nocturnal or woodland dwellers. In common with other insects, there is often a resting stage during larval development, broken mainly by changes in temperature; the majority of adults therefore emerge at the same time, optimizing successful mating. Mating efficiency is also increased by the use of sex pheromones (chemical attractants), which may be used to control beetles by attracting them to a site where they can be destroyed. Work in the USSR has shown that the great spruce bark beetle *(Dendroctonus micans),* found recently in Britain, disperses from the 'home tree' when the temperature reaches 15°C, otherwise inbreeding occurs amongst members of the same brood without dispersal (Harris and Harris 1988c). Such limiting factors control both breeding efficiency and distribution. The body structure of male and female beetles may be different; for example, the female glow-worm is much larger than the male and has no wings. Glow-worms are therefore less able than species with winged females to disperse widely and rapidly.

Beetles eat a wide range of food and can be classified into predators, scavengers, parasites and herbivores. *Predators and scavengers* are active beetles with both adults and larvae eating a wide range of foods. They occur in woodland but are not dependent on the tree species that form the cover. As with some moths and butterflies, the habitat and microclimate provided in the various forest stages are the factors determining the presence of beetles. The predators include ground beetles (Carabidae) and tiger beetles (Cicindelidae). The latter are found in the south of England and are on the edge of their range in Britain. Ground beetles are more widespread and are found in the early stages of coppicing when the ground is clear but numbers fall as the ground vegetation increases. Ground scarification after felling was found to increase the numbers of ground beetles (Parry 1986). Clear felling, because it exposes the ground, therefore benefits these species, which also use rides and roads. Ladybirds are also predators and their larvae can be found feeding on aphids in a wide range of woodland sites.

Scavengers include dung and carrion beetles and their presence is dictated by available food in the form of dung or carcasses. A succession of beetle species exploit these 'foods'; those using recent carcasses differ from those exploiting later stages of decay (Evans 1977). *Parasites* live on live hosts and so are restricted to situations where their hosts occur.

Herbivores are found on a wide range of plant foods, including fungi. Some are general feeders on the leaves of both broadleaves and conifers, whilst others are more specialized, feeding on either broadleaves or conifers. Specialized feeders, adapted to digesting cellulose, burrow in timber. Beetles also eat pollen, seeds and fruits. Many are pests, attacking crops, timber and food products.

There are many herbivorous beetles that eat a wide range of plant food, but each occupies a specialized habitat. Different stages of forest succession favour different species. Beetle succession in Scots pine has been discussed by Hunter (1977). Weevils in the late successional stages of birch colonization are more generalized feeders than those on herbs in the earlier stages. Conversely, leaf beetles become more specialized as succession proceeds (Brown and Hyman 1986). Decaying timber also attracts a succession of species, each of which is restricted to a particular stage of decay (Evans 1977). Many of these are pests of timber. Other pest species attack seeds, fruits, roots and leaves. A specialized group feed on fungi growing on old and dying trees.

Table 14.2 Beetle succession in dead timber

State of tree	Beetles found attacking
Recently dead	Bark beetles and weevils
Bark separating	Wood borers (Longhorn beetles)
Tree breaking up	Ground and click beetles

It is often thought that only old large trees are attacked by this last group of beetles but this is not correct (see table 14.2). Most of them can be found attacking dead trees of all ages and as the timber decays the species involved become more general feeders, using both conifers and broadleaves (Hunter 1977). Table 14.3 summarizes approximately the activities of the large number of beetles that occur in Britain, many of which occur in woodland. It is intended as a guide only.

Conservation measures for beetles

Forest management designed to produce sustained yield will provide varied conditions within woodland for a succession of harmless species. The trees themselves, depending on the geographical situation and the microclimates provided by them,

Table 14.3 Categorization of British beetle species

Type of beetle		Number of species
Harmless beetles		
Predators and scavengers		1800
Water beetles		300
Beetles feeding on fungi		400
Beetles feeding on rotting wood		120
	Total	2620
Beetles causing damage		
Beetles feeding on leaves/needles		800
Beetles boring in wood/bark		130
Beetles feeding on sap		95
Beetles feeding on roots		60
Beetles forming galls		30
	Total	1115

will also provide habitats. Unfortunately, the problem with many of the beetles feeding on trees in woodland is one of control rather than conservation. Control can be exercised in various ways: prevention, such as the prompt removal of felled timber (especially conifers with bark on); chemical control with insecticides; the use of artificial pheromones as attractants; the use of viruses; the release of parasites; and removing bark from felled timber.

Commercial forests normally contain little in the way of dead and dying overmature trees. However, a considerable amount of dead and diseased material can safely be left in a growing crop; especially if crown thinning is employed, suppressed and damaged trees, not competing with final crop trees, need not be removed. Old or ringbarked trees can also be left in the growing crop with benefits to other species as well as insects.

The preservation of beetle communities, such as those described for pasture woodland by Harding and Rose (1986), are in the province of pure conservation. These communities are a mixture of (1) species with a general distribution that can also be found in managed, plantation forest, (2) species with a northern distribution that can be found in plantations in the north but are confined to ancient or pasture woodland on the southern edge of their range and (3) southern species on the edge of their range in Britain. The last group may occur in woodlands in the south but at the northern edge of their range they are confined to pasture woodlands (Harding and Rose 1986). Many of these southern species are rare because summer temperatures are not high enough for them to disperse readily. This complex mixture of species is a reflection of the position of Britain as a meeting place of climatic zones.

BUGS

About 1650 species of bug occur in the British Isles and most of them feed on plants; their mouth parts are adapted to sucking out sap. (Note that we use the term 'bug' in this more specific sense, rather than in the general sense, common in the USA and elsewhere, to mean any kind of insect.) Some feed on other soft-bodied animals. Many bugs are therefore pests, causing considerable damage to crops. As with other insects, more species occur in the south of Britain than in the north.

Bugs can be divided into two main groups depending on their wing structure. The fore wings, when present, in the first group are partially thicker and tougher nearer the body but membranous nearer the tip. These bugs are known as the Heteroptera as the two pairs of wings are different, and include the water bugs we have already mentioned when discussing aquatic insects. The second group are known as the Homoptera as there is no difference in the structure of the fore and hind wings. The familiar aphids are members of this group. As with other insects, flight provides an efficient means of dispersal, except in the few wingless forms. Another characteristic of many bugs is the ability to reproduce without a sexual stage. The process is called parthenogenesis. In plant aphids this leads to very rapid population increases, resulting in heavy damage to crops.

Heteropteran bugs

The heteropteran bugs are on the whole not as important economically as the homopterans and few are potential pests. Apart from the water bugs (the Corixids and pond skaters, which are found in a variety of pond types), the members of this group are terrestrial. The majority of these are found in the southern counties, preferring a continental type of climate that enables them to breed rapidly in the summer. As with other insects, damp winters exercise a high toll on numbers and cool summers reduce breeding success (Chinery 1979). A wide variety of plants are eaten, both by adults and larvae, and include grasses and flowers. Some heteropterans feed on trees and are specific to their hosts; examples are the pine flat bug *(Aradus cinnamomeus),* which feeds on pine sap, and the hawthorn shield bug *(Acanthosoma haemorrhoidale),* which feeds on hawthorn fruits and leaves. This group also contains predators, such as the assassin bugs that feed on other insects by sucking out the body fluids. Most heteropteran bugs will benefit by any measures taken to encourage butterflies, as a wide range of flowers, grasses and shrubs is then available. In common with other insect species, the impact of such measures in Britain would be greatest in the south of the country.

Homopteran bugs

The homopteran bugs are divided into two main groups: the first includes the cicadas and hoppers, which cause no appreciable damage, and the second the Psyllids, whiteflies, scale insects and aphids, which are all pest species.

The cicadas are southern species and the only British representative *(Cicadetta montana)* is found in the New Forest. The cool summers of the 1970s and 1980s have reduced the numbers of this species. The hoppers include the well known frog hoppers or 'cuckoo spit' insects that feed on grasses, as well as the leaf hoppers that feed on a wide range of plant material, including grasses, shrubs and trees; though particular species may be restricted to a few food plants.

The Psyllids, or plant lice, are a specialized group of small insects, and each species is confined to its own host plant. Whiteflies, so called from their white waxy appearance, are mainly southern species and cause damage, especially to greenhouse crops. The aphids form the largest group and it contains many pest species; these not only damage the crops themselves but also carry viruses from one plant to another as they suck sap from infected plants and inject it into uninfected plants. Many aphids produce a sweet secretion known as 'honey dew' on which ants, and sometimes butterflies, feed. Aphids breed very rapidly and as a result have evolved forms specific to their hosts. Some aphids produce galls on their hosts. The adelgid aphids are only found on conifers and some adelgid species are restricted to one tree genus, whilst others are more general feeders and can be found on many exotic coniferous trees in both forests and gardens. It can safely be said that aphids require little conservation but they do provide an important food source for many other animal species, especially birds.

Conservation measures for bugs

The bugs are a very successful group in general and for many species the need is for control rather than conservation. As we mentioned above, measures adopted to encourage butterflies will benefit the harmless bug species by providing a wide range of grasses and plants as food and suitable microclimates within the forest. The opportunities for conserving rare species are mainly in the south. The presence of species feeding on shrubs and trees will be determined by the type of woodland.

FLIES

Over 5000 species of fly are found in Britain. They feed on a very wide range of foods, including liquids; many, such as the mosquitoes, are blood-suckers. There are also harmless species that feed on nectar. Although many flies are undesirable species as they spread disease, the role played by the more useful species should not be forgotten. Pollination of many plants is effected by flies feeding on flowers; and predators, such as the robber flies, feed on harmful insects.

Fly larvae also use a wide range of foods and, in addition to predators and parasites, feed on decaying material (house flies), dung (dung flies), living animals (warble flies), plant roots (leatherjackets, which are the larvae of crane flies), fungi (the fungus gnats), and other plant material, including humus. All these species are not dependent on the type of trees present but rather on an abundant food source and suitable conditions provided within woodland. Many prefer the sunlit open areas as adult insects but require cooler and damper conditions in dense woodland for their larvae.

The gall midges, of which there are about 600 species, and the leaf miners, of which there are about 100 species, feed on trees and shrubs. Many of them are specific to the species that their larvae feed on. The amount of damage caused by these species is not significant compared with other insects. Their presence in a woodland is dictated by the tree species present and its geographical situation. A number of fly species depend upon decaying wood. The composition of fly communities in old pasture woodland follows a similar pattern to that of beetles described in an earlier section (Harding and Rose 1986).

Lacewing and alder flies (Neuroptera) are not true flies (Diptera) but they can be found in and near woods. Both their larvae and adults are predators, aphids being a common prey. Snake flies lay eggs in bark crevices, so smooth-barked trees and young trees offer few sites. The bark of conifers provides suitable places for eggs to be laid and so do some broadleaved species in the later stages of the rotation as the trees mature.

Conservation measures for flies

Little is known about many species and the requirements of adults and larvae differ. Many larvae feed on humus and in shady conditions, whilst the adults fly and feed in sunlit areas. Species feeding on rotting vegetation or humus will find sources under trees, whilst clearings, such as rides and cleared areas, will provide sunlit areas and feeding grounds for adult flies. The management of woodland for other insects will benefit the majority of fly species using woodland that are not dependent on the trees present.

Larvae feeding on the leaves of trees and shrubs will require the presence of the right species but succession of the woodland itself also provides an increasing variety of niches. For example, the species diversity and abundance of some leaf miners increases after coppicing, through to high forest (NCC 1987d). For many species of flies, however, control is more important than conservation because of their ability to breed rapidly when conditions are favourable.

ANTS, BEES, WASPS AND SAWFLIES

The order containing ants, bees, wasps and sawflies is divided into two main groups. The ants, bees and wasps have the familiar 'wasp waist' whilst the sawflies do not.

The majority of ants are heath or short grassland species and are warmth-loving; direct sunshine in May and June being more important to them than the overall temperature (Brian 1977). Most of the 42 British species are confined to the south of England. Woodlands can provide habitats for species requiring open conditions on unshaded ride sides and roads and on the woodland edge. Cleared areas, especially if bare ground is exposed, provide suitable conditions for a time for many species. Wood ants (*Formica* spp.) tolerate shade but build their nests in situations where they will receive maximum sunshine; they are to be found in open woods, especially of pine and larch on light soils. In Wales and Shropshire, the wood ant *Formica rufa* has colonized the edge of conifer plantations at altitudes up to 200 m, and *Formica lugubris*, a northern species that prefers cooler conditions and is more tolerant of shade, has colonized the edge of spruce plantations above 305 m (Hughes 1972). Wood ants are considered to be important in many countries to the control of the numbers of moth and sawfly larvae in pine, spruce and oak. Unmanaged broadleaved woods with a dense canopy are too shady for wood ants in the summer (Hughes 1972).

Most of the 469 species of bees and wasps are also confined in Britain to the southern parts of the country; many are on the edges of their range (Else et al. 1979). The hornet *Vespa crabo,* for example, is confined to southern England and has become scarcer in recent years as a result of climatic change. Bees and some wasps are important flower pollinators but most wasps are predators; some species are social and breed in colonies whilst others are solitary. Several of the solitary species require suitable soils in which to make nesting burrows, which they excavate themselves. Disturbance is therefore important for these species; for example, banks with exposed earth caused by erosion or excavation, roadside cuttings, bare ground, paths, windblown tree roots holding a vertical soil surface, and old walls are all used. Other species make use of cut plant stems. All bees and wasps need sunny and open areas; within forest areas, roadside banks, clear falls and wide rides provide suitable sheltered habitats. Ride management and selective weed control on restocked areas can provide the range and succession of flower species for honey and pollen collection.

Some wasp species are parasitic on other insects, so they are only present if their host species are also present. Species that parasitize wood-burrowing larvae have long ovipositors that enable them to reach their victims; *Rhyssa* is a well known example, its larva feeding on that of the great wood wasp *(Urocerus gigas),* which attacks conifers. Other wasp species form galls on various trees and shrubs and many of these are specific to their hosts, which may be either conifers or broadleaves. Oak hosts a number of species whose larvae all have different requirements and feed on different parts of the tree.Welch (1995) reports that exotic oaks are colonized by native oak gall wasps.

Sawfly caterpillars feed on a range of conifers and broadleaves as well as on a number of shrubs, and many species cause damage. Several species are confined to a specific host, whilst others are more general feeders. Some species are confined to either conifers or broadleaves. Coniferous species include the wood wasps, which bore into conifers and lay their eggs in the wood; usually the trees are dead or dying and damage is not significant, although felled timber is damaged if it is left lying about.

Conservation measures for ants bees, wasps and sawflies

As with other insect groups, ants, bees and wasps can be divided into those using habitats provided by a woodland and those dependent upon the trees themselves. For the first, the trees forming the woodland are not important. Sheltered sunny areas, especially south-facing ones, can be provided on roads, wide rides, clearings such as ride junctions, groups being regenerated and clear cuts. Nesting sites for the social bees and wasps can be provided by leaving hollow trees, and dead trees provide nesting material for wasps. Disturbance, which creates areas of bare soil for the solitary species excavating their own burrows, is important in keeping areas free of vegetation. The restricted use of chemicals for weed control will provide a wide range of flowers for pollen and nectar. Hand weeding of brambles and docks provides nesting sites in the cut stems for some species.

The gall wasps and sawflies are dependent on the presence of their host plant species, so their occurrence in a woodland is dictated by the tree species present. For many of these species, control is more important than conservation.

GRASSHOPPERS
Britain's 30 species of grasshoppers and crickets can hardly be regarded as true woodland insects but sunny, open areas in woodland, such as rides, banks and felled areas, provide conditions suitable for many of them. Some prefer longer grass than others. They are mainly southern species, the rarer forms being confined in Britain to the southern counties.

SPIDERS
About 600 species of spiders are found in Britain but the majority of these occur in open situations and are southern in distribution. Distribution patterns are similar to insect groups. A Nature Conservation Review (1977) lists 59 species found only in woodland but states that it is not certain 'that any species is confined to a particular type of woodland' and 'a northern spider which occurs only in Highland pine woods is really more typical of upland than woodland'. It would appear, therefore, that it is the degree of shade and woodland structure that is important in determining the presence of spiders. Spiders either hunt their prey or build webs (orb webs or sheet webs) in which to catch it. It has been shown that the diversity of species of both sheet and orb spiders increases with succession as the woodland structure develops (NCC 1987d). Short rotations and coppicing prevent colonization by species arriving late in succession.

Conservation measures for spiders
There appear to be no records of spider populations and species in plantation woodlands but management itself, by varying the structure, must provide suitable habitats for many species.

SNAILS AND SLUGS
Few of the 116 British land snails and slugs occur in the uplands where soils are acid

and conditions cool. Most are to be found in the south of Britain on non-acid soils and they include 21 introduced species. Some southern species are at the edge of their range in Britain but were formerly more widespread (Kerney and Stubbs 1980). Of the 95 'native' species, 49 use woodland as their main habitat but in a study quoted by Peterken (1981) 'no species was absolutely confined to woodland but many were found mainly in a series of relatively sheltered, undisturbed, semi-natural habitats'. Retention of adequate cover, irrespective of vegetation type, appears therefore to be the most important factor. A wide range of foods, both living and dead vegetable matter and decaying animal matter, is eaten by most species. Woodlands with good herb and litter layers, together with dead branches lying on the woodland floor, provide the damp conditions needed by most species and those situated on the more basic soils sustain the greatest numbers. Damp areas near ponds, ditches and other water sources also provide a suitable habitat. Snails and slugs dislike short vegetation and any habitat disturbance which causes drying out, though snails are able to survive periods of desiccation. Most species only feed at night or in humid conditions. The greatest diversity of species occurs in situations with a diverse structure and where tree and shrub cover has not been totally removed; for example, in primary woodland (Kerney and Stubbs 1980). As well as the land snails, 74 species of snails and mussels are to be found in freshwater in Britain, so ponds within woodland can also provide a habitat for some species.

Conservation measures for snails and slugs

As with spiders, there is no information on slug and snail species in managed woodland of any kind but the presence of a herbaceous layer providing damp conditions is important for many species. Slugs and snails can, however, cause considerable damage to young, naturally regenerated seedlings. Control is not usually undertaken in woodland, though metaldehyde is used in nurseries.

OTHER INVERTEBRATES

Woodland also provides a damp habitat for centipedes (mainly predators), millipedes, woodlice and other soil and litter animals. Fallen logs, brash and stones all provide damp hiding places for many of these species, as humidity and soil moisture are more important to them than the type of woodland (Ellenberg 1988). A study in China (Zhang et al. 1987) of soil invertebrates in temperate mountain forests showed that all groups studied were present in all forest types, with the exception of warmth-loving genera, which were restricted to the lower slopes. Total numbers of individual species decreased with altitude. The majority of soil invertebrates were present between the litter layer on the soil surface and just below the soil surface but worms and springtails were found at greater depths in the soil. Springtails, beetles and mites were more abundant in coniferous forest, with nematode worms the most abundant group in broadleaved forest.

GENERAL PRINCIPLES FOR THE CONSERVATION OF INSECTS AND OTHER INVERTEBRATES

It is evident from the above accounts that insects and other invertebrates exploit a wide range of habitats, and that individual species either exploit a wide range of foods or are specific to the trees and shrubs that their larvae feed on. It is also clear that the majority of species are found in the south of the country, and that many of them are on the edge of their ranges (either at their northern or western limits) because they need more continental conditions with hotter summers than normally occur in Britain. Insects in particular are very sensitive to small climate changes compared with other animal groups.

The insect species of a particular woodland will therefore depend on several factors:

1 *The geographical position,* which will determine the insect population found within it. Southern and lowland woods that are able to grow both conifers and broadleaves have the potential to hold a wide range of species, as most insects are more limited by temperature and sunshine than are other animal groups and the physiological parameters within which they live are more restricting. The rarest species are confined to the southern counties and are on the northern edge of their range in Britain. Other local species confined to the south and east in Britain have a northern distribution elsewhere but require a continental climate with hot summers. In lowland areas a wide range of tree species can be grown, thus increasing the conservation potential of managed woodland. Conversely, northern species of insects do not extend into southern Britain but many are now found in coniferous plantations in the north. Little work has been undertaken to record and monitor the spread of northern insect species, other than pest species, in coniferous plantations of both native and exotic tree species. Insects with a general distribution, many of which can be found in all kinds of plantations, have also received little attention.

2 For both insects and other invertebrates, *a succession of various stages,* from clear felling to mature woodland and including the presence of dead and dying trees. Trees do not need to be old or overmature to attract most species of insect, the wood of dead, dying and diseased trees being attacked from the pole stage onwards. There are, however, some species that will only occur in logs that have been felled for some time and others that feed on fungi growing on old wood. Some use fallen trees as protection and as a damp habitat. The latter is important for many other invertebrate species.

3 *The presence of other food sources* required at other stages of the life cycle, such as flowers supplying nectar to many adult insects. Such food sources are important as they enable the adult insects to survive for a longer period than would be possible without them and thus maximize reproduction.

4 *The range of plant, tree and shrub species,* including both conifers and

broadleaves, available for larval food. Many of the insects feeding on trees and shrubs, on foliage, seeds, wood or roots, cause damage and some of them become pests under certain conditions. Control and not conservation is then required. The use of chemical control requires careful application because of effects on other insects. Biological control using specific parasites and viruses has less of an impact, but destruction of the pest species is not so immediate.

In general, conservation for insects falls into two main divisions. The first includes provision for those that are not dependent upon the species of tree forming the woodland but that use the microclimates and shelter provided within it. The majority of other invertebrate species also fall into this category. The second approach to conservation provides for insects feeding on trees and shrubs, their presence being dependent on the tree and shrub species present. Most shrub food plants are woodland edge species, so the trees composing the woodland are not of particular importance.

For insects not dependent on the trees composing the woodland, but on the habitats and foods found within it, there is considerable scope to widen the range of habitat available in all types of woodland. Such species can be encouraged by ordinary forest management, as it can provide a continuous succession of varying conditions within the woodland and can maximize other factors on which a particular species depends. For example, favourable conditions can be created by ride management, to encourage the appropriate flowering plants for larva (to feed upon) and adults (to collect pollen and nectar), and by making large enough groups for regeneration to provide sunlit 'glades'.

Roads, rides (preferably curving to avoid 'wind tunnels'), woodland edge and sunny dry banks, all provide conditions suiting many insects. Morning sunshine is particularly important for basking as it enables insects to 'warm up' for the day. The value of sheltered, sunlit areas for many insects must therefore be borne in mind. On flat ground, east/west rides provide maximum sunlight but on sloping ground north/south rides are better (Fry and Lonsdale 1991). In hilly woodland areas, the provision of open ground for insect conservation needs to be concentrated on south-facing slopes. Once insects have 'warmed up', other parts of the wood can then be used. Clear falls provide habitat for open-ground species. If not smothered by lop and top, herbaceous growth on restocked sites produces a range of plants and grasses attractive to insects. Windrowing or burning lop and top encourages the regrowth of herbaceous vegetation. Methods of ride management have been developed to maintain meadow species important to a wide variety of insects and the wide mown rides provided for pheasants provide conservation benefits with no extra cost. Other forest operations, such as the extraction of thinnings, create bare ground which gives certain flowers and grasses an opportunity to colonize and also provides nesting places for some insects, such as solitary bees. Thinnings, large logs and lop and top, left on the ground, all provide insect habitats.

Different stages of the rotation will produce further habitats, and the opening out of trees by thinning allows sunlight to penetrate and to raise the low temperatures typical of dense plantations. Light penetration into woodland is important for insects; in the summer when adult insects are flying, even a broadleaved canopy intercepts over 70 per cent of the sunlight. The value of 'skylighting' for pheasants (that is, the creation of small patches within broadleaved woodland to receive full sunlight) has been shown to attract many species of butterfly. Group regeneration will provide a similar effect. Regular crown thinning will assist sunlight penetration and the irregular intensity of light reaching the ground benefits many insects.

The provision of water in the form of ponds benefits a wide range of aquatic insects. Ponds for insects should receive maximum sunlight but the value of some overhanging trees as resting places should not be overlooked. Such trees also provide a source of food for aquatic species in the form of insect food falling into the water and plant food from leaves. Streams and ponds already present also need variation in light and shade, so some opening out may be needed in an existing wood, and open unplanted areas at water edges could be provided.

Consideration should also be given to reintroducing species into suitable areas, particularly where they have been found in the past. Advice should be sought and licences obtained. The usual argument against such introductions is the effect on the genetics of the native population. However, the inherent variability of most animal (and plant) species in response to prevailing conditions cannot be underestimated. The Dutch race of the large copper, reintroduced in Cambridgeshire, has 'diverged considerably from the ancestral Dutch population and many of the changes (especially in the hind wing pattern) are in the direction of the extinct British race' (Berry 1977).

The introduction of non-native species is not usually wise, except where predators can be released to control an exotic pest. The dangers of introducing non-native species are well illustrated by the case of the gipsy moth in the USA: after its introduction at the beginning of the twentieth century, the moth escaped into the wild and is now a major pest, causing severe defoliation on many tree species.

SUMMARY

Part II of this book has demonstrated the adaptability of most species of animals and plants that inhabit woodland. It shows that the constantly changing conditions occurring in managed, productive forest offer opportunities for extending conservation. It also illustrates that for many species of animals and plants, the habitat provided by the woodland structure and its management is paramount, rather than the tree species forming the woodland. The provision of ecotones (transitional conditions between habitats), formed both by internal edges within the forest or woodland, or particularly by the external edge, will sustain a high number of species. Thus, ancient semi-natural woodland should by no means be regarded as the only, or indeed the primary, woodland conservation resource: managed forests have an increasingly important part to play.

Bibliography

Academia Sinica et al. 1978. *Flora,* 7. Science Press.

Aichele, D. 1984. *Wildflowers.* Octopus.

Alexander, I. and Watling, R. 1987. Macrofungi of Sitka spruce in Scotland. In: *Sitka Spruce,* edited by D. M. Henderson and R. Faulkner. *Proceedings of The Royal Society of Edinburgh,* 93B.

Allen, D. E. 1978. *The Naturalist in Britain. Penguin.*

Anderson, M. A. 1979. The development of plant habitats under exotic forest crops. In: *Ecology and Design in Amenity Land Management,* edited by S. Wright and P. Buckley. Wye College Symposium.

Anderson, M. A. 1987. The effects of forest plantations on some lowland soils. *Forestry,* 60.

Anderson, M. A. and Carter, C. I. 1987. Shaping ride sides to benefit wild plants and butterflies. In: *Wildlife Management in Forests,* edited by D. C. Jardine. Lancaster discussion meeting; Institute of Chartered Foresters.

Anderson, M. A. and Carter, C. I. 1988. Duration of daylight length and level of energy input to sunlit ridesides. *Forest Research 1988.* Forestry Commission, HMSO.

Anderson, M. L. 1967. *A History of Scottish Forestry.* Nelson.

Art, H. W. 1989. 'This Mountain was Exceeding Good Land': A Journal, vol. 6, Centre for Environmental Studies, New England, USA.

Aune, E. I. 1977. Scandinavian pine forests and their relationship to Scottish pinewoods. In: *Native Pinewoods in Scotland,* edited by R. G. H. Bunce and J. N. R. Jeffers. Aviemore Symposium; Institute of Terrestrial Ecology.

Avery, M. and Leslie, R. 1990. *Birds and Forestry.* T & A Poyser, London.

Baker, C. A., Moxey, P. A. and Oxford, P. M. 1978. Woodland Continuity and Change in Epping Forest. *Field Studies,* 4.

Ballasina, D. 1984. *Amphibians of Europe.* David & Charles.

Bamford, R. 1987. Forestry and ferns. *Quarterly Journal of Forestry,* 81.

Bamford, R. 1991. Nest box usage in high elevation plantations. *Quarterly Journal of Forestry,* 85(3).

Barry, R. G. and Chorley, R. J. 1976. *Atmosphere, Weather and Climate.* Methuen.

Belovsky, I. 1983. Foraging models and territory size. *Nature,* 305.

Berry, R. J. 1977. *Inheritance and Natural History.* Collins.

Berry, R. J. 1981. Town Mouse, Country Mouse. *Mammal Review,* 11.

Bevan D. 1987. Forest Insects. *Forestry Commission Handbook 1*, HMSO.

Bibby, C. 1987. Effects of management of commercial conifer plantations on birds. In: *Environmental Aspects of Plantation Forestry in Wales*, edited by J. E. G. Good. ITE Symposium no. 22; Institute of Terrestrial Ecology.

Bibby, C. 1988. Management in commercial conifers for birds. In: *Wildlife Management in Forests*, edited by D. C. Jardine. Lancaster discussion meeting; Institute of Chartered Foresters.

Blandford, P. R. S. 1987. Biology of the polecat: A literature review. *Mammal Review*, 17.

Bolund, L. 1987. *Nestboxes for the Birds of Great Britain and Europe*. Sainsbury.

Bon, M. 1987. *The Mushrooms and Toadstools of Britain and North-west Europe*. Hodder & Stoughton.

Brambell, F. W. R. 1974. Voles and field mice. *Forestry Commission Record 90*. HMSO .

Brian, M. V. 1977. *Ants*. Collins.

Brian, M. V. 1979. *Ant Hill*. Faber and Faber.

Brown, A. H. F. 1981. The recovery of ground vegetation in coppice wood: The significance of buried seed. In: *Forest and Woodland Ecology*, edited by F. T. Last and A. S. Gardiner. ITE Symposium no. 8; Institute of Terrestrial Ecology.

Brown, A. H. F. 1987. The effects of tree mixtures on soil and tree growth and the role of buried seed in the dynamics of woodland vegetation. In: *Report of the first meeting of The Uneven Aged Silviculture Group*, edited by D. R. Helliwell (members' report).

Brown, U. K. and Hyman, P. S. 1986. Successional communities of plants and phytophagous Coleoptera. *Journal of Ecology*, 74.

Bunce, R. G. H. 1977. The range of variation within the pinewoods. In: *Native Pinewoods of Scotland*, edited by R. G. H. Bunce and J. N. R. Jeffers. ITE Aviemore Symposium; Institute of Terrestrial Ecology.

Bunce, R. H. G. 1982. *A Field Key for Classifying British Woodland Vegetation*, part 1, Institute of Terrestrial Ecology.

Bunn, D. S., Warburton, A. B. and Wilson, R. D. 1982. *The Barn Owl*. Poyser.

Burnett, J. H. 1964. *The Vegetation of Scotland*. Oliver & Boyd.

Buse, A. 1980. The distribution of beetles in some upland habitat types. In: *1979 Annual Report*. Institute of Terrestrial Ecology.

Cadbury, C. J. and Everett, M. 1987. *RSPB Conservation Review*. Royal Society for the Protection of Birds.

Cannell, M. G. R., Grace, J. and Booth, A. 1989. Possible impacts of climatic warming on trees and forests in the United Kingdom. *Forestry*, 62(4).

Carter, C. I. and Anderson, M. A. 1987. Enhancement of lowland forest ridesides and roadsides to benefit wild plants and butterflies. *Forestry Commission Research Information Note 126*. Forestry Commission, Alice Holt.

Carter, D. 1982. *Butterflies and Moths*. Pan.

Chabot, B. F. and Mooney, H. A. 1985. *Physiological Ecology of North American Plant Communities and Short Succession*. Chapman & Hall.

Chapman, N. D. 1978. *Fallow Deer*. British Deer Society Publication No. 1.

Chapman and Crawford 1985. Growth and regeneration in Britain's most northerly natural woodland. *Botanical Society of Edinburgh*, 43.

Chapman, N., Harris, S. and Stanford, A. 1994. Reeves Muntjac in Britain. *Mammal Review*, 24(3).

Champness, S. S. and Moss, K. 1948. The population of buried viable seeds. *Journal of Ecology*, 36.

Changbai Mountain Research Station 1980. *An Enumeration of the Plants of Changbai Mountain.* Academia Sinica.

Charles, W. J. 1981. Abundance of field voles in conifer plantations. In: *Forest and Woodland Ecology,* edited by F. T. Last and A. S. Gardiner. ITE Symposium no. 8; Institute of Terrestrial Ecology.

Cheeseman, C. L., Little, T. W. A., Mallinson, P.J., Page R. J. C., Wilesmith, J. W. and Pritchard, D. G. 1985. Population ecology and prevalence of tuberculosis in Badgers in an area of Staffordshire. *Mammal Review,* 15.

Chelmik, D., Hammond, C., Moore, N. and Stubbs, A. 1980. *The Conservation of Dragonflies.* Nature Conservancy Council.

Chinery, M. 1979. *Field Guide to the Insects of Britain.* Collins.

Christenson, P. E. and Kimber, P. C. 1975. Effects of prescribed burning on the flora and fauna of south-west Australian forest. In: *Managing Terrestrial Ecosystems. Proceedings of the Ecological Society of Australia,* 9.

Churchfield, J. S. 1979. A note on the diet of the European water shrew. *Mammal Society Notes,* 38. Mammal Society.

Clapham, A. R., Tutin, T. G. and Warburg, E. F. 1952. *Flora of the British Isles.* Cambridge University Press.

Clements, E. D., Neal, E. G. and Yalden, D. W. 1988. The National Badger Sett Survey. *Mammal Review,* 18.

Collier, R. 1988. From mountain to ashwood. *Bird Watching.*

Collin, N. N. 1987. The Bradford-Hutt continuous cover forestry system. In: *Report of the first meeting of The Uneven Aged Silviculture Group,* edited by D. R. Helliwell (members' report).

Conner, E. F., Faeth, S. H., Simberloff, D. and Opler, P. A. 1980. Taxonomic isolation and the accumulation of herbivorous insects: A comparison of introduced and native trees. *Ecological Entomology,* 5.

Coombe, S. E. 1957. The spectral composition of shade light in woodland. *Journal of Ecology,* 45.

Corbet, G. B. and Southern, H. N. 1977. *The Handbook of British Mammals.* Blackwell Scientific Publications.

Corbet, P. S., Longfield, C. and Moore, N. 1960. *Dragonflies.* Collins.

Corke, D. 1972. The distribution of *Apodemus flavicollis* in Britain. *Mammal Review,* 7.

Cotton, M. J. and Griffiths, D. A. 1967. Observations on temperature conditions in vole nests. *Mammal Society Notes,* 15.

Country Life 1988. Editorial: Nature reserves: Public or private?

Cramp, S. et al. 1977 et seq. *The Birds of the Western Palearctic,* volumes 1-5. Oxford University Press.

Cresswell, W. J. and Harris, S. 1988. Foraging behaviour and home-range utilisation in a suburban badger population. *Mammal Review,* 18.

Crooke, M. 1979. The development of populations of insects. In: *The Ecology of Even Aged Forest Plantations,* edited by Ford, E. D., Malcolm, D. C. and Atterson, J. (IUFRO). Institute of Terrestrial Ecology.

Currie, F. A. and Bamford, R. 1981. Bird populations in sample pre-thicket forest plantations. *Quarterly Journal of Forestry,* 75(1).

Dansie, O. 1970. *Muntjac.* British Deer Society publication no. 2.

Darvill, T. 1986. *The Archaeology of the Uplands.* The Council for British Archaeology.

Davies, R. J. and Pepper, H. W. 1987. Protecting trees from field voles. *Arboricultural*

Research Note 74. Forestry Commission, Alice Holt.

Day, M. G. 1968. Food habits of British stoats and weasels. *Journal of Zoology,* 155.

Degn, J. 1974. Feeding activity in the red squirrel. *Mammal Society Notes,* 29.

Delmas R. J., Scencio, J-M. and Legrand, M. 1980. Polar ice evidence that atmospheric CO_2 20 000 BP was 50% of present. *Nature,* 284.

Dennis, R. H. 1987. Boxes for goldeneye: A success story. *RSPB Conservation Review.* Royal Society for the Protection of Birds.

Dent, S. and Spellerberg, I. F. 1987. Habitats of the lizards *Lacerta agilis* and *L. vivipara* on forest ride verges in Britain. *Biological Conservation,* 42.

De Schaunsee, R. M. 1984. *The Birds of China.* Oxford University Press.

DFS 1949. *Native Trees of Canada.* Dominion Forest Service, Ottawa.

Dickman, C. R. 1980. Estimation of population density in the common shrew from a conifer plantation. *Journal of Zoology,* 192.

Dimbleby, G. W. 1952. Soil regeneration on the north-east Yorkshire moors. *Journal of Ecology,* 40.

Dimbleby, G. W. 1965. Post glacial changes in soil profiles. *Proceedings of The Royal Society,* B161.

Dimbleby, G. W. 1978. *Plants and Archaeology.* Granada.

Dimbleby, G. W. and Gill, J. 1957. The occurrence of podsols under deciduous woodland in the New Forest. *Forestry,* 28.

Disney, R. H. L. 1986. Assessments using invertebrates. In: *Wildlife Conservation Evaluation,* edited by M. Usher. Chapman & Hall.

Don, B. A. C. 1979. Gut analysis of small mammals during a sawfly outbreak. *Mammal Society Notes,* 38.

Don, B. A. C. 1983. Home range characteristics and correlates in tree squirrels. *Mammal Review,* 13.

Dunwell, M. R. and Killingsley, A. 1969. The distribution of badger setts in relation to the geology of the Chilterns. *Mammal Society Notes,* 18.

Eldridge, M. J. 1968. Live trapping. *Mammal Society Notes,* 16.

Eldridge, M. J. 1969. Observations on food eaten by wood mice and voles in a hedge. *Mammal Society Notes,* 18.

Eldridge, M. J. 1971. Some observations on the dispersal of small mammals in hedgerows. *Mammal Society Notes,* 23.

Ellenberg, H. 1988. *Vegetation Ecology of Central Europe.* Cambridge University Press.

Ellenberg, H. and Klotzli, F. 1972. *Waldgesellschaften und Waldstandorte der Schweiz.* Zurich.

Else, G., Felton, J. and Stubbs, A. 1979. *The Conservation of Bees and Wasps.* Nature Conservancy Council.

Evans, H. 1977. *The Life of Beetles.* Allen & Unwin.

Evans, H. F. 1987. Sitka spruce insects. In: *Sitka Spruce,* edited by D. M. Henderson and R. Faulkner. *Proceedings of the Royal Society of Edinburgh.* 93B.

Everett, M. 1995. Death threat. *Bird Watching,* September.

Eyre, R. 1984. *Vegetation and Soils.* London.

Fairburn, W. 1968. Climatic zonation in the British Isles. *Forestry,* 41.

FAO 1982. *Forests in China.* FAO.

Fearson, K. and Weiss, N. D. 1987. Improved growth rates within tree shelters. *Quarterly Journal of Forestry,* 81.

Finegan, B. 1984. Forest Succession. *Nature,* 312.

Hunter, M. L. 1990. *Wildlife, Forests, and Forestry.* Prentice Hall, New Jersey, USA.

Huntley, B. and Birks, J. B. 1983. *An Atlas of Past and Present Pollen Maps of Europe. 0-13000 Years Ago.* Cambridge University Press.

Hurrell, H. G. 1968. Pine martens. *Forestry Commission Record, 64.* HMSO.

Hurrell, E. and McIntosh, G. 1984. The Mammal Society dormouse survey 1975- 1979. *Mammal Review, 14.*

Hvass, H. 1978. *Reptiles and Amphibians.* Blandford.

Hyams, E. 1979. *The Story of England's Flora.* Kestrel.

Insley, H. 1977. An estimate of the population density of the red fox in the New Forest. *Mammal Society Notes, 35.*

Institute of Terrestrial Ecology. 1979. *Annual Report.*

IUCN 1980. *World Conservation Strategy.* International Union for the Conservation of Nature.

Jackson, J. 1974. Feeding habits of deer. *Mammal Review, 4.*

Jackson, J. 1977. A note on the food of muntjac. *Mammal Society Notes, 4.*

Jackson, J. 1994. The edible or fat dormouse in Britain. *Quarterly Journal of Forestry, 88(2).*

Jahns, H. M. 1983. *Ferns, Mosses and Lichens.* Collins.

James, N. D. G. 1981. *A History of British Forestry.* Basil Blackwell.

Jardine, D. C. 1988. Wildlife management in forests. Lancaster Symposium; Institute of Chartered Foresters.

Jefferies, D. J. 1974. Earthworms in the diet of the red fox. *Mammal Society Notes, 28.*

Jenkins, D. 1986. Trees and wildlife in the Scottish uplands. ITE Symposium no. 17; Institute of Terrestrial Ecology.

Jennings, T. J. 1975. Notes on the burrowing systems of wood mice. *Mammal Society Notes, 31.*

Johnson, H. 1984. *Encyclopaedia of Trees.* Mitchell Beazley.

Jones, E. W. 1974. Introduction. In: *The British Oak,* edited by M. G. Morris and F. H. Perring. Classey.

Joyce, C. 1988. Global warming could wipe out wildlife. *New Scientist.*

Judes, U. 1982. Bibliography of wood mouse studies. *Mammal Review, 12.*

Kallio, M., Dykstra, D. P. and Binkley, C. S. 1987. *The Global Forest Sector.* Wiley.

Karnil, A. C., Krebs, J. R. and Pulliam, H. R. 1987. *Foraging Behaviour.* Plenum.

Kauppi, P. 1987. Forests and the changing chemical composition of the atmosphere. In: The *Global Forest Sector,* edited by M. Kallio, D. Dykstra and C. Binkley. Wiley.

Keble Martin, W. 1965. *The Concise British Flora in Colour.* Ebury Press/ Michael Joseph.

Kennedy, C. E. J. and Southwood, T. R. E. 1984. The number of species of insects associated with British trees: A re-analysis. *Journal of Animal Ecology, 53.*

Kenward, R. E. 1983. The cause of damage by red and grey squirrels. *Mammal Review, 13.*

Kenward, R. E. and Parish, T. 1986. Bark stripping by grey squirrels. *Journal of Zoology, 210.*

Kenward, R. E., Parish, T., Holm, J. and Harris, E. H. M. 1988a. Grey squirrel bark stripping 1. *Quarterly Journal of Forestry, 82(1).*

Kenward, R. E., Parish, T. and Dyle, F. I. B. 1988b. Grey squirrel bark stripping 2. *Quarterly Journal of Forestry, 82(2).*

Kenward, R. E., Dutton, J. C. F., Parish T., Doyle, F. I. B., Walls, S. S. and Robertson, P. A. 1996. Damage by grey squirrels 1 & 2. *Quarterly Journal of Forestry, 90 (1/2).*

Kerney, M. and Stubbs, A. 1980. *The Conservation of Snails, Slugs and Freshwater Mussels.* Nature Conservancy Council.

Fitter, R. 1959. *The Ark in our Midst.* Collins.

Flint, V., Boehme, R. L., Kostin, Y. V. and Kuznetsov, A. A. 1984. *Birds of the USSR.* Princeton University Press.

Flower, R.J. and Batterbee, R. W. 1983. Diatom evidence for recent acidification of two Scottish lochs. *Nature, 305.*

Flowerdew, J. R. 1976. The effects of a local increase in food supply on the distribution of woodland mice in oaks. *Mammal Society Notes, 32.*

Forbes, H. and Lance, A. N. 1976. The contents of fox scats. *Mammal Society Notes, 32.*

Ford, E. B. 1945. *Butterflies.* Collins.

Ford, H. 1987. Bird communities on habitat islands in England. *Bird Study, 34.*

Ford, E. D., Malcolm, D. C. and Atterson, J. (eds) 1979. Ecology of Even Aged Forest Plantations (IUFRO). Institute of Terrestrial Ecology.

Forestry Authority. 1994a. *Native Pinewoods.* Forestry Practice Guide 7. Forestry Commission, Edinburgh.

Forestry Authority. 1994b. *Caledonian Pinewood Inventory.* Leaflet, The Forestry Authority Scotland, Glasgow.

Forestry Commission 1987. *Facts and Figures, 1986-87.* Edinburgh.

Forestry Commission 1988a. *Forests and Water: Guidelines.* Edinburgh.

Forestry Commission 1988b. *Facts and Figures, 1987-1988.* Edinburgh.

Frank, L. G. 1979. Selective predation and seasonal variation in the diet of the fox. *Mammal Society Notes, 38.*

Frazer, D. 1983. *Reptiles and Amphibians.* Collins.

French, D. D., Jenkins, D. and Conroy, J. W. H. 1986. Guidelines for managing woods in Aberdeenshire for song birds. In: *Trees and Wildlife in the Scottish Uplands,* edited by D. Jenkins. ITE Symposium no. 17; Institute of Terrestrial Ecology .

Fry, R. and Lonsdale, D. 1991. Habitat conservation for insects. *Amateur Entomologist, 21.*

Fuller, R. J. and Langslow, D. R. 1986. Ornithological evaluation. In: *Wildlife Conservation Evaluation,* edited by M. Usher. Chapman & Hall.

Fuller, R. J. and Taylor, K. 1984. Breeding birds and woodland management in some Lincolnshire limewoods. In: *Research Reports Digest 8.* Nature Conservancy Council.

Fuller, R. J. 1995. *Bird Life of Woodland and Forest.* Cambridge University Press.

Garwood, N., Janos, D. P. and Brokaw, N. 1979. Earthquake caused landslides. *Science, 205.*

Gill, R. M. A. 1992. A review of damage by mammals in north temperate forests 1 & 2. *Forestry, 65(2/3).*

Gilmour, J. and Walters, M. 1954. *Wild Flowers.* Collins.

Glue, D. 197S. Harvest mice as barn owl prey in the British Isles. *Mammal Review, 5.*

Godwin, H. 1956. *History of the British Flora.* Cambridge University Press.

Godwin, H. 1975. History of the natural forests of Britain. *Philosophical Transactions of the Royal Society, 271.*

Goldsmith, B. 1983. Ecological effects of upland afforestation. In: *Conservation in Perspective,* edited by A. Warren and B. Goldsmith. Wiley.

Good, J. E. G. 1987. Environmental Aspects of Plantation Forestry in Wales. ITE Symposium no. 22; Institute of Terrestrial Ecology.

Gorman, M. L. and Robertson, J. 1981. *Mammal Society Notes, 43.*

Grant, W. and Mitchell, B. 1981. Notes on the performance of red deer in woodland habitat. *Mammal Society Notes, 42.*

Gurnell, J. 1979. Woodland mice. *Forest Record 118.* Forestry Commission, HMSO.

Gurnell, J. 1983. Squirrel numbers and the abundance of tree seeds. *Mammal Review, 18.*

Gurnell, J. 1987. *Squirrels.* Christopher Helm.

Gurnell, J. and Pepper, H. 1988. Perspectives on the management of red and grey squirrels. In: *Wildlife Management in Forests,* edited by D. C. Jardine. Lancaster discussion meeting; Institute of Chartered Foresters.

Gurnell, J. and Pepper, H. 1993. A critical look at conserving the red squirrel. *Mammal Review,* 23(3/4).

Hale Mason, E. 1974. *The Biology of Lichens.* Edward Arnold.

Hambler, C. and Speight, M. R. 1995. Seeing the wood for the trees. *Tree News,* Autumn. The Tree Council, London.

Harde, K. N. *Beetles.* Octopus.

Harding, P. T. and Rose, F. 1986. *Pasture Woodlands in Lowland Britain.* Institute of Terrestrial Ecology.

Harris, E. H. M. 1983. *Forestry and Conservation.* Royal Forestry Society, Tring.

Harris, E. H. M. 1985. Forestry in Austria: The Society Visit, Sept. 1983. *Quarterly Journal of Forestry,* 79.

Harris, E. H. M. 1986. The case for Sycamore. *Quarterly Journal of Forestry,* 80.

Harris, E. H. M. and Harris, J. A. 1981. *The Guinness Book of Trees.* Guinness.

Harris, E. H. M. and Harris, J. A. 1987. *A Glimpse of Forestry and the Countryside of China.* Royal Forestry Society, Tring.

Harris, E. H. M. and Harris, J. A. 1988a. The Society Visit to Poland, September 1987. *Quarterly Journal of Forestry,* 82(3).

Harris, E. H. M. and Harris, J. A. 1988b. The Society visit to Hungary, Autumn 1987. *Quarterly Journal of Forestry,* 82(2).

Harris, E. H. M. and Harris, J. A. 1988c. A visit to Russia. *Quarterly Journal of Forestry,* 82(4).

Harris, E. H. M. and Harris, J. A. 1990. Society study tour, Eastern USA. *Quarterly Journal of Forestry,* 84(2).

Harris, J. A. 1960. Notes on the Capercaillie. *Scottish Birds,* 1.

Harris, J. A. 1983. *Birds and Coniferous Plantations.* Royal Forestry Society, Tring.

Harris, J. A. 1984. Spruce and wildlife: Some thoughts on the Austrian tour. *Quarterly Journal of Forestry,* 78.

Harris, J. A. 1987. Origins of British sika. *Deer: Journal of the British Deer Society,* 7.

Harris, M. J. and Kent, M. 1987. Ecological benefits of the Bradford-Hutt system of commercial forestry, 1 and 2. *Quarterly Journal of Forestry,* 81.

Harris, S. 1977. The food and suburban foxes. *Mammal Review,* 7.

Harris, S. 1979. History, distribution, status and habitat requirements of the harvest mouse. *Mammal Review,* 9.

Harrison Matthews, L. 19S2. *British Mammals.* Collins.

Hart, C. 1995. Alternative silvicultural systems to clear cutting in Britain: a review. *Forestry Commission Bulletin* 115. HMSO.

Healing, T. D. 1980. The dispersal of bank voles and mice in dry stone dykes. *Mammal Notes,* 40.

Heimburger, C. C. 1934. *Forest Type Studies in the Adirondack Region,* Memoir 165, Cornell, Ithaca.

Heinselman, M. L. 1973. Fire in the virgin forests of the Boundary Waters Canoe area, Minnesota. in: *The Ecological Role of Fire in Natural Conifer Forests of Western and Northern America,* edited H. E. Wright and M. L. Heinselman. Quaternary Research, 3, Academic Press.

Henderson, D. M. and Faulkner, R. 1987. Sitka spruce. *Proceedings of the Royal Society of Edinburgh,* 93B.

Henry, B. A. M. 1981. Distribution patterns of roe deer. *Mammal Society Notes,* 52.

Hewson, R. 1972. Changes in the number of stoats. *Mammal Society Notes,* 25.

Hewson, R. 1977. Browsing by mountain hares. *Mammal Society Notes,* 34.

Hewson, R. and Healing, T. D. 1971. The stoat and its prey. *Mammal Society Notes,* 22.

Hewson, R. and Kolb, H. 1975. The food of foxes in Scottish forests. *Mammal Society Notes,* 30.

Higgins, L. G. and Riley, N. D. 1983. *A Field Guide to the Butterflies of Britain and Europe.* Collins.

Hill, M. O., Bunce, R. G. H. and Shaw, M. W. 1975. Indicator Species Analysis: a divisive polythetic method of classification and its application to a survey of native pinewoods in Scotland. *Journal of Ecology,* 63.

Hill, M. O. 1979. The development of a flora in even aged plantations. In: *The Ecology of Even Aged Forest Plantations,* edited by E. D. Ford, D. C. Malcolm and J. Atterson (IUFRO). Institute of Terrestrial Ecology.

Hill, M. O. 1986. Ground flora and succession in commercial forests. In: *Trees and Wildlife in the Scottish Uplands,* edited by D. Jenkins. ITE Symposium no. 17; Institute of Terrestrial Ecology.

Hill, M. O. 1987a. Opportunities for vegetation management in plantation forests. In: *Environmental Aspects of Plantation Forestry in Wales,* edited by J. E. G. Good. ITE Symposium no. 22; Institute of Terrestrial Ecology.

Hill, M. O. 1987b. Quoted by K. Kirby at Nature Conservancy Council Conference, Oxford.

Hill, M. O. and Jones, E. W. 1978. Vegetation changes resulting from afforestation of rough grazings in Caeo Forest. *Journal of Ecology,* 66.

Hill, M. O. and Stevens, P. A. 1981. The density of viable seed in the soils of forest plantations in upland Britain. *Journal of Ecology,* 69.

HMSO. 1994a. *Biodiversity: The UK Action Plan.*

HMSO. 1994b. *Sustainable Forestry: The UK Action Plan.*

HMSO. 1995. *Biodiversity: The UK Action Plan (revised).*

Hodkinson, I. D. and Hughes, M. K. 1982. *Insect Herbivory.* Chapman & Hall.

Hoodless, A. and Morris, P. A. 1993. An estimate of population density of the fat dormouse. *Communications from the Mammal Society,* 66.

Hora, B. 1981. *Encyclopaedia of Trees.* Oxford University Press.

Hora, B. 1959 and 1972. Quoted in *The British Oak,* edited by M. G. Morris and F. H. Perring. Classey.

Hornung, M., Stevens, P. A. and Reynolds, B. 1987. The effects of forestry on soils, soil water and surface chemistry. In: *Environmental Aspects of Plantation Forestry in Wales,* edited by J. E. G. Good. ITE Symposium 22; Institute of Terrestrial Ecology.

How, F. C. 1984. *A Dictionary of the Families and Genera of Chinese Seed Plants.* Science Press.

Horwood, M. T. and Masters, E. H. 1970. Sika deer. *British Deer Society,* publication no. 2.

Hosey, G. R. 1981. Foods of the roe deer. *Mammal Society Notes,* 42.

Howard, R. W. and Bradbury, K. 1979. Feeding by regurgitation in the badger. *Mammal Society Notes,* 38.

Hubbard, C. E. 1954. *Grasses.* Penguin.

Hughes, I. G. 1972. Habitat preferences in wood ants. *Journal of the Entomological Society.*

Hunter, F. A. 1977. Ecology of pinewood beetles. In: *Native Pinewoods of Scotland,* edited by R. G. H. Bunce and J. N. R. Jeffers. Aviemore Symposium; Institute of Terrestrial Ecology.

King, C. M. 1973. A system for trapping and handling live weasels in the field. *Journal of Zoology,* 171.

Kingsolver, J. G. 1982. Butterfly engineering. *Scientific American.*

Kinloch, B. B., Westfall, R. D. and Forrest, G. I. 1987. Caledonian Scots pine: Origins and genetic structure. *New Phytologist,* 104.

Kirby, K. 1988. Changes in ground flora under plantations on ancient woodland sites. *Forestry,* 4.

Knystautas, A. 1987. *The Natural History of the USSR.* Century Press.

Koiriukstis, L. A. 1968. Solar energy in mixed stands. *Forestry,* 41.

Kolb, H. H. 1994. Rabbit populations in Scotland since the introduction of myxomatosis. *Mammal Review,* 25(1).

Konig, E. and Gossow, H. 1979. Even-aged plantations as a habitat for deer. In: *Ecology of Even-Aged Forest Plantations,* edited by E. D. Ford, D. C. Malcolm and J. Atterson (IUFRO). Institute of Terrestrial Ecology.

Kormanik, P. P. 1979. Biological means of improving nutrient uptake in trees. In: *Ecology of Even-Aged Forest Plantations,* edited E. D. Ford, D. C. Malcolm and J. Atterson (IUFRO). Institute of Terrestrial Ecology.

Krebs, J. R. 1986. Are lagomorphs similar to other mammals in population ecology? *Mammal Review,* 16.

Kuchler, A. W. 1964. Potential natural vegetation of the conterminous United States. *Special Publication no. 36.* American Geographical Society.

Kuusela, K. 1987. Forest production in boreal countries. *Newsletter, 25.* International Union of Societies of Foresters.

LaMarche, V. C., Greybill, D. A., Fritts, H. C. and Rose, M. R. 1984. Increasing atmospheric CO_2: Tree ring evidence for growth enhancement in natural vegetation. *Science,* 225.

Lack, P. C. 1988. Hedge intersection and breeding bird distribution in farmland. *Bird Study,* 35.

Lance, A. N. 1973. The number of wood mice on improved and unimproved blanket bog. *Mammal Notes,* 27.

Langley, P. J. W. and Yalden, D. W. 1977. The decline of the rarer carnivores in Gt. Britain during the 19th century. *Mammal Review,* 7.

Larsen, J. A. 1980. *The Boreal Ecosystem.* Academic Press.

Last, F. T. and Gardiner, A. S. 1981. *Forest and Woodland Ecology.* ITE Symposium no. 8; Institute of Terrestrial Ecology.

Last, F. T. et al. 1986. Whither forestry? The scene in AD 2025. In: *Trees and Wildlife in the Scottish Uplands,* edited by D. Jenkins. ITE Symposium no. 17; Institute of Terrestrial Ecology.

Leeks, G. J. L. and Roberts, G. 1987. The effects of forestry on upland streams. In: *Environmental Aspects of Plantation Forestry in Wales,* edited by J. E. G. Good. ITE Symposium no. 22; Institute of Terrestrial Ecology.

Legg, C. J. 1986. Impact of upland afforestation on vegetation. In: *Afforestation and Nature Conservation Interactions.* Timber Growers United Kingdom.

Leslie, R. 1981. The birds of N.E. England forests. *Quarterly Journal of Forestry,* 75.

Lever, C. 1977. *The Naturalised Animals of the British Isles.* Hutchinson.

Little, B. and Davison, M. 1992. Merlins using crows' nests in Kielder Forest, Northumberland. *Bird Study,* 39.

Linnaird, W. 1982. *Welsh Woods and Forests.* National Museum of Wales.

Lloyd, H. G. 1983. Past and present distribution of red and grey squirrels. *Mammal Review,* 13.

316

Lockie, J. D. 1966. *Territory in Small Carnivores.* Symposium no. 18; London Zoological Society.

Lorrain-Smith, R. 1973. Systems of management for timber production and wildlife conservation in lowland forestry. *Quarterly Journal of Forestry,* 67.

Low, A. J. 1986. 1986 Use of broadleaved species in upland forests. *Forestry Commission Leaflet no. 88.* HMSO.

Lowe, P. D. 1983. Values and institutions in the history of British nature conservation. In: *Conservation in Perspective,* edited by A. Warren and F. B. Goldsmith. Wiley.

Lowe, V. P. W. 1975. Pinewoods as habitats for mammals. In: *Native Pinewoods of Scotland,* edited by R. G. H. Bunce, and J. N. R. Jeffers. Aviemore Symposium; Institute of Terrestrial Ecology.

Lu, H. and Shens, H. 1984. Status of the black muntjac. *Mammal Review,* 14.

Lyneborg, L. 1976. *Moths.* Blandford.

Mabberley, D. J. 1987. *The Plant Book.* Cambridge University Press.

Macdonald, D. W. 1976. Food preference in the red fox. *Mammal Review,* 6.

Macdonald, D. W. 1987. *Running with the Fox.* Guild.

MacKinnon, K. 1978. Competition between red and grey squirrels. *Mammal Review,* 8.

McVean, D. N. and Ratcliffe, D. A. 1962. *Plant Communities of the Scottish Highlands.* HMSO.

Mallorie, H. C. and Flowerdew, J. R. 1994. Woodland small mammal population ecology in Britain. *Mammal Review,* 24(1).

Maloney, B. K. 1984. Disease and the elm decline. *Nature,* 312.

Mammal News. 1995. No. 97. Mammal Society.

Manley, G. 1953. *Climate and the British Scene.* Collins.

Marquiss, M., Newton, I. and Radcliffe, D. 1978. The decline of the raven in relation to afforestation in southern Scotland and northern England. *Journal of Applied Ecology,* 15.

Mason, P. A. 1981. Toadstools and trees. In: *Forest and Woodland Ecology,* edited by F. T. Last and A. S. Gardiner. ITE Symposium no. 8; Institute of Terrestrial Ecology.

Mason, P. A. and Last, F. T. 1986. Are the occurrences of sheathing mycorrhizal fungi in new and regenerating forests and woodlands in Scotland predictable? In: *Trees and Wildlife in the Scottish Uplands,* edited by D. Jenkins. ITE Symposium no. 17; Institute of Terrestrial Ecology.

Mason, C. F. and MacDonald, S. M. 1982. The input of terrestrial invertebrates from tree canopies to a stream. *Freshwater Biology* 12.

Mason, C. F., MacDonald, S. M. and Hussey, A. 1984. Structure, management and conservation value of the riparian woody plant community. *Biological Conservation* 29.

Matthews, J. D. 1989. *Silvicultural Systems.* Oxford University Press.

May, R. M. 1979a. Patterns in the abundance of parasites on plants. *Nature,* 281.

May, R. M. 1979b. Nutrient retention in tropical rain forests. *Nature,* 282.

Mellanby, K. 1981. *Farming and Wildlife.* Collins.

Miles, J. 1979. *Vegetation Dynamics.* Chapman & Hall.

Miles, J. 1981. Effects of trees on soils. In: *Forest and Woodland Ecology,* edited by F. T. Last and A. S. Gardiner. ITE Symposium no. 8; Institute of Terrestrial Ecology.

Miles, J. 1986. What are the effects of trees on soils? In: *Trees and Wildlife in the Scottish Uplands,* edited by D. Jenkins. ITE Symposium no. 17; Institute of Terrestrial Ecology.

Miller, H. G. 1986. Effects on soil and water. In: *Afforestation and Nature Conservation Interactions.* Timber Growers United Kingdom.

Mills, D. H. 1980. The management of forest streams. *Forestry Commission Leaflet no. 78.* HMSO.

Mills, D. H. 1986. Effects of afforestation on salmon and trout rivers and suggestions for their control. In: *Afforestation and Nature Conservation Interactions*. Timber Growers United Kingdom.

Milsom, T. P. 1987. Aerial insect hunting by hobbies in relation to weather. *Bird Study*, 34.

Mitchell, A. 1974. *A Field Guide to the Trees of Britain and Northern Europe*. Collins.

Mitchell, P. L. 1989. *Ecological Effects of Forestry Practice in Long Established Woodland and Their Implication for Nature Conservation*. Occasional Paper no. 39. Oxford Forestry Institute.

Mitchell Beazley 1981. *The Forest Realm*. Mitchell Beazley.

Moffat, B. 1987. A curious assemblage of seeds from Waltham Abbey. *Transactions of the Essex Society for Archaeology and History*, 18.

Moller, H. 1983. Foods and foraging behaviour of red and grey squirrels. *Mammal Review*, 13.

Montgomery, W. I. 1979. Seasonal variation in numbers of *Apodemus sylvaticus, A. flavicollis* and *Clethrionomys glareolus*. *Journal of Zoology*, 188.

Montgomery, W. I. 1980. The use of arboreal runways by woodland rodents. *Mammal Review*, 10.

Moore, P. D. 1977. Ancient distribution of lime trees in Britain. *Nature*, 268.

Moore, P. D. 1981. The spruce invasion. *Nature*.

Moore, P. D. 1982. Seeds of thought for plant conservationists. *Nature*, 303.

Moore, P. D. 1984a. Forest history from pollen. *Nature*, 311.

Moore, P. D. 1984b. Hampstead heath clues to the historical decline of elms. *Nature*, 312.

Moore, P. D. 1987. Tree boundaries on the move. *Nature*, 326.

Morris, M. G. 1974. Oak as a habitat for insect life. In: *The British Oak*, edited by M. G. Morris and F. H. Perring. Classey.

Morris, M. G. and Perring, F. H. 1974. *The British Oak*. Classey.

Morrison, B. R. S. 1988. The effects of forest management on the biology of water courses. In: *Wildlife Management in Forests*, edited by D. Jardine. Lancaster discussion meeting; Institute of Chartered Foresters.

Morton Boyd, J. 1987. Commercial forests and woods: The nature conservation baseline. *Forestry*, 60.

Murray, J. S. 1979. The development of populations of pests and pathogens.In: *The Ecology of Even-Aged Forest Plantations*, edited by E. D. Ford, D. C. Malcolm and J. Atterson (IUFRO). Institute of Terrestrial Ecology.

Natural World. Summer 1994. Royal Society for Nature Conservation.

Natural World. Spring / Summer 1996. Royal Society for Nature Conservation.

NCC 1981. *The Conservation of Butterflies*. Nature Conservancy Council.

NCC 1983a. *Eighth Annual Report 1981-82*.

NCC 1983b. *The Ecology and Conservation of Amphibian and Reptile Species Endangered in Britain*.

NCC 1984. *Nature Conservation in Great Britain*.

NCC 1986. *Nature Conservation and Afforestation in Britain*.

NCC 1987a. *Corporate Plan 1987-1988*.

NCC 1987b. *Birds, Bogs and Forestry*.

NCC 1987c. The golden eagle in a changing landscape. *Topical Issues*.

NCC 1987d. *Oxford Meeting*.

NCC 1988. *Thirteenth Annual Report*.

Neal, E. G. 1955. *The Badger*. Forestry Commission Leaflet.

Neal, E. G. 1972. National badger survey. *Mammal Review*, 2.

318

Newton, I. 1979. *Population Ecology of Raptors.* Poyser.

Newton, I. 1983. Birds and forestry. In: *Forestry and Conservation,* edited by E. H. M. Harris, Royal Forestry Society, Tring.

Newton, I. 1986a. Principles underlying bird numbers in Scottish woodlands. In: *Trees and Wildlife in the Scottish Uplands,* edited by D. Jenkins. ITE Symposium no. 17; Institute of Terrestrial Ecology.

Newton, I. 1986b. Birds and forestry. In: *Afforestation and Nature Conservation Interactions.* Timber Growers United Kingdom.

Newton, I. and Moss, D. 1977. Breeding birds of Scottish pinewoods. In: *Native Pinewoods of Scotland,* edited by R. G. H. Bunce and J. N. R. Jeffers. Aviemore Symposium; Institute of Terrestrial Ecology.

Newton, I. et al. 1981. Distribution and breeding of red kites in relation to land use in Wales. *Journal of Applied Ecology,* 18.

O'Conner, T. P. 1986. The garden dormouse from Roman York. *Mammal Society Notes,* 53.

Odum, S. 1965. Germination of ancient seeds. *Dansk Botanisk Arkit,* 24.

Orchel, J. 1992. *Forest Merlins in Scotland.* The Hawk and Owl Trust.

Ormerod, S. J., Mawle, G. W. and Edwards, R. W. 1987. The influence of forest on aquatic fauna. In: *Environmental Aspects of Plantation Forestry in Wales,* edited by J. E. G. Good. ITE Symposium no. 22; Institute of Terrestrial Ecology.

O'Sullivan, P. J. 1983. The distribution of the pine marten in the Republic of Ireland. *Mammal Review,* 13.

Ovington, J. D. 1953. Studies of the development of woodland conditions under different trees. *Journal of Ecology,* 41.

Ovington, J. D. 1954. Studies of the development of woodland conditions under different trees [continued]. *Journal of Ecology,* 42.

Ozanne, C. 1996. Wildlife. *Timber Grower,* 139.

Parker, D. E. 1988. The climate of England and Wales during the past 150 years. *Journal of The Royal Agricultural Society of England.*

Parry, W. H. 1986. Effects of afforestation on invertebrate conservation. In: *Afforestation and Nature Conservation Interactions.* Timber Growers United Kingdom.

Peachey, C. 1986. Quoted by Young 1986.

Pearsall, W. H. 1959. *Mountains and Moorlands.* Collins.

Perring, F. 1983. The voluntary movement. In: *Conservation in Perspective,* edited by A. Warren and F. B. Goldsmith. Wiley.

Perring, F. and Walters, S. M. 1962. *Atlas of the British Flora.* Botanical Society of the British Isles. Nelson.

Peterken, G. F. 1981. *Woodland Conservation and Management.* Chapman & Hall .

Peterken, G. F. 1986. The status of native woods in the Scottish uplands. In: *Trees and Wildlife in the Scottish Uplands,* edited by D. Jenkins. ITE Symposium no. 17; Institute of Terrestrial Ecology.

Peterken, G. F. 1987. Natural features in the management of upland conifer forests. In: *Sitka Spruce,* edited by D. M. Henderson and R. Faulkner. *Proceedings of the Royal Society of Edinburgh,* 93B.

Peterken, G. F. and Game, M. 1984. Historical factors affecting the number and distribution of vascular plant species in the woods of central Lincolnshire. *Journal of Ecology,* 72.

Peterken, G. F., Ausherman, D., Buchenau, M. and Forman, R. T. T. 1992. Old growth conservation within British upland conifer plantations. *Forestry,* 65 (2).

Peterson, R. T. 1947. *A Field Guide to the Birds.* Houghton Mifflin.

319

Peterson, R. T. 1954. *A Field Guide to the Birds of Britain and Europe.* Collins.

Peterson, R. T. 1961. *A Field Guide to Western Birds.* Houghton Mifflin.

Peterson, R. T. and McKenny, M. 1968. *A Field Guide to Wildflowers.* Houghton Mifflin.

Petty, S. J. 1987. The design and use of nest boxes for tawny owls in upland forests. *Quarterly Journal of Forestry,* 81.

Petty, S. J. 1988. The management of raptors in upland forests. In: *Wildlife Management in Forests,* edited by D. C. Jardine. Lancaster discussion meeting; Institute of Chartered Foresters.

Petty, S. J. 1989. Goshawks: Their status, requirements and management. *Forestry Commission Bulletin* 81. HMSO.

Petty, S. J. and Anderson, D. 1986. Breeding by hen harriers on restocked sites in upland forests. *Bird Study,* 33.

Petty, S. and Avery, M. 1990. *Forest Bird Communities.* Occasional Paper 26. Forestry Commission.

Pickell, S. T. A. and White, P. S. 1985. *Patch Dynamics and the Effect of Disturbance.* Academic Press.

Pigott, C. D. 1988a. The growth of lime in an experimental plantation and its influence on soil development and vegetation. *Quarterly Journal of Forestry,* 83.

Pigott, C. D. 1988b. The ecology and silviculture of limes. In: *Report of the National Hardwoods Programme and Second Meeting of The Uneven-Aged Sylviculture Group,* edited by P. Savill. Oxford Forestry Institute.

Pigozzi, G. 1988. Diet of the European badger in the Maremma Natural Park, Central Italy. *Mammal Review,* 18.

Pinder, N. 1979. Faunal re-introductions. *University College London, Discussion Paper, 23.*

Pollard, E. 1979. The expansion in range of the white admiral. *ITE Annual Report 1978.* Institute of Terrestrial Ecology.

Pollard, E. 1980. The butterfly monitoring scheme. In: *ITE Annual Report, 1979.* Institute of Terrestrial Ecology.

Pollard, E. 1981. Population studies of woodland butterflies. In: *Forest and Woodland Ecology,* edited by F. T. Last and A. S. Gardiner. ITE Symposium no. 8; Institute of Terrestrial Ecology.

Pollard, E. 1987. *Research Reports Digest no. 10.* Nature Conservancy Council.

Polunin, O. and Walters, M. 1985. *A Guide to the Vegetation of Britain and Europe.* Oxford University Press.

Poole, T. B. 1970. Polecats. *Forestry Commission Record 76.* HMSO.

Poore, M. E. D. 1955. The use of phytosociological methods in ecological investigations. *Journal of Ecology,* 43.

Poore, M. E. D. and McVean, D. N. 1957. A new approach to Scottish mountain vegetation. *Journal of Ecology,* 45.

Potter, M. 1988. Exotic broadleaves. In: *Broadleaves – Changing Horizons,* edited by M. Potter. Herriot-Watt discussion meeting, Institute of Chartered Foresters.

Pyatt, D. G., Anderson, A. R. and Ray, D. 1988. *Deep Peats: Report on Forest Research.* Forestry Commission, HMSO London.

Racey, P. A. and Stebbings, R. E. 1972. Bats in Britain. *Oryx,* 11.

Racey, P. A. and Swift, -. 1986. Bats and buildings. *Veterinary Record,* 118.

Rackham, J. 1979. Quoted in Yalden 1982.

Rackham, O. 1974. The oak tree in historic times. In: *The British Oak,* edited by M. G. Morris and F. H. Perring. Classey.

Rackham, O. 1977. Quoted in Peterken 1981.

Rackham, O. 1981. *Trees and Woodland in the British Landscape.* Dent.

Ramsbottom, J. 1953. *Mushrooms and Toadstools.* Collins.

Ranger, J. and Nys, C. 1996. Biomass and nutrient content of extensively and intensively managed coppice stands. *Forestry,* 69(2).

Ratcliffe, D. R. 1977. *A Nature Conservation Review.* Cambridge University Press.

Ratcliffe, D. R. 1980. *The Peregrine.* Poyser.

Ratcliffe, D. R. 1984. Tree nesting by peregrines in Britain and Ireland. *Bird Study,* 31.

Ratcliffe, D. R. 1986. The effects of afforestation on the wildlife of open habitats. In: *Trees and Wildlife in the Scottish Uplands,* edited by D. Jenkins. ITE Symposium no. 17; Institute of Terrestrial Ecology.

Ratcliffe, P. R. 1987. The Management of Red Deer in Upland Forests. *Forestry Commission Bulletin 71.* HMSO.

Ratcliffe, P. R. 1988. The management of red deer populations resident in upland forests. In: *Wildlife Management in Forests,* edited by D. C. Jardine. Lancaster discussion meeting; Institute of Chartered Foresters.

Ratcliffe, P. R. and Petty, S. J. 1986. The management of commercial forests for wildlife. In: *Trees and Wildlife in the Scottish Uplands,* edited by D. Jenkins. ITE Symposium no. 17; Institute of Terrestrial Ecology.

Ravenscroft, N. O. M. 1989. The status and habitat of the nightjar in coastal Suffolk. *Bird Study,* 36.

Reid, R. and Wilson, G. 1986. *Agroforestry in Australia and New Zealand.* Capitol Press, Victoria.

Richards, C. G. J. et al. 1984. The food of the common dormouse in S. Devon. *Mammal Review,* 14.

Richardson, D. H. S. *The Biology of Mosses.* Blackwell Scientific Publications.

Richens, R. H. 1983. *The Elm.* Cambridge University Press.

Roberts, G. 1996. Conserving moths in woodland. *Quarterly Journal of Forestry,* 90(1).

Roberts, G. and Warren, M. 1994. Butterfly conservation; new life for old woods. *Quarterly Journal of Forestry,* 88(3).

Robertson, P. A. 1988. Pheasant management in small broadleaved woodlands. In: *Wildlife Management in Forests,* edited by D. C. Jardine. Lancaster discussion meeting; Institute of Chartered Foresters.

Robinson, T. C. 1968. Butterflies in woodlands. *Forestry Commission Record 65.* HMSO.

Roche, L. and Haddock, P. G. 1987. Sitka spruce in N. America with special reference to its role in British forestry. In: *Sitka Spruce,* edited by D. M. Henderson and R. Faulkner. *Proceedings of The Royal Society of Edinburgh,* 93B.

Rochelle, J. A. and Bunnell, F. L. 1979. Plantation management and vertebrate wildlife. In: *Ecology of Even-Aged Forest Plantations,* edited by E. D. Ford, D. C. Malcolm and J. Atterson (IUFRO). Institute of Terrestrial Ecology.

Roden, D. 1968. Woodland and its management in the medieval Chilterns. *Forestry,* 41.

Rodwell, J. S. 1991. *British Plant Communities. Woodlands and Scrub, Vol. 1.* Cambridge University Press.

Rogers, H. H., Thomas, J. F. and Bingham, G. E. 1983. Response of agronomic and forest species to elevated atmospheric carbon dioxide. *Science,* 220.

Rose, F. 1974. The epiphytes of oak. In: *The British Oak,* edited by M. G. Morris and F. H. Perring. Classey.

Rose, C. I. and Hawksworth, D. L. 1981. Lichens' recolonisation of London's cleaner air. *Nature,* 289.

Ross J. and Tittensor, A. M. 1986. The influence of myxomatosis in regulating rabbit numbers. *Mammal Review,* 16.

Rowe, J. 1983. Squirrel management. *Mammal Review,* 13.

Rowe, J. S. and Scotter, G. W. 1973. Fire in the boreal forest. In: *The Ecological Role of Fire in Natural Conifer Forests of Western and Northern America,* edited by H. E. Wright and M. L. Heinselman. *Quarternary Research,* 3. Academic Press.

Royal Forestry Society. 1984. The Dukeries and Lincolnshire. *Quarterly Journal of Forestry,* 78.

Royal Forestry Society. 1985. South Downs. *Quarterly Journal of Forestry,* 79.

Royal Forestry Society. 1987. Lake District, 1987. *Quarterly Journal of Forestry,* 81.

Royal Society. 1988. Seminar: Acid rain and British forests.

RSPB. 1987. *Conservation Review.* Royal Society for the Protection of Birds.

Russell, E. J. 1957. *The World of the Soil.* Collins.

Salisbury, E. J. 1961. *Weeds and Aliens.* Collins.

Savill, P. S. and Evans, J. 1986. *Plantation Silviculture in Temperate Regions.* Clarendon Press.

Sawford, B. 1987. *The Butterflies of Hertfordshire.* Castle Mead.

Schauer, T. 1978. *Wild Flowers.* Collins.

Scott, D. and Clarke, R. 1993. Taking to the trees. *Bird Watching,* 1992.

Seitz, R. 1986. Siberian fire as 'nuclear winter' guide. *Nature,* 323.

Sharrock, J. T. R. 1976. *The Atlas of Breeding Birds in Britain and Ireland.* Poyser.

Shaw, G. and Livingstone, J. 1991. Goldfinches and other birds eating Sitka spruce seed. *British Trust for Ornithology News,* 174.

Shawyer, C. 1987. The barn owl in the British Isles: its past, present and future. Report of the Hawk Trust.

Simberloff, D. 1986. Design of nature reserves. In: *Wildlife Conservation Evaluation,* edited by M. Usher. Chapman & Hall.

Smal, C. M. and Fairley, J. S. 1980. Food of woodmice and bank voles. *Mammal Society Notes,* 40.

Smart, N. and Andrews, J. 1985. *Birds and Broadleaves Handbook.* Royal Society for the Protection of Birds.

Southern, H. N. 1964. *The Handbook of British Mammals.* Blackwell Scientific Publications.

Southwood, T. R. E. 1961. The number of species of insects associated with various trees. *Journal of Animal Ecology,* 30.

Spellerberg, I. 1988a. Management of Forest Habitats for Reptiles. In: *Wildlife Management in Forests,* edited by D. C. Jardine. Lancaster discussion meeting; Institute of Chartered Foresters.

Spellerberg, I. 1988b. Ecology and management of reptile populations in forests. *Quarterly Journal of Forestry,* 82.

Spencer, J. A. 1995. Restructuring in Britain's forests. *Quarterly Journal of Forestry,* 89(3).

Sprugel, D. and Bormann, E. H. 1981. Natural disturbance and the steady state in high altitude balsam fir forests. *Science,* 211.

Stace, C. 1995. *New Flora of the British Isles.* Cambridge University Press.

Staines, B. W. 1974. A review of factors affecting deer dispersal and their relevance to management. *Mammal Review,* 4.

Staines, B. W. 1983. The conservation and management of mammals in commercial plantations. In: *Forestry and Conservation,* edited by E. H. M. Harris, Royal Forestry Society, Tring.

Staines, B. W. 1986. Mammals in Scottish upland woods. In: *Trees and Wildlife in the Scottish Uplands.* ITE Symposium no. 17; Institute of Terrestrial Ecology.

Staines, B. W. and Crisp, J. M. 1978. Observations on food quality in Scottish red deer. *Mammal Review*, 3.

Staines, B. W., Petty, S. J. and Ratcliffe, P. R. 1987. Sitka spruce forests as a habitat for birds and mammals. In: *Sitka Spruce*, edited by D. M. Henderson and R. Faulkner. *Proceedings of The Royal Society of Edinburgh*, 93B.

Staines, B. W. and Welch, D. 1987. Conflict between roe and red deer in Scottish woodlands. *Mammals as Pests*. Mammal Society symposium.

Steel, C. 1989. Butterfly recording in Sheephouse wood 1986. *Research Report Digest no. 12*. Nature Conservancy Council.

Steele, R. C. 1975. Wildlife Conservation in Woodlands. *Forestry Commission Booklet 29*. HMSO.

Stern, R. C. 1982. The use of sycamore in British forestry. In: *Broadleaves in Britain*, edited by D. C. Malcolm, J. Evans and P. N. Edwards. Institute of Chartered Foresters.

Stern, R. C. 1989. Sycamore in Wessex forests. *Forestry* 62(4).

Steven, H. M. and Carlisle, A. 1959. *The Native Pinewoods of Scotland*. Oliver & Boyd, Edinburgh and London.

Stowe, T. J. 1987. Management of sessile oakwood for pied flycatchers. *RSPB Conservation Review*. The Royal Society for the Protection of Birds.

Strong, D. R. 1974a. The insects of British trees: community equilibration in ecological time. *Annals of the Missouri Botanical Garden*, 61.

Strong, D. R. 1974b. Non-asymptotic species richness models and the insects of British trees. *Proceedings of the National Academy of Sciences, USA*.

Strong, D. R. 1979. Biogeographic dynamics of insect-host plant communities. *Annual Review of Entomology*, 24.

Strong, D. R. and Levin, S. 1979. (Quoted in May 1979a). *American Naturalist*, 114.

Strong, D. R., McCoy, E. D. and Rey, J. R. 1977. Time and the number of herbivore species. *Ecology*, 58.

Summerhayes, V. S. 1957. *Wild Orchids of Britain*. Collins.

Sumption, K. J. and Flowerdew, J. R. 1985. Ecological effects of the decline of the rabbit. *Mammal Review*, 15.

Sutherland, W. J. and Watkinson, A. R. 1986. Do plants evolve differently? *Nature*, 320.

Swindel, B. F., Conde, L. F. and Smith, J. E. 1983. Plant cover and biomass response to clear cutting: Site preparation and planting in *P. elliottii* flatwoods. *Science*, 219.

Sylva Sinica. 1985. Volumes I and 2. Chinese Forestry Institute, Beijing.

Tansley, A. G. 1949. *The British Isles and their Vegetation*. Cambridge University Press.

Tapper, S. C. 1976a. Population fluctuations in mammals. *Mammal Review*, 6.

Tapper, S. C. 1976b. Diet of weasels and stoats. *Mammal Society Notes*, 32.

Tapper, S. C. et al. 1982. Effect of mammalian predators on partridge populations. *Mammal Review*, 12.

Tasker, M. L. 1989. *North-East Scotland Bird Report 1988*. North-East Scotland Bird Club.

Tate, P. 1989. *The Nightjar*. Shire Natural History.

Taylor, C. 1994. Report on the activities of the native pinewood managers. *Scottish Forestry*, 48(2).

Thomas, J. A. 1974. Quoted by Young 1986.

Thomas, J. A. 1980. The extinction of the large blue and the conservation of the black hair-streak butterflies (a contrast of failure and success). *ITE Annual Report 1979*. Institute of Terrestrial Ecology.

Thompson, M. 1983. The feeding habits of bats. *Mammal Society Newsletter*, 55.

Thompson, A. 1986. Anomalies in the estimation of small mammal abundance in conifer plantations. *Mammal Notes*, 52.

Thornton, P. S. 1988. Density and distribution of badgers in south-west England. In: *Proceedings of a Symposium on the Badger. Mammal Review*, 18.

Timber Growers United Kingdom. 1985. *The Forestry and Woodland Code*. London.

Timber Growers United Kingdom. 1986. *Afforestation and Nature Conservation Interactions*. London.

Tittensor, A. M. 1970. Red squirrel dreys. *Journal of Zoology*, 162.

Tonkin, J. M. 1983. Activity pattern of the red squirrel. *Mammal Review*, 13.

Trippensee, R. E. 1948. *Wildlife Management*. McGraw-Hill.

Trout, R. C. et al. 1978. A review of studies on captive harvest mice. *Mammal Review*, 4.

Trout, R. C., Tapper, S. C. and Harradine, J. 1986. Recent trends in the rabbit population in Britain. *Mammal Review*, 16.

Tuley, G. 1985. The growth of young oak trees in shelters. *Forestry*, 58.

Tumajanov, I.I. 1971. Changes in the Great Caucasus forest vegetation during the Pleistocene and Holocene. In: *Plant life of SW Asia*. Botanical Society of Edinburgh.

Turrill, W. B. 1948. *British Plant Life*. Collins.

USDA 1949. *Trees*. United States Department of Agriculture. Washington.

Usher, M. 1986. *Wildlife Conservation Evaluation*. Chapman & Hall.

van den Brink, 1967. *A Field Guide to the Mammals of Britain and Europe*. Collins.

Velander, K. A. 1983. *Pine Marten Survey of Scotland, England and Wales* 1980-82. Vincent Wildlife Trust.

Velander, K. A. 1989. A study of pine-marten ecology in Inverness-shire. *Research Reports Digest no. 12*. Nature Conservancy Council.

Viereck, L. A. 1973. Wildlife in the taiga of Alaska. In: *The Ecological Role of Fire in Natural Conifer Forests of Western and Northern America*, edited by H. E. Wright and M. L. Heinselman. *Quaternary Research*, 3. Academic Press.

Von Droste, B. 1987. International co-operation in the Man and Biosphere programme of UNESCO. In: *The Temperate Forest Ecosystem*, edited by Yang Hanxi, Wang Zhan, J. N. R. Jeffers and P. A. Ward. ITE Symposium no. 20; Institute of Terrestrial Ecology.

Von Droste, B. and Gregg, W. P. 1985. *Biosphere Reserves*. Parks, 10.

Walker, C. 1987. Sitka spruce Mycorrhizas? In: *Sitka Spruce*, edited by D. M. Henderson and R. Faulkner. *Proceedings of The Royal Society of Edinburgh*, 93B.

Walker, D. R. G. 1972. Observations on a collection of weasels from estates in S.W. Hertfordshire. *Journal of Zoology*, 166.

Walter, H. 1973. *Vegetation of the Earth in Relation to Climate and the Ecophysiological Conditions*. English Universities Press.

Walton, K. C. 1968. The distribution of the polecat. *Journal of Zoology*, 155.

Wang, C. 1961. *The Forests of China*. Publication no. 5. The Maria Moors Cabot Foundation, Harvard.

Ward, L. 1995. Red squirrel conservation. *Quarterly Journal of Forestry*, 89(2).

Ward, L. K. and Spalding, D. F. 1981. Trees and the phytophagous insects data bank. In: *Forest and Woodland Ecology*, edited by F. T. Last and A. S. Gardiner. ITE Symposium no. 8; Institute of Terrestrial Ecology.

Waring, P. and Haggett, G. 1991. Coppice woodland habitats. In Fry and Lonsdale 1991, *Habitat Conservation for Insects. Amateur Entomologist*, 21.

Warren, M. S. 1979. The ecology of the wood white butterfly. *ITE Annual Report 1979*. Institute of Terrestrial Ecology.

Warren, A. and Goldsmith, F. B. 1983. *Conservation in Perspective.* Wiley.

Watling, R. 1974. Macrofungi in the oakwoods of Britain. In: *The British Oak,* edited by M. G. Morris and F. H. Perring. Classey.

Watling, R. 1986. A 150 years of paddock stools. *Botanical Society of Edinburgh,* 45.

Watson, A. and Hewson, R. 1973. Population densities of mountain hares on western Scottish moors and Irish moors and on Scottish hills. *Journal of Zoology,* 170.

Watson, A. and Staines, B. W. 1978. Differences in the quality of wintering areas used by male and female red deer. *Mammal Society Notes,* 37.

Watt, A. D. 1986. The ecology of the pine beauty moth in commercial woods in Scotland. In: *Trees and Wildlife in the Scottish Uplands,* edited by D. Jenkins. ITE Symposium no. 17; Institute of Terrestrial Ecology.

Watt, K. E. F. 1987. An alternative explanation for widespread tree mortality. *International Union of Societies of Foresters, Newsletter,* 25.

Wedgewood, M. 1990. Barn owls. *Country Life,* 8 Feb.

Welch, D., Chambers, H. G., Scott, D. and Staines, B. W. 1988. Roe deer browsing on spring flush growth of Sitka spruce. *Scottish Forestry,* 42.

Welch, R. C. 1981. Insects on exotic broadleaved trees of the Fagaceae. In: *Forest and Woodland Ecology,* edited by F. T. Last and A. S. Gardiner. ITE Symposium no. 8; Institute of Terrestrial Ecology.

Welch, R. C. 1986. What do we know about insects in Scottish woods? In: *Trees and Wildlife in the Scottish Uplands,* edited by D. Jenkins. ITE Symposium no. 17; Institute of Terrestrial Ecology.

Welch, R. C. 1995. Introduced oaks and their galls at Bicton, Devon. *Quarterly Journal of Forestry,* 89(2).

Whalley, P. 1980. *Butterfly Watching.* Severn House.

Whitbread, A. M. 1990. How relevant is the identification of ancient woodland to woodland conservation in the upland counties of Britain? *Quarterly Journal of Forestry,* 84(1).

White, E. J. and Smith, R. I. 1982. *Climatological Maps of Great Britain.* Institute of Terrestrial Ecology.

Whitehead, P. F. 1984. Observations on the biota of the middle Lena river, Yakotia, USSR. *Quarterly Journal of Forestry,* 78.

Whitmore, T. C. 1975. *Tropical Rain Forests of the Far East.* Clarendon Press.

Whitmore, T. C. 1985. Forest Succession. *Nature,* 315.

Whitmore, T. C. and Prance, G. T. 1987. *Biogeography and Quaternary History in Tropical America.* Oxford Science Publications.

Wilkinson, W. 1987. Montane wood mice. *Journal of Zoology,* 212.

Williams, C. B. 1958. *Insect Migration.* Collins.

Williamson, D. R. and Lane, P. B. 1989. The use of herbicides in the forest. *Field Book No 8.* Forestry Commission, HMSO, London.

Winter, T. G. 1979. New host plant records of Lepidoptera associated with conifer afforestation. *Entomologist's Gazette,* 25.

Winter, T. G. 1983. *A Catalogue of Phytophagus Insects and Mites on Trees in Great Britain.* Forestry Commission Booklet 53. Edinburgh.

Witherby, H. F. et al. 1943-4. *The Handbook of British Birds,* vols 1-5. Witherby, London.

Woillard, G. 1979. Abrupt end of the last interglacial s.s. in north-east France. *Nature,* 281.

Wray, S. and Harris, S. 1994. Brown hares in commercial forestry in Great Britain. *Quarterly Journal of Forestry,* 88(3).

Wright, H. E. and Heinselman, M. L. 1973. The Ecological role of Fire in the Natural Conifer Forest of Western and Northern America. *Quaternary Research, 3.*

Wroot, A. J. 1985. Foraging in the European hedgehog. *Mammal Review,* 15.

Yalden, D. W. 1969. The food of the hedgehog. *Mammal Society Bulletin,* 32.

Yalden, D. W. 1971. A population of the yellow necked mouse. *Mammal Notes,* 52.

Yalden, D W. 1982. Origins of British mammals. *Mammal Review,* 12.

Yalden, D. W. 1986. Opportunities for re-introducing British mammals. *Mammal Review,* 16.

Young, M. R. 1986. The effect of commercial forestry on woodland Lepidoptra. In: *Trees and Wildlife in the Scottish Uplands,* edited by D. Jenkins. ITE Symposium no. 17; Institute of Terrestrial Ecology.

Zeuner, F. E. 1952. *Dating the Past.* Methuen.

Zhang, Y. et al. 1987. Review of the study of soil animals during the past six years (1979-85) on Changbai mountain. In: *The Temperate Forest Ecosystem,* edited by Yang, H., Wang, Z., Jeffers, J. N. R. and Ward, P. A. ITE Symposium no. 20; Institute of Terrestrial Ecology.

Zobel, B. J., Van Wyk, G. and Stahl, P. 1987. *Growing Exotic Forests.* Wiley.

Zvelebil, M. 1982. Post glacial foraging in the forests of Europe. *Scientific American.*

Zvelebil, M. 1994. Plant use in the Mesolithic and its role in transition to farming. *Proceedings of the Prehistoric Society,* 60.

Index

herb Paris 154,158,167
herb robert 156
heritage value 142
holly 98,143,206,286
home range and territory 178
honey dew 77,84,280,288,298
honeysuckle 200,286,289,291
hornbeam 14,29,45,53,76,77,80,89,100, 110,222
hornet 300
Howard, Eliot 34
humidity 163,171-3,269,302
humus 7,12,23
    mor 17,154,161-3
    mull 163
    raw *see above* mor
Hungary 20
hurdles 57
hurricane 47

Ice Age, 17,21,22,71,72,89,92,175, 213, 266,294
ice storm 47
increment 51,53,54,136
India 13
indicator species analysis 151
Industrial Revolution 30
inherited resistance 192
insecticides *see* pesticides
insects 271,274,304
    control of 271
    saproxylic 58
    successions 272,303
Institute of Terrestrial Ecology 6,7,34, 38,151
Inter-governmental Panel on Forests 146
Inverliever 30
Ireland 24,175
iron 28
    ships 30
    smelting 28,29
ivy 206,286
IUCN 39

Japan 14
Joint Nature Conservation Committee 39
juniper 143,149

kangaroo 176
Kerb granules 64
Kielder Forest 141
knotgrass 283
Korea 14,20,47

ladies' smock 284
ladybird *see* beetles
Lake District 28,57,92,155,165
land use 95,245
Land Use Survey Group 34
Land Utilisation Survey 34
landscape 141
landscape enhancement 142
landslides 47
larch 17,24,30,45,47,62,82,88, 92,95,97,100,118,119,154-5,156,161,169, 176,194,195,196,213,265,300
    European 91
    hybrid 91
    Japanese 91
larvae, insect 181,186,187,198,199,274,275
leaf-hopper *see* bugs
leaf miner 299
latitude, influence of 5,110,177
law, forest 28
leaching 11,13,28,149
lead mining 28
lichens 22,58,85,168,172-3
    succession of 172
light 60,97,161-2,171-2,288,304-5
light demanders *see* pioneer species
lily of the valley 155,158
lime 15,22,25,54,72,75,76,77- 9,82,88,89,110,150,158,222,287
limestone 76,82,84,128,129,155,285
liming 103
liverworts 85,171
lizard
    common 254,261,266
    sand 254,261,266
lop and top 304
Lusitanian flora 22

maidenhair tree *see* ginkgo
Malaysia 13
Mammal Society 133,183,201
Man and the Biosphere Programme 38

replanting conditions 141
reserves
    and island biogeography 43
    fragility of 44
    naturalness of 41
    representativeness of 43
    size of 43
    typicalness of 43
    uniqueness of 44
rest harrow 283
restocked areas 181,190,201,209,218,220,
226,227,228,231,234,236,237,241,263,266,
276,288,300,304
restocking 59
restructuring 141
rhododendron 102,129,143,208
rides and roads 122,127,128,130,133,165,
181,183,186,191,205,209,260,261,263,264,
276,282,288,289,295,299,301,304
    wide 127,128,130,288,304
ride junctions 130,165,288
ringbarking 25,129,208,222,223,224,229,
296
Rio Principles 143
Rio Summit 40,143
rock rose 283,284
Romans 28,82,83,179
rotation 51,57,68,166
    age 54
    coppice 57,58,75
    extended 106,124,166,241,245,266
    long 141
    second 111,141,227
    short 58,68,95,101,132,141,164,234,
    243,263,266
rowan 92,100,143,149,193
Royal Navy 30
Royal Society for the Protection of Birds
see RSPB
RSNC 35
RSPB 34,37,100,133,144
RUSSIA 22,95,135,213, 9.6

Sahara desert 154
sallow 287,291
sanicle 156,167
Santiago Principles 145
sawflies 300

larch 67
saxifrage, golden 155
scabious 284
Scandinavia 135,155,186
scarification 133,294
Scotland 19,22,29,30,57,62,74,84,92, 96,
105,109,136,150,155,157,158,159,166,168,
182,186,213,227
Scottish Natural Heritage 39
scouring of water courses 105,266,275
scrub 1,5,57,130,280
secondary species
45,62,90,100,102,103,111,112,118,140,224
sedge 158,281
    pale 158
    pill 156
    remote 158
    wood 154
sedimentation 105,241,266,275
seed
    bank 163
    buried 162-3
    crop 210
    dormant 163
    viability of 163
selection systems 66,112,122,124,139,266
    see also under forest/forestry
self-heal 155
shakes
    ring 86
    star 86
sheep 30,122
    walks 101
shelterwood 56,224
shelterwood system 56
ship building 29,59,75
shrews 187,188
    common 181-2
    pygmy 181-2
    water 182
silting *see* sedimentation
Silva, Evelyns's 30
silverweed 284
Sites of Special Scientific Interest
32,37,39,123
sloe *see* blackthorn
slowworm 254,263-4,266
slugs 301-2